Iterative Learning Control for Multi-agent Systems Coordination

Iterative Learning Control for Multi-agent Systems Coordination

Shiping Yang
Jian-Xin Xu
Xuefang Li
National University of Singapore
Dong Shen
Beijing University of Chemical Technology, P.R. China

Library of Congress Cataloging-in-Publication Data
Names: Yang, Shiping, 1987– author. | Xu, Jian-Xin, author. | Li, Xuefang,
    1985– author. | Shen, Dong, 1982– author.
Title: Iterative learning control for multi-agent systems coordination / by
    Shiping Yang, Jian-Xin Xu, Xuefang Li, Dong Shen.
Description: Singapore : John Wiley & Sons, Inc., 2017. | Includes bibliographical references and index.
Identifiers: LCCN 2016052027 (print) | LCCN 2016056133 (ebook) | ISBN 9781119189046 (hardback) |
    ISBN 9781119189060 (pdf) | ISBN 9781119189077 (epub)
Subjects: LCSH: Intelligent control systems. | Multiagent systems. | Machine learning. | Iterative methods
    (Mathematics) | BISAC: TECHNOLOGY & ENGINEERING / Robotics.
Classification: LCC TJ217.5 .Y36 2017 (print) | LCC TJ217.5 (ebook) | DDC 629.8/9–dc23
LC record available at https://lccn.loc.gov/2016052027
A catalog record for this book is available from the British Library.

Set in 10/12pt Warnock by SPi Global, Pondicherry, India

Cover design: Wiley
Cover image: © Michael Mann/Gettyimages

Printed in Singapore by C.O.S. Printers Pte Ltd

10 9 8 7 6 5 4 3 2 1

Contents

Preface

The coordination and control problems of multi-agent systems (MAS) have been extensively studied by the control community due to the broad practical applications, for example, the formation control problem, search and rescue by multiple aerial vehicles, synchronization, sensor fusion, distributed optimization, the economic dispatch problem in power systems, and so on. Meanwhile, many industry processes require both repetitive executions and coordination among several independent entities. This observation motivates the research of multi-agent coordination from an iterative learning control (ILC) perspective. This book is dedicated to the application of iterative learning control to multi-agent coordination problems.

In order to study multi-agent coordination by ILC, an extra dimension, the iteration domain, is introduced into the problem. A challenging issue in controlling multi-agent systems by ILC is the non-perfect repeating characteristics of MAS. The inherent nature of MAS such as heterogeneity, information sharing, sparse and intermittent communication, imperfect initial conditions, and inconsistent target tracking trajectories increases the complexity of the problem. Due to these factors, controller design becomes a challenging problem. This book provides detailed guidelines for the design of learning controllers under various coordination conditions, in a systematic manner. The main content can be classified into two parts following the two main frameworks of ILC, namely, contraction-mapping (CM) and composite energy function (CEF) approaches. Chapters 2–7 apply the CM approach, while Chapters 8–10 apply the CEF approach. Each chapter studies the coordination problem under certain conditions. For example, Chapter 2 assumes a fixed communication topology, Chapter 3 assumes a switching topology, and Chapter 4 addresses the initial state error problem in multi-agent coordination. In a sense, each chapter addresses a unique coordination control problem for MAS. Chapters 2–10 discuss continuous-time systems. In Chapter 11 we present a generalized iterative learning algorithm to solve an optimal power dispatch problem in a smart grid by utilizing discrete-time consensus algorithms.

This book is self contained and intensive. Prior knowledge of ILC and MAS is not required. Chapter 1 provides a rudimentary introduction to the two areas. Two minimal examples of ILC are presented in Chapter 1, and a short review of some terminologies in graph theory is provided in Appendix A. Readers can skip the preliminary parts if they are familiar with the domain. We present detailed convergence proofs for each controller as we believe that understanding the theoretical derivations can benefit readers in two ways. On the one hand, it helps readers appreciate the controller design. On the other hand, the control design and analysis techniques can be transferred to other domains to facilitate further exploration in various control applications. Specifically for industrial experts and practitioners, we provide detailed illustrative examples in each chapter to show how those control algorithms are implemented. The examples demonstrate the effectiveness of the learning controllers and can be modified to handle practical problems.

1

Introduction

1.1 Introduction to Iterative Learning Control

Iterative learning control (ILC), as an effective control strategy, is designed to improve current control performance for unpredictable systems by fully utilizing past control experience. Specifically, ILC is designed for systems that complete tasks over a fixed time interval and perform them repeatedly. The underlying philosophy mimics the human learning process that "practice makes perfect." By synthesizing control inputs from previous control inputs and tracking errors, the controller is able to learn from past experience and improve current tracking performance. ILC was initially developed by Arimoto *et al.* (1984), and has been widely explored by the control community since then (Moore, 1993; Bien and Xu, 1998; Chen and Wen, 1999; Longman, 2000; Norrlof and Gunnarsson, 2002; Xu and Tan, 2003; Bristow *et al.*, 2006; Moore *et al.*, 2006; Ahn *et al.*, 2007a; Rogers *et al.*, 2007; Ahn *et al.*, 2007b; Xu *et al.*, 2008; Wang *et al.*, 2009, 2014).

Figure 1.1 shows the schematic diagram of an ILC system, where the subscript i denotes the iteration index and y_d denotes the reference trajectory. Based on the input signal, u_i, at the ith iteration, as well as the tracking error $e_i = y_d - y_i$, the input u_{i+1} for the next iteration, namely the $(i + 1)$th iteration, is constructed. Meanwhile, the input signal u_{i+1} will also be stored into memory for use in the $(i + 2)$th iteration. It is important to note that in Figure 1.1, a closed feedback loop is formed in the iteration domain rather than the time domain. Compared to other control methods such as proportional-integral-derivative (PID) control and sliding mode control, there are a number of distinctive features about ILC. First, ILC is designed to handle repetitive control tasks, while other control techniques don't typically take advantage of task repetition—under a repeatable control environment, repeating the same feedback would yield the same control performance. In contrast, by incorporating learning, ILC is able to improve the control performance iteratively. Second, the control objective is different. ILC aims at achieving perfect tracking over the whole operational interval. Most control methods aim to achieve asymptotic convergence in tracking accuracy over time. Third, ILC is a feedforward control method if viewed in the time domain. The plant shown in Figure 1.1 is a generalized plant, that is, it can actually include a feedback loop. ILC can be used to further improve the performance of the generalized plant. As such, the generalized plant could be made stable in the time domain, which is helpful in guaranteeing transient response while learning takes place. Last but not

Iterative Learning Control for Multi-agent Systems Coordination, First Edition.
Shiping Yang, Jian-Xin Xu, Xuefang Li, and Dong Shen.
© 2017 John Wiley & Sons Singapore Pte. Ltd. Published 2017 by John Wiley & Sons Singapore Pte. Ltd.

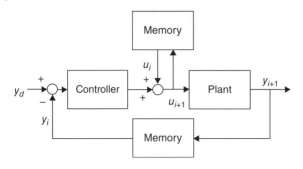

Figure 1.1 The framework of ILC.

least, ILC is a partially model-free control method. As long as an appropriate learning gain is chosen, perfect tracking can be achieved without using a perfect plant model.

Generally speaking, there are two main frameworks for ILC, namely contraction-mapping (CM)-based and composite energy function (CEF)-based approaches. A CM-based iterative learning controller has a very simple structure and is easy to implement. A correction term in the controller is constructed from the output tracking error; to ensure convergence, an appropriate learning gain is selected based on system gradient information in place of an accurate dynamic model. As a partially model-free control method, CM-based ILC is applicable to non-affine-in-input systems. These features are highly desirable in practice as there are plenty of data available in industry processes but there is a shortage of accurate system models. CM-based ILC has been adopted in many applications, for example X-Y tables, chemical batch reactors, laser cutting systems, motor control, water heating systems, freeway traffic control, wafer manufacturing, and so on (Ahn *et al.*, 2007a). A limitation of CM-based ILC is that it is only applicable to global Lipschitz continuous (GLC) systems. The GLC condition is required by ILC in order to form a contractive mapping, and rule out the finite escape time phenomenon. In comparison, CEF-based ILC, a complementary approach to CM-based ILC, applies a Lyapunov-like method to design learning rules. CEF is an effective method to handle locally Lipschitz continuous (LLC) systems, because system dynamics is used in the design of learning and feedback mechanisms. It is, however, worthwhile pointing out that in CM-based ILC, the learning mechanism only requires output signals, while in CEF-based ILC, full state information is usually required. CEF-based ILC has been applied in satellite trajectory keeping (Ahn *et al.*, 2010) and robotic manipulator control (Tayebi, 2004; Tayebi and Islam, 2006; Sun *et al.*, 2006).

This book follows the two main frameworks and investigates the multi-agent coordination problem using ILC. To illustrate the underlying idea and properties of ILC, we start with a simple ILC system.

Consider the following linear time-invariant dynamics:

$$\dot{x}_i(t) = ax_i(t) + u_i(t), \ t \in [0, T], \tag{1.1}$$

where i is the iteration index, a is an unknown constant parameter, and T is the trial length. Let the target trajectory be $x_d(t)$, which is generated by

$$\dot{x}_d(t) = ax_d(t) + u_d(t), \ t \in [0, T], \tag{1.2}$$

with $u_d(t)$ is the desired control signal. The control objective is to tune $u_i(t)$ such that without any prior knowledge about the parameter a, the tracking error

$e_i(t) \triangleq x_d(t) - x_i(t)$ can converge to zero as the iteration number increases, that is, $\lim_{i \to \infty} e_i(t) = 0$ for $t \in [0, T]$.

We perform the ILC controller design and convergence analysis for this simple control problem under the frameworks of both CM-based and CEF-based approaches, in order to illustrate the basic concepts in ILC and analysis techniques. To restrict our discussion, the following assumptions are imposed on the dynamical system (1.1).

Assumption 1.1   The identical initialization condition holds for all iterations, that is, $x_i(0) = x_d(0)$, $\forall i \in \mathbb{N}$.

Assumption 1.2   For $\forall x_d(t), t \in [0, T]$, there exists a $u_d(t), t \in [0, T]$ such that $u_i(t) \to u_d(t)$ implies $x_i(t) \to x_d(t), t \in [0, T]$.

1.1.1 Contraction-Mapping Approach

Under the framework of CM-based methodology, we apply the following D-type updating law to solve the trajectory tracking problem:

$$u_{i+1} = u_i + \gamma \dot{e}_i, \tag{1.3}$$

where $\gamma > 0$ is the learning gain to be determined. Our objective is to show that the ILC law (1.3) can converge to the desired u_d, which implies the convergence of the tracking error $e_i(t), t \in [0, T]$ as i increases.

Define $\Delta u_i = u_d - u_i$. First we can derive the relation

$$\begin{aligned} \Delta u_{i+1} &= u_d - u_{i+1} \\ &= u_d - u_i - \gamma \dot{e}_i \\ &= \Delta u_i - \gamma \dot{e}_i. \end{aligned} \tag{1.4}$$

Furthermore, the state error dynamics is given by

$$\begin{aligned} \dot{e}_i &= \dot{x}_d - \dot{x}_i \\ &= (ax_d + u_d) - (ax_i + u_i) \\ &= ae_i + \Delta u_i. \end{aligned} \tag{1.5}$$

Combining (1.4) and (1.5) gives:

$$\begin{aligned} \Delta u_{i+1} &= \Delta u_i - \gamma \dot{e}_i \\ &= (1 - \gamma)\Delta u_i - a\gamma e_i. \end{aligned} \tag{1.6}$$

Integrating both sides of the state error dynamics and using Assumption 1.1 yields

$$\begin{aligned} e_i(t) &= e_i(0) + \int_0^t e^{a(t-\tau)} \Delta u_i(\tau) d\tau \\ &= \int_0^t e^{a(t-\tau)} \Delta u_i(\tau) d\tau. \end{aligned} \tag{1.7}$$

Then, substituting (1.7) into (1.6), we obtain

$$\Delta u_{i+1} = (1 - \gamma)\Delta u_i - a\gamma \int_0^t e^{a(t-\tau)} \Delta u_i(\tau) d\tau. \tag{1.8}$$

Taking λ-norm on both sides of (1.8) gives

$$|\Delta u_{i+1}|_\lambda \le |1 - \gamma||\Delta u_i|_\lambda + a\gamma \frac{1 - e^{-(\lambda-a)T}}{\lambda - a}|\Delta u_i|_\lambda,$$

$$\triangleq \rho_1 |\Delta u_i|_\lambda, \tag{1.9}$$

where $\rho_1 \triangleq |1 - \gamma| + a\gamma \frac{1-e^{-(\lambda-a)T}}{\lambda-a}$, and the λ-norm is defined as

$$|\Delta u_{i+1}|_\lambda = \sup_{t\in[0,T]} e^{-\lambda t}|\Delta u_{i+1}(t)|.$$

The λ-norm is just a time weighted norm and is used to simplify the derivation. It will be formally defined in Section 1.4.

If $|1 - \gamma| < 1$ in (1.9), it is possible to choose a sufficiently large $\lambda > a$ such that $\rho_1 < 1$. Therefore, (1.9) implies that $\lim_{t\to\infty} |\Delta u_i|_\lambda = 0$, namely $\lim_{t\to\infty} u_i(t) = u_d(t)$, $t \in [0, T]$.

1.1.2 Composite Energy Function Approach

In this subsection, the ILC controller will be developed and analyzed under the framework of CEF-based approach. First of all, the error dynamics of the system (1.1) can be expressed as follows:

$$\dot{e}_i(t) = -ax_i + \dot{x}_d - u_i, \tag{1.10}$$

where x_d is the target trajectory.

Let k be a positive constant. By applying the control law

$$u_i = -ke_i + \dot{x}_d - \hat{a}_i(t)x_i \tag{1.11}$$

and the parametric updating law $\forall t \in [0, T]$,

$$\hat{a}_i(t) = \hat{a}_{i-1}(t) + x_i e_i, \quad \hat{a}_{-1}(t) = 0, \tag{1.12}$$

we can obtain the convergence of the tracking error e_i as i tends to infinity.

In order to facilitate the convergence analysis of the proposed ILC scheme, we introduce the following CEF:

$$E_i(t) = \frac{1}{2}e_i^2(t) + \frac{1}{2}\int_0^t \phi_i^2(\tau)d\tau, \tag{1.13}$$

where $\phi_i(t) \triangleq \hat{a}_i - a$ is the estimation error of the unknown parameter a.

The difference of E_i is

$$\Delta E_i(t) = E_i - E_{i-1}$$

$$= \frac{1}{2}e_i^2 + \frac{1}{2}\int_0^t (\phi_i^2 - \phi_{i-1}^2)d\tau - \frac{1}{2}e_{i-1}^2. \tag{1.14}$$

By using the identical initialization condition as in Assumption 1.1, the error dynamics (1.10), and the control law (1.11), the first term on the right hand side of (1.14) can be calculated as

$$\frac{1}{2}e_i^2 = \int_0^t e_i \dot{e}_i d\tau$$

$$= \int_0^t e_i(-\dot{x}_d + ax_i + u_i)d\tau$$

$$= \int_0^t (-\phi_i x_i e_i - ke_i^2)d\tau. \tag{1.15}$$

In addition, the second term on the right hand side of (1.14) can be expressed as

$$\frac{1}{2}\int_0^t (\phi_i^2 - \phi_{i-1}^2)d\tau = \frac{1}{2}\int_0^t (\hat{a}_{i-1} - \hat{a}_i)(2a - 2\hat{a}_i + \hat{a}_i - \hat{a}_{i-1})d\tau$$

$$= \int_0^t (\phi_i x_i e_i - \frac{1}{2}x_i^2 e_i^2)d\tau, \tag{1.16}$$

where the updating law (1.12) is applied. Clearly, $\phi_i x_i e_i$ appears in (1.15) and (1.16) with opposite signs. Combining (1.14), (1.15), and (1.16) yields

$$\Delta E_i(t) = -k\int_0^t e_i^2 d\tau - \frac{1}{2}\int_0^t x_i^2 e_i^2 d\tau - \frac{1}{2}e_{i-1}^2$$

$$\leq -\frac{1}{2}e_{i-1}^2 < 0. \tag{1.17}$$

The function E_i is a monotonically decreasing sequence, hence is bounded if E_0 is bounded.

Now, let us show the boundedness of E_0. For the linear plant (1.1) or in general GLC plants, there will be no finite escape time, thus E_0 is bounded. For local Lipschitz continuous plants, ILC designed under CEF guarantees there is no finite escape time (see Xu and Tan, 2003, chap. 7), thus E_0 is bounded. Hence, the boundedness of $E_0(t)$ over $[0, T]$ is obtained.

Consider a finite sum of ΔE_i,

$$\sum_{j=1}^{i} \Delta E_j = \sum_{j=1}^{i}(E_j - E_{j-1}) = E_i - E_0, \tag{1.18}$$

and apply the inequality (1.17); we have:

$$E_i(t) = E_0(t) + \sum_{j=1}^{i} \Delta E_j$$

$$\leq E_0(t) - \frac{1}{2}\sum_{j=1}^{i} e_{j-1}^2. \tag{1.19}$$

Because of the positiveness of E_i and boundedness of E_0, $e_i(t)$ converges to zero in a pointwise fashion as i tends to infinity.

1.2   Introduction to MAS Coordination

In the past several decades, MAS coordination and control problems have attracted considerable attention from many researchers of various backgrounds due to their potential applications and cross-disciplinary nature. *Consensus* in particular is an important class of MAS coordination and control problems (Cao *et al.*, 2013). According to Olfati-Saber *et al.* (2007), in networks of agents (or dynamic systems), consensus means to reach an agreement regarding certain quantities of interest that are associated with all agents. Depending on the specific application, these quantities could be velocity, position, temperature, orientation, and so on. In a consensus realization, the control action of an agent is generated based on the information received or measured from its neighborhood.

Since the control law is a kind of distributed algorithm, it is more robust and scalable compared to centralized control algorithms.

The three main components in MAS coordination are the agent model, the information sharing topology, and the control algorithm or consensus algorithm.

Agent models range from simple single integrator model to complex nonlinear models. Consensus results on single integrators are reported by Jadbabaie *et al.* (2003), Olfati-Saber and Murray (2004), Moreau (2005), Ren *et al.* (2007), and Olfati-Saber *et al.* (2007). Double integrators are investigated in Xie and Wang (2005), Hong *et al.* (2006), Ren (2008a), and Zhang and Tian (2009). Results on linear agent models can be found in Xiang *et al.* (2009), Ma and Zhang (2010), Li *et al.* (2010), Huang (2011), and Wieland *et al.* (2011). Since the Lagrangian system can be used to model many practical systems, consensus has been extensively studied by means of the Lagrangian system. Some representative works are reported by Hou *et al.* (2009), Chen and Lewis (2011), Mei *et al.* (2011), and Zhang *et al.* (2014).

Information sharing among agents is one of the indispensable components for consensus seeking. Information sharing can be realized by direct measurement from on board sensors or communication through wireless networks. The information sharing mechanism is usually modeled by a graph. For simplicity in the early stages of consensus algorithm development, the communication graph is assumed to be fixed. However, a consensus algorithm that is robust or adaptive to topology variations is more desirable, since many practical conditions can be modeled as time-varying communications, for example asynchronous updating, or communication link failures and creations. As communication among agents is an important topic in the MAS literature, various communication assumptions and consensus results have been investigated by researchers (Moreau, 2005; Hatano and Mesbahi, 2005; Tahbaz-Salehi and Jadbabaie, 2008; Zhang and Tian, 2009). An excellent survey can be found in Fang and Antsaklis (2006). Since graph theory is seldom used in control theory and applications, a brief introduction to the topic is given in Appendix A.

A consensus algorithm is a very simple local coordination rule which can result in very complex and useful behaviors at the group level. For instance, it is widely observed that by adopting such a strategy, a school of fish can improve the chance of survival under the sea (Moyle and Cech, 2003). Many interesting coordination problems have been formulated and solved under the framework of consensus, for example distributed sensor fusion (Olfati-Saber *et al.*, 2007), satellite alignment (Ren and Beard, 2008), multi-agent formation (Ren *et al.*, 2007), synchronization of coupled oscillators (Ren, 2008b), and optimal dispatch in power systems (Yang *et al.*, 2013). The consensus problem is usually studied in the infinite time horizon, that is, the consensus is reached as time tends to infinity. However, some finite-time convergence algorithms are available (Cortex, 2006; Wang and Hong, 2008; Khoo *et al.*, 2009; Wang and Xiao, 2010; Li *et al.*, 2011). In the existing literature, most consensus algorithms are model based. By incorporating ILC into consensus algorithms, the prior information requirement from a plant model can be significantly reduced. This advantage will be shown throughout this book.

1.3 Motivation and Overview

In practice, there are many tasks requiring both repetitive executions and coordination among several independent entities. For example, it is useful for a group of satellites to orbit the earth in formation for positioning or monitoring purposes (Ahn *et al.*, 2010). Each satellite orbiting the earth is a repeated task, and the formation task fits perfectly in the ILC framework. Another example is the cooperative transportation of a heavy load by multiple mobile robots (Bai and Wen, 2010; Yufka *et al.*, 2010). In such kinds of task implementation, the robots have to maneuver in formation from the very beginning to the destination. The economic dispatch problem in power systems (Xu and Yang, 2013; Yang *et al.*, 2013) and formation control for ground vehicles with nonholonomic constraints (Xu *et al.*, 2011) also fall in this category. These observations motivate the study of multi-agent coordination control from the perspective of ILC.

As discussed in the previous subsection, the consensus tracking problem is an important multi-agent coordination problem, and many other coordination problems can be formulated and solved in this framework, such as the formation, cooperative search, area coverage, and synchronization problems. We chose consensus tracking as the main topic in this book. Here we briefly describe a prototype consensus tracking problem and illustrate the concepts of distributed tracking error which are used throughout the book. In the problem formulation, there is a single leader that follows a prescribed trajectory, and the leader's behavior is not affected by others in the network. There are many followers, and they can communicate with each other and with the leader agent. However, they may not know which one the leader is. Due to communication limitations, a follower is only able to communicate with its near neighbors. The control task is to design an appropriate *local* controller such that all the followers can track the leader's trajectory. A local controller means that an agent is only allowed to utilize local information. To illustrate these concepts, Figure 1.2 shows an example of a communication network. (Please see Appendix A for a revision of graph theory.) Each node in the graph represents an agent (agents will be modeled by dynamic systems in later chapters). Edges in the graph show the information flow. For instance, there is an edge starting from agent 2 and ending at agent 1, which means agent 1 is able to obtain information from agent 2. In this example there are two edges ending at agent 1. This implies that agent 1 can utilize the information received from agents 0 and 2. Let x_i denote the variable of interest for the ith agent, for instance, position, velocity, orientation, temperature, pressure, and so on. The distributed error ξ_1 for agent 1 is defined as

$$\xi_1 = (x_0 - x_1) + (x_2 - x_1).$$

The distributed error ξ_1 will be used to construct the distributed learning rule.

Figure 1.2 Example of a network.

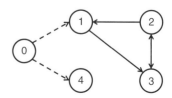

With this problem in mind, the main content of the book is summarized below.

1) In Chapter 2, a general consensus tracking problem is formulated for a group of global Lipschitz continuous systems. It is assumed that the communication is fixed and connected, and the perfect identical initialization condition (*iic*) constraint is satisfied as well. A D-type ILC rule is proposed for the systems to achieve perfect consensus tracking. By adoption of a graph dependent matrix norm, a local convergence condition is devised at the agent level. In addition, optimal learning gain design methods are developed for both directed and undirected graphs such that the λ-norm of tracking error converges at the fastest rate.

2) In Chapter 3, we investigate the robustness of the D-type learning rule against communication variations. It turns out that the controller is insensitive to iteration-varying topology. In the most general case, the learning controller is still convergent when the communication topology is uniformly strongly connected over the iteration domain.

3) In Chapter 4, the PD-type learning rule is proposed to deal with imperfect initialization conditions as it is difficult to ensure perfect initial conditions for all agents due to sparse information communication—hence only a few of the follower agents know the desired initial state. The new learning rule offers two main features. On the one hand, it can ensure controller convergence. On the other hand, the learning gain can be used to tune the final tracking performance.

4) In Chapter 5, a novel input sharing learning controller is developed. In the existing literature, when designing the learning controller, only the tracking error is incorporated in the control signal generation. However, if the follower agents can share their experience gained during the process, this may accelerate the learning speed. Using this idea, the new controller is developed for each agent by sharing its learned control input with its neighbors.

5) In Chapter 6, we apply the learning controller to a formation problem. The formation contains two geometric configurations. The two configurations are related by a high-order internal model (HOIM). Usually the ILC control task is fixed. The most challenging part of this class of problem is how to handle changes in configuration. By incorporating the HOIM into the learning controller, it is shown that, surprisingly, the agents are still able to learn from different tasks.

6) In Chapter 7, by combining the Lyapunov analysis method and contraction-mapping analysis, we explore the applicability of the P-type learning rule to several classes of local Lipschitz systems. Several sufficient convergence conditions in terms of Lyapunov criteria are derived. In particular, the P-type learning rule can be applied to a Lyapunov stable system with quadratic Lyapunov functions, an exponentially stable system, a system with bounded drift terms, and a uniformly bounded energy bounded state system under control saturation. The results greatly complement the existing literature. By using the results of this chapter, we can immediately extend the results in Chapters 2–5 to more general nonlinear systems.

7) In Chapter 8, the composite energy function method is utilized to design an adaptive learning rule to deal with local Lipschitz systems that can be modeled by system dynamics that are linear in parameters. With the help of a special parameterization method, the leader's trajectory can be treated as an iteration-invariant parameter

that all the followers can learn from local measurements. In addition, the initial rectifying action is applied to reduce the effect of imperfect initialization conditions. The method works for high-order systems as well.

8) Chapter 9 addresses the consensus problem of nonlinear multi-agent system (MAS) with state constraints. A novel type of barrier Lyapunov function (BLF) is adopted to deal with the bounded constraints. An ILC strategy is introduced to estimate the unknown parameter and basic control signal. To address the consensus problem comprehensively from both theoretical and practical viewpoints, five control schemes are designed in turn: the original adaptive scheme, a projection-based scheme, a smooth function based scheme as well as its alternative, and a dead-zone like scheme. The consensus convergence and constraints guarantee are strictly proved for each control scheme by using the barrier composite energy function (BCEF) approach.

9) Lagrangian systems have wide applications in practice. For example, industry robotic manipulators can be modeled as a Lagrangian system. In Chapter 10, we develop a set of distributed learning rules to synchronize networked Lagrangian systems. In the controller design, we fully utilize the inherent features of Lagrangian systems, and the controller works under a directed acyclic graph.

10) In Chapter 11, we focus our attention on discrete-time system and present a generalized iterative learning algorithm to solve an optimal power generation problem in a smart grid. Usually the optimal power dispatch problem is solved by centralized methods. Noticing that the optimal solution is achieved when the incremental costs for all power generators are equal, if we consider the incremental cost as the variable of interest, it may be possible to devise a distributed algorithm. Following this idea and by virtue of the distributed nature of the consensus algorithm, a hierarchical two-level algorithm is developed. The new learning algorithm is able to find the optimal solution, as well as taking power generator constraints and power line loss into account.

1.4 Common Notations in This Book

The set of real numbers is denoted by $\mathbb{R}$, and the set of complex numbers is denoted by $\mathbb{C}$. The set of natural numbers is denoted by $\mathbb{N}$, and $i \in \mathbb{N}$ is the number of iteration. For any $z \in \mathbb{C}$, $\mathfrak{R}(z)$ denotes its real part. For a given vector $\mathbf{x} = [x_1, x_2, \cdots, x_n]^T \in \mathbb{R}^n$, $|\mathbf{x}|$ denotes any l_p vector norm, where $1 \leq p \leq \infty$. In particular, $|\mathbf{x}|_1 = \sum_{k=1}^{n} |x_k|$, $|\mathbf{x}|_2 = \sqrt{\mathbf{x}^T \mathbf{x}}$, and $|\mathbf{x}|_\infty = \max_{k=1,\ldots,n} |x_k|$. For any matrix $A \in \mathbb{R}^{n \times n}$, $|A|$ is the induced matrix norm. $\rho(A)$ is its spectral radius. Moreover, $\otimes$ denotes the Kronecker product, and I_m is the $m \times m$ identity matrix.

Let $C^m[0, T]$ denote a set consisting of all functions whose mth derivatives are continuous on the finite-time interval $[0, T]$. For any function $\mathbf{f}(\cdot) \in C[0, T]$, the supremum norm is defined as $\|\mathbf{f}\| = \sup_{t \in [0,T]} |\mathbf{f}(t)|$. Let λ be a positive constant, the time weighted norm (λ-norm) is defined as $\|\mathbf{f}\|_\lambda = \sup_{t \in [0,T]} e^{-\lambda t} |\mathbf{f}(t)|$.

2

Optimal Iterative Learning Control for Multi-agent Consensus Tracking

2.1 Introduction

The idea of using ILC for multi-agent coordination first appears in Ahn and Chen (2009), where the multi-agent formation control problem is studied for a group of global Lipschitz nonlinear systems, in which the communication graph is identical to the formation structure. When the tree-like formation is considered, perfect formation control can be achieved. Liu and Jia (2012) improve the control performance in Ahn *et al.* (2010). The formation structure can be independent of the communication topology, and time-varying communication is assumed in Liu and Jia (2012). The convergence condition is specified at the group level by a matrix norm inequality, and the learning gain can be designed by solving a set of linear matrix inequalities (LMIs). It is not clear under what conditions the set of LMIs admit a solution, and there is no insight as to how the communication topologies relate to the convergence condition. In Meng and Jia (2012), the idea of terminal ILC (Xu *et al.*, 1999) is brought into the consensus problem. A finite-time consensus problem is formulated for discrete-time linear systems in the ILC framework. It is shown that all agents reach consensus at the terminal time as the iteration number tends to infinity. In Meng *et al.* (2012), the authors extend the terminal consensus problem in their previous work to track a time-varying reference trajectory over the entire finite-time interval. A unified ILC algorithm is developed for both discrete-time and continuous-time linear agents. Necessary and sufficient conditions in the form of spectral radii are derived to ensure the convergence properties. Shi *et al.* (2014) develop a learning controller for second-order MAS to execute formation control using a similar approach.

In this chapter, we study the consensus tracking problem for a group of time-varying nonlinear dynamic agents, where the nonlinear terms satisfy the global Lipschitz continuous condition. For simplicity, a number of conditions in the problem formulation are idealized, for example the communication graph is assumed to be fixed, the initial conditions are assumed to be perfect, and so on. These conditions will be relaxed in subsequent chapters.

Two-dimensional theory (Chow and Fang, 1998) is an effective tool for analyzing ILC learning rules, and Meng *et al.* (2012) successfully apply this method to the multi-agent coordination problem. However it is only applicable to linear systems. As the system dynamics here are nonlinear, it is not possible to apply two-dimensional system theory in order to analyze and design the controller. To overcome this difficulty, we propose the concept of a graph dependent matrix norm. As a result of the newly defined

Iterative Learning Control for Multi-agent Systems Coordination, First Edition.
Shiping Yang, Jian-Xin Xu, Xuefang Li, and Dong Shen.

matrix norm, we are able to obtain the convergence results for global Lipschitz nonlinear systems. Usually the convergence condition is specified at the group level in the form of a matrix norm inequality, and the learning gain is designed by solving a set of LMIs. Owing to the graph dependent matrix norm, the convergence condition can be expressed at the individual agent level in the form of spectral radius inequalities, which are related to the eigenvalues associated with the communication graph. This shows that these eigenvalues play crucial roles in the convergence condition. In addition, the results are less conservative than the matrix norm inequality since the spectral radius of a matrix is less or equal to its matrix norm. Furthermore, by using the graph dependent matrix norm and λ-norm analysis, the learning controller design can be extended to heterogeneous systems. The convergence condition obtained motivates us to consider optimal learning gain designs which can impose the tightest bounding functions for actual tracking errors.

The rest of this chapter is organized as follows. In Section 2.2, notations and some useful results are introduced. Next, the consensus tracking problem for heterogeneous agents is formulated. Then, learning control laws are developed in Section 2.3, for both homogeneous and heterogeneous agents. Next, optimal learning design methods are proposed in Section 2.4, where optimal designs for undirected and directed graphs are explored. Then, an illustrative example for heterogeneous agents under a fixed directed graph is given in Section 2.5 to demonstrate the efficacy of the proposed algorithms. Finally, we draw conclusions for the chapter in Section 2.6.

2.2 Preliminaries and Problem Description

2.2.1 Preliminaries

Graph theory (Biggs, 1994) is an instrumental tool for describing the communication topology among agents in MAS. The basic terminologies and some properties of algebraic graph theory are revisited in Appendix A, in which the communication topology is modeled by the graph defined there. In particular, the vertex set $\mathcal{V}$ represents the agent index and the edge set $\mathcal{E}$ describes the information flow among agents.

For simplicity, a 0–1 weighting is adopted in the graph adjacency matrix $\mathcal{A}$. However, any positive weighted adjacency matrix preserves the convergence results. The strength of the weights can be interpreted as the reliability of information in the communication channels. In addition, positive weights represent collaboration among agents, whereas negative weights represent competition among agents. For example, Altafini (2013) shows that consensus can be reached on signed networks but the consensus values have opposite signs. If the controller designer has the freedom to select the weightings in the adjacency matrix, Xiao and Boyd (2004) demonstrate that some of the edges may take negative weights in order to achieve the fastest convergence rate in a linear average algorithm. Although interesting, negative weighting is outside the scope of this book.

The following propositions and lemma lay the foundations for the convergence analysis in the main results.

Proposition 2.1   For any given matrix $M \in \mathbb{R}^{n \times n}$ satisfying $\rho(M) < 1$, there exists at least one matrix norm $| \cdot |_S$ such that $\lim_{k \to \infty} \left(|M|_S \right)^k = 0$.

Proposition 2.1 is an extension of Lemma 5.6.10 in Horn and Johnson (1985). The proof is given in Appendix B.2 as the idea in the proof will be used to prove Theorem 2.1 and illustrate the graph dependent matrix norm.

Proposition 2.2 (Horn and Johnson, 1985, pp. 297) For any matrix norm $| \cdot |_S$, there exists at least one compatible vector norm $| \cdot |_s$, and for any $M \in \mathbb{R}^{n \times n}$ and $\mathbf{x} \in \mathbb{R}^n$, $|M\mathbf{x}|_s \leq |M|_S |\mathbf{x}|_s$.

The following Propositions (2.3, 2.4), and Lemma (2.1) will be utilized in the optimal learning gain designs.

Proposition 2.3 (Xu and Tan, 2002b) Denoting the compact set $\mathcal{I} = [\alpha_1, \alpha_2]$, where $0 < \alpha_1 < \alpha_2 < +\infty$, the cost function

$$J = \min_{\gamma \in \mathbb{R}} \max_{d \in \mathcal{I}} |1 - d\gamma|$$

reaches its minimum value $\frac{\alpha_2 - \alpha_1}{\alpha_2 + \alpha_1}$ when $\gamma^* = \frac{2}{\alpha_2 + \alpha_1}$.

Proposition 2.4 maximum modulus theorem (Zhou and Doyle, 1998) Let $f(z)$ be a continuous complex-value function defined on a compact set $\mathcal{Z}$, and analytic on the interior of $\mathcal{Z}$, then $|f(z)|$ cannot attain the maximum in the interior of $\mathcal{Z}$ unless $f(z)$ is a constant.

By using Proposition 2.4, Lemma 2.1 is proved in Appendix B.3.

Lemma 2.1 When $\gamma^* = \alpha_1 / \alpha_2^2$, the following min-max problem reaches its optimal value

$$\min_{\gamma \in \mathbb{R}} \max_{\alpha_1 < a < \sqrt{a^2 + b^2} < \alpha_2} |1 - \gamma(a + jb)| = \frac{\sqrt{\alpha_2^2 - \alpha_1^2}}{\alpha_2}.$$

2.2.2 Problem Description

Consider a group of N heterogeneous time-varying dynamic agents that work in a repeatable control environment. Their interaction topology is depicted by graph $\mathcal{G} = (\mathcal{V}, \mathcal{E}, \mathcal{A})$, which is iteration-invariant. At the ith iteration, the dynamics of the jth agent take the following form:

$$\begin{cases} \dot{\mathbf{x}}_{i,j}(t) = \mathbf{f}_j(t, \mathbf{x}_{i,j}(t)) + B_j(t)\mathbf{u}_{i,j}(t) \\ \mathbf{y}_{i,j}(t) = C_j(t)\mathbf{x}_{i,j}(t) \end{cases}, \forall t \in [0, T], \ \forall j \in \mathcal{V}, \tag{2.1}$$

with initial condition $\mathbf{x}_{i,j}(0)$. Here $\mathbf{x}_{i,j}(t) \in \mathbb{R}^{n_j}$ is the state vector, $\mathbf{y}_{i,j}(t) \in \mathbb{R}^m$ is the output vector, $\mathbf{u}_{i,j}(t) \in \mathbb{R}^{p_j}$ is the control input. For any $j = 1, 2, \ldots, N$, the unknown nonlinear function $\mathbf{f}_j(\cdot, \cdot)$ satisfies the global Lipschitz continuous condition with respect to $\mathbf{x}$ uniformly in $t, \forall t \in [0, T]$. In addition, the time-varying matrices $B_j(t)$ and $C_j(t)$ satisfy that $B_j(t) \in C^1[0, T]$ and $C_j(t) \in C^1[0, T]$.

The desired consensus tracking trajectory is denoted by $\mathbf{y}_d(t) \in C^1[0, T]$. Meanwhile, the state of each agent is not measurable. The only information available is the output signal of each agent.

Unlike the traditional tracking problem in ILC, in which each agent should know the desired trajectory, here $\mathbf{y}_d(t)$ is only accessible to a subset of agents. We can think of the desired trajectory as a (virtual) leader, and index it by vertex 0 in the graph representation. Thus, the complete information flow can be described by another graph $\overline{\mathcal{G}} = (\mathcal{V} \cup \{0\}, \overline{\mathcal{E}}, \overline{\mathcal{A}})$, where $\overline{\mathcal{E}}$ is the edge set and $\overline{\mathcal{A}}$ is the weighted adjacency matrix of $\overline{\mathcal{G}}$.

Let $\xi_{i,j}(t)$ denote the distributed information measured or received by the jth agent at the ith iteration. More specifically,

$$\xi_{i,j}(t) = \sum_{k \in \mathcal{N}_j} a_{j,k}(\mathbf{y}_{i,k}(t) - \mathbf{y}_{i,j}(t)) + d_j(\mathbf{y}_d(t) - \mathbf{y}_{i,j}(t)), \tag{2.2}$$

where $a_{j,k}$ is the (j,k)th entry in the adjacency matrix $\mathcal{A}$, $\mathcal{N}_j$ is the neighborhood set of the jth agent, $\mathbf{y}_{i,j}(t)$ is the output of the jth agent at the ith iteration, $d_j = 1$ if agent j can access the desired trajectory, that is, there is an edge from the virtual leader to the jth agent or $(0,j) \in \overline{\mathcal{E}}$, and $d_j = 0$ otherwise. The tracking error is defined as $\mathbf{e}_{i,j}(t) \triangleq \mathbf{y}_d(t) - \mathbf{y}_{i,j}(t)$.

The control objective is to design an appropriate iterative learning law such that the output from each agent converges to the desired trajectory $\mathbf{y}_d(t)$ when only some of the agents know the desired trajectory.

To simplify the analysis, the following assumptions are used.

Assumption 2.1   For any $j = 1, 2, \ldots, N$, the unknown nonlinear term $\mathbf{f}_j(t, \mathbf{x})$ satisfies

$$|\mathbf{f}_j(t, \mathbf{x}_1) - \mathbf{f}_j(t, \mathbf{x}_2)| \le l_j |\mathbf{x}_1 - \mathbf{x}_2|, \text{ for any } \mathbf{x}_1, \mathbf{x}_2 \in \mathbb{R}^{n_j},$$

where l_j is a positive constant.

Remark 2.1   In the existing literature, CM-based ILC is only applicable to global Lipschitz systems. Extension to local Lipschitz systems remains an open question. Two possible research directions are available. If the nonlinear terms can be linearly parameterized, CEF-based ILC (Xu and Tan, 2003) can be applied to overcome the global Lipschitz assumption. The other method makes use of the stability properties of system dynamics. By combining Lyapunov and CM analysis methods, it is possible to extend CM-based ILC to certain classes of local Lipschitz continuous systems. This kind of methodology will be explored in Chapter 7.

Assumption 2.2 $C_j(t)B_j(t)$ is of full row rank for all $t \in [0, T]$.

Remark 2.2   The requirement that $C_j(t)B_j(t)$ is of full row rank for any $j = 1, 2, \ldots, N$ and any $t \in [0, T]$ can be relaxed if using the higher order derivatives of $\xi_{i,j}(t)$ (if they exist) in the learning updating law. The proof technique will be very similar. If the higher order derivatives do not exist, some smooth approximations of these higher order derivatives can be applied.

Assumption 2.3   The communication graph $\overline{\mathcal{G}}$ contains a spanning tree with the (virtual) leader being the root.

Remark 2.3   Assumption 2.3 is a necessary communication requirement for the solvability of the consensus tracking problem. If there is an isolated agent, it is impossible for that agent to follow the leader's trajectory as it does not even know the control objective. It is noted that the original communication graph $\mathcal{G}$ does not necessarily contain a spanning tree. By selecting a (virtual) leader and its communications carefully, the proposed updating law can still work in such a situation.

Furthermore, the following identical initialization condition (*iic*) is needed.

Assumption 2.4   The systems are reset to the same initial state after each execution, and $\mathbf{e}_{i,j}(0) = 0$ for any $j = 1, 2, \ldots, N, i \in \mathbb{N}$.

Remark 2.4   The *iic* is a standard assumption in ILC design to ensure perfect tracking performance. It is possible to remove this condition with a sacrifice in tracking performance, but this requires either extra system information or additional control mechanisms, for instance the initial state learning rule (Chen *et al.*, 1999) and initial rectifying action (Sun and Wang, 2002). Note that without perfect initial conditions, perfect tracking can never be achieved. More discussions on various initial conditions in the learning context can be found in Park *et al.* (1999), Xu and Yan (2005), Chi *et al.* (2008) and references therein. Note that only the output of each agent is required to start from the same initial value as $\mathbf{y}_d(0)$. For example, in many applications the state of the system includes both position and velocity, whereas the output is just the velocity information. In such a situation, it is very natural to assume that the output has zero initial velocity as the desired trajectory. The initial condition problem will be explored further in Chapters 4 and 8.

2.3   Main Results

In the consensus literature, the consensus problem is usually studied for a group of identical agents. In contrast, the problem formulation presented in systems (2.1) is very general, with all the parameters being agent dependent.

For simplicity, the learning law is first designed for MAS with identical agents. Then the results will be extended to heterogeneous systems (2.1).

2.3.1   Controller Design for Homogeneous Agents

Assume that in (2.1), each agent has identical dynamics, that is, $\mathbf{f}_j(t, \mathbf{x}) = \mathbf{f}(t, \mathbf{x})$, $C_j(t) = C(t)$, and $B_j(t) = B(t)$ for all $j = 1, 2, \ldots, N$.

The following D-type updating law is used to solve the consensus tracking problem:

$$\mathbf{u}_{i+1,j}(t) = \mathbf{u}_{i,j}(t) + \Gamma(t)\dot{\xi}_{i,j}(t), \quad \mathbf{u}_{0,j}(t) \equiv 0, \tag{2.3}$$

where $\Gamma(t) \in C^1[0, T]$ is a time-varying learning gain matrix to be designed.

Remark 2.5   The updating law (2.3) sets zero initial condition for $\mathbf{u}_{0,j}(t)$ for simplicity. Feedback laws can be used to construct $\mathbf{u}_{0,j}(t)$ such that the systems are stable in the time domain, which may be helpful for the transient performance in the learning process.

The distributed measurement in (2.2) can be rewritten in terms of the tracking errors as

$$\xi_{i,j}(t) = \sum_{k \in N_j} a_{j,k}(\mathbf{e}_{i,j}(t) - \mathbf{e}_{i,k}(t)) + d_j \mathbf{e}_{i,j}(t). \tag{2.4}$$

Define three column stack vectors in the ith iteration $\mathbf{x}_i(t) = [\mathbf{x}_{i,1}(t)^T, \mathbf{x}_{i,2}(t)^T, \dots,$ $\mathbf{x}_{i,N}(t)^T]^T$, $\mathbf{e}_i(t) = [\mathbf{e}_{i,1}(t)^T, \mathbf{e}_{i,2}(t)^T, \dots, \mathbf{e}_{i,N}(t)^T]^T$, and $\xi_i(t) = [\xi_{i,1}(t)^T, \xi_{i,2}(t)^T, \dots,$ $\xi_{i,N}(t)^T]^T$. Consequently, (2.4) can be written in a compact form,

$$\xi_i(t) = ((L + D) \otimes I_m)\mathbf{e}_i(t), \tag{2.5}$$

where L is the Laplacian matrix of graph $\mathcal{G}$, and $D \triangleq \operatorname{diag}(d_1, d_2, \dots, d_N)$.

By using (2.5), the updating law (2.3) can be rewritten in terms of the tracking errors:

$$\mathbf{u}_{i+1}(t) = \mathbf{u}_i(t) + ((L + D) \otimes \Gamma(t))\dot{\mathbf{e}}_i(t). \tag{2.6}$$

For convenience, we define $\lambda_j, j = 1, 2, \dots, N$ as the jth eigenvalue of $L + D$.

The following theorem summarizes the convergence properties of the consensus algorithms (2.6).

Theorem 2.1 Assume that Assumptions 2.1–2.4 hold for the time-varying nonlinear systems (2.1) with the systems' parameters being identical. If the learning gain matrix $\Gamma(t)$ satisfies the following condition:

$$\max_{j=1,2..N} \max_{t \in [0, T]} \rho\left(I_m - \lambda_j \cdot C(t) \cdot B(t) \cdot \Gamma(t)\right) \leq \rho_0 < 1, \tag{2.7}$$

for some $\rho_0 \in (0, 1)$, then there exists a positive constant λ such that

$$\|\mathbf{e}_{i+1}\|_\lambda \leq \rho_0 \|\mathbf{e}_i\|_\lambda, \forall j = 1, 2, \dots, N, \tag{2.8}$$

which indicates that $\lim_{i \to \infty} \mathbf{y}_{i,j}(t) = \mathbf{y}_d(t)$ for all $t \in [0, T], j = 1, 2, \dots, N$.

Proof: The tracking error of the jth agent between two consecutive iterations can be expressed as

$$\begin{aligned} \mathbf{e}_{i+1,j}(t) &= \mathbf{y}_d(t) - \mathbf{y}_{i+1,j}(t) \\ &= \mathbf{e}_{i,j}(t) - (\mathbf{y}_{i+1,j}(t) - \mathbf{y}_{i,j}(t)), \end{aligned}$$

which can be written as the compact form,

$$\mathbf{e}_{i+1}(t) = \mathbf{e}_i(t) - (I_N \otimes C(t))(\mathbf{x}_{i+1}(t) - \mathbf{x}_i(t)). \tag{2.9}$$

The state difference $\mathbf{x}_{i+1}(t) - \mathbf{x}_i(t)$ can be calculated by integrating the system dynamics (2.1) along the time domain:

$$\mathbf{x}_{i+1}(t) - \mathbf{x}_i(t) \tag{2.10}$$

$$= \mathbf{x}_{i+1}(0) - \mathbf{x}_i(0) + \int_0^t \left(\bar{\mathbf{f}}(\tau, \mathbf{x}_{i+1}) - \bar{\mathbf{f}}(\tau, \mathbf{x}_i) + (I_N \otimes B(t))(\mathbf{u}_{i+1}(\tau) - \mathbf{u}_i(\tau))\right) d\tau$$

where $\bar{\mathbf{f}}(t, \mathbf{x}_i) \triangleq [\mathbf{f}(t, \mathbf{x}_{i,1})^T, \mathbf{f}(t, \mathbf{x}_{i,2})^T, \dots, \mathbf{f}(t, \mathbf{x}_{i,N})^T]^T$.

According to Assumption 2.4, and using the updating law (2.6), it yields

$$
\mathbf{x}_{i+1}(t) - \mathbf{x}_i(t)
$$
$$
= \int_0^t \left(\bar{\mathbf{f}}(\tau, \mathbf{x}_{i+1}) - \bar{\mathbf{f}}(\tau, \mathbf{x}_i) + (L+D) \otimes (B(\tau)\Gamma(\tau))\dot{\mathbf{e}}_i(\tau) \right) d\tau. \tag{2.11}
$$

Apply integration by parts to the last term in (2.11), noting Assumption 2.4, gives

$$
\int_0^t (L+D) \otimes (B(\tau)\Gamma(\tau))\dot{\mathbf{e}}_i(\tau)\, d\tau
$$
$$
= (L+D) \otimes (B(t)\Gamma(t))\mathbf{e}_i(t) + \int_0^t (L+D) \otimes \left(\frac{d}{d\tau} B(\tau)\Gamma(\tau) \right) \mathbf{e}_i(\tau)\, d\tau. \tag{2.12}
$$

Substituting (2.12) into (2.11), we can obtain

$$
\mathbf{x}_{i+1}(t) - \mathbf{x}_i(t) = (L+D) \otimes B(t)\Gamma(t)\mathbf{e}_i(t) + \int_0^t [\bar{\mathbf{f}}(\tau, \mathbf{x}_{i+1}) - \bar{\mathbf{f}}(\tau, \mathbf{x}_i)]\, d\tau
$$
$$
+ \int_0^t \left((L+D) \otimes \left(\frac{d}{d\tau} B(\tau)\Gamma(\tau) \right) \mathbf{e}_i(\tau) \right) d\tau. \tag{2.13}
$$

Then, substituting (2.13) in (2.9) yields

$$
\mathbf{e}_{i+1}(t) = (I_{mN} - (L+D) \otimes C(t)B(t)\Gamma(t))\mathbf{e}_i(t) - (I_N \otimes C(t)) \left(\int_0^t [\bar{\mathbf{f}}(\tau, \mathbf{x}_{i+1}) \right.
$$
$$
\left. -\bar{\mathbf{f}}(\tau, \mathbf{x}_i)]\, d\tau + \int_0^t \left((L+D) \otimes \left(\frac{d}{d\tau} B(\tau)\Gamma(\tau) \right) \mathbf{e}_i(\tau) \right) d\tau \right). \tag{2.14}
$$

For simple presentation, the following constants are used in the sequel:

$$
b_1 \triangleq \| I_N \otimes C(t) \|,
$$
$$
b_2 \triangleq \left\| (L+D) \otimes \left(\frac{d}{dt} B(t)\Gamma(t) \right) \right\|,
$$
$$
b_3 \triangleq \| (L+D) \otimes B(t)\Gamma(t) \|.
$$

Taking the norm on both sides of (2.14), and noticing the Lipschitz condition in Assumption 2.1, we have

$$
|\mathbf{e}_{i+1}(t)| \leq \| I_{mN} - (L+D) \otimes C(t)B(t)\Gamma(t) \|\, |\mathbf{e}_i(t)|
$$
$$
+ b_1 \bar{k}_f \int_0^t |\mathbf{x}_{i+1}(\tau) - \mathbf{x}_i(\tau)|\, d\tau + b_1 b_2 \int_0^t |\mathbf{e}_i(\tau)|\, d\tau, \tag{2.15}
$$

where $\bar{k}_f$ is the global Lipschitz constant of $\bar{\mathbf{f}}(\cdot, \cdot)$.

Furthermore, taking the λ-norm on both sides of (2.15) yields

$$
\| \mathbf{e}_{i+1} \|_\lambda \leq \| I_{mN} - (L+D) \otimes C(t)B(t)\Gamma(t) \|\, \| \mathbf{e}_i \|_\lambda
$$
$$
+ b_1 \bar{k}_f \| \mathbf{x}_{i+1} - \mathbf{x}_i \|_\lambda \frac{1 - e^{-\lambda T}}{\lambda} + b_1 b_2 \| \mathbf{e}_i \|_\lambda \frac{1 - e^{-\lambda T}}{\lambda}. \tag{2.16}
$$

To derive the convergence property of $\| \mathbf{e}_i \|_\lambda$ along the iteration axis, it suffices to explore the relation between $\| \mathbf{x}_{i+1} - \mathbf{x}_i \|_\lambda$ and $\| \mathbf{e}_i \|_\lambda$.

Taking the norm on both sides of (2.13), and applying the Lipschitz condition for system nonlinearity $\bar{\mathbf{f}}(\cdot, \cdot)$, it can be shown that

$$|\mathbf{x}_{i+1}(t) - \mathbf{x}_i(t)| \leq b_3 |\mathbf{e}_i(t)| + \bar{k}_f \int_0^t |\mathbf{x}_{i+1}(\tau) - \mathbf{x}_i(\tau)|\, d\tau + b_2 \int_0^t |\mathbf{e}_i(\tau)|\, d\tau. \quad (2.17)$$

Next, taking the λ-norm on both sides of (2.17) yields

$$\|\mathbf{x}_{i+1} - \mathbf{x}_i\|_\lambda \leq b_3 \|\mathbf{e}_i\|_\lambda + \bar{k}_f \frac{1 - e^{-\lambda T}}{\lambda} \|\mathbf{x}_{i+1} - \mathbf{x}_i\|_\lambda + b_2 \frac{1 - e^{-\lambda T}}{\lambda} \|\mathbf{e}_i\|_\lambda. \quad (2.18)$$

Rearrange $\|\mathbf{x}_{i+1} - \mathbf{x}_i\|_\lambda$ and $\|\mathbf{e}_i\|_\lambda$ in (2.18),

$$\|\mathbf{x}_{i+1} - \mathbf{x}_i\|_\lambda \leq (b_3 + b_2 \frac{1 - e^{-\lambda T}}{\lambda})(1 - \bar{k}_f \frac{1 - e^{-\lambda T}}{\lambda})^{-1} \|\mathbf{e}_i\|_\lambda. \quad (2.19)$$

Substituting (2.19) in (2.16), $\|\mathbf{e}_{i+1}\|_\lambda$ becomes

$$\|\mathbf{e}_{i+1}\|_\lambda \leq \|I_{mN} - (L + D) \otimes C(t)B(t)\Gamma(t)\| \|\mathbf{e}_i\|_\lambda + \mathcal{O}(\lambda^{-1}) \|\mathbf{e}_i\|_\lambda, \quad (2.20)$$

where

$$\mathcal{O}(\lambda^{-1}) = b_1 \bar{k}_f (b_3 + b_2 \frac{1 - e^{-\lambda T}}{\lambda})(1 - \bar{k}_f \frac{1 - e^{-\lambda T}}{\lambda})^{-1} \frac{1 - e^{-\lambda T}}{\lambda} + b_1 b_2 \frac{1 - e^{-\lambda T}}{\lambda}.$$

As $\mathcal{O}(\lambda^{-1})$ is in the same order with λ^{-1}, hence, it can be made negligibly small by choosing a sufficiently large λ. As a result if

$$\|I_{mN} - (L + D) \otimes C(t)B(t)\Gamma(t)\| \leq \rho_0 < 1, \quad (2.21)$$

then there exists ρ_1 with $0 < \rho_0 < \rho_1 < 1$ such that $\|\mathbf{e}_{i+1}\|_\lambda \leq \rho_1 \|\mathbf{e}_i\|_\lambda$.

Define $M(t) \triangleq I_{mN} - (L + D) \otimes C(t)B(t)\Gamma(t)$. Based on Proposition 2.1, it is sufficient to design a suitable $\Gamma(t)$ such that $\rho(M(t)) \leq \rho_0$, for all $t \in [0, T]$, then the consensus tracking is fulfilled by (2.6). This is because when $\rho(M(t)) \leq \rho_0$ for all $t \in [0, T]$, we can always find an appropriate matrix norm such that $\|M\| \leq \rho_0$.

Note that $M(t) \in \mathbb{R}^{mN \times mN}$, and the condition $\rho(M(t)) \leq \rho_0, \forall t \in [0, T]$ is specified at the group level since it contains all the agents' dynamics and complete information about the communication topology. It does not make much difference in the homogeneous case. However, we will see later that it becomes more complex in the heterogeneous case. Next, let us derive the convergence condition at the agent level. For simplicity in what follows, the time argument is dropped when no confusion arises.

Following the concepts in Appendix B.2, $L + D$ can be decomposed in the following form:

$$\Delta = U^*(L + D)U,$$

where Δ is an upper triangular matrix with diagonal entries being the eigenvalues of $L + D$, U is an associated unitary matrix, and $*$ denotes the conjugate transpose.

Let the matrix norm operation in the above development be defined as below:

$$|\cdot| \triangleq |[(QU^*) \otimes I_m](\cdot)[(UQ^{-1}) \otimes I_m]|,$$

where Q is a constant matrix defined in Appendix B.2. Based on Proposition 2.2, there always exists a corresponding vector norm which is compatible to the previously defined matrix norm. Hence, all the derivations here remain valid.

Then, we have:

$$
\begin{aligned}
|M| \\
&= \left| [(QU^*) \otimes I_m][I_{mN} - (L+D) \otimes CB\Gamma][(UQ^{-1}) \otimes I_m] \right| \\
&= \left| I_{mN} - [(QU^*)(L+D)(UQ^{-1})] \otimes CB\Gamma \right| \\
&= \left| I_{mN} - (Q\Delta Q^{-1}) \otimes CB\Gamma \right| \\
&= \left| \begin{bmatrix}
I_m - \lambda_1 CB\Gamma & (\star) & (\star) & (\star) \\
0 & I_m - \lambda_2 CB\Gamma & (\star) & (\star) \\
\vdots & \vdots & \ddots & \vdots \\
0 & 0 & \cdots & I_m - \lambda_N CB\Gamma
\end{bmatrix} \right|,
\end{aligned}
$$

(2.22)

where $(\star)$ can be made arbitrarily small by choosing sufficiently large α in Q, as these terms are in the order of $\mathcal{O}(\alpha^{-k})$ where $k = 1, 2, \ldots, N - 1$.

By Assumption 2.3, the leader has a path to any follower agent; thus $L + D$ is nonsingular and all the eigenvalues have positive real parts (Ren and Beard, 2008). Together with (2.22), we can conclude that if

$$
\max_{j=1,2..N} \max_{t \in [0,T]} \rho \left(I_m - \lambda_j C(t)B(t)\Gamma(t) \right) \leq \rho_0,
$$

then the inequality (2.21) holds, that is, the tracking error $\mathbf{e}_i(t)$ converges to zero as i tends to infinity. ∎

Remark 2.6 In the proof of Theorem 2.1, the constant λ is only an analysis tool, and it can be chosen to be arbitrarily large. This is because λ is not used in the implementation, and it does not affect the actual control performance.

Remark 2.7 The matrix norm defined in the proof depends on the communication graph $\bar{\mathcal{G}}$ as U is calculated from $L + D$; that is why we call it the graph dependent norm. Such a norm enables us to derive a simpler convergence condition, and reveals the insight of the relation between the communication topology and the convergence property.

The convergence condition (2.7) is specified at the agent level, and it has the same form as the traditional D-type ILC convergence condition. The influence of communication on the convergence condition is reflected through the eigenvalues $\lambda_j, j = 1, 2, \ldots, N$, which are associated with the graph $\bar{\mathcal{G}}$. Hence, the local controller design is decoupled from other agents' dynamics. This motivates us to consider heterogeneous agents. Surprisingly, a similar convergence condition applies to heterogeneous agents as well. Detailed treatment is given in Section 2.3.2.

Remark 2.8 In Assumption 2.2, CB is assumed to be of full row rank. We can select $\Gamma = \gamma (CB)^T (CB(CB)^T)^{-1}$, then the convergence condition becomes much simpler:

$$
\max_{j=1,2..N} |1 - \gamma \lambda_j| < 1.
$$

(2.23)

This expression motivates us to consider the optimal learning gain design in the sense that $\|\mathbf{e}_i\|_\lambda$ converges at the fastest rate, which is explored in Section 2.4.

2.3.2 Controller Design for Heterogeneous Agents

Heterogeneity is the nature of MAS. Even between the same types of agent, they may have the similar structures, but it is unlikely that they share identical parameters. The consensus tracking problem for heterogeneous agents is thus of more practical use, but of course more challenging.

Consider the model in (2.1). Notice that, unlike the homogeneous systems case, all the agent dynamics are different from each other now. As the target is for output consensus tracking, the outputs from all agents should have the same dimension, whereas the state dimensions may not necessarily be the same.

Now we adopt the following learning rule:

$$\mathbf{u}_{i+1,j} = \mathbf{u}_{i,j} + \Gamma_j \dot{\xi}_{i,j}, \tag{2.24}$$

where Γ_j is the agent dependent learning gain matrix to be designed. The controller (2.24) is similar to (2.3) in the homogeneous case, except that the learning gain Γ_j is agent dependent.

Writing (2.24) in a compact form, we have

$$\mathbf{u}_{i+1} = \mathbf{u}_i + \Gamma((L+D) \otimes I_m)\dot{\mathbf{e}}_i(t), \tag{2.25}$$

where $\Gamma = \text{diag}(\Gamma_1, \Gamma_2, \dots, \Gamma_N)$. Due to the heterogeneity, the structure of (2.25) is rather different from the one in the homogeneous case. Following a similar development to the previous section, we eventually obtain

$$\|\mathbf{e}_{i+1}\|_\lambda \le \|I_{mN} - CB\Gamma((L+D) \otimes I_m)\| \, \|\mathbf{e}_i\|_\lambda + \mathcal{O}(\lambda^{-1})\|\mathbf{e}_i\|_\lambda, \tag{2.26}$$

where $B = \text{diag}(B_1, B_2, \dots, B_N)$, and $C = \text{diag}(C_1, C_2, \dots, C_N)$. If the agent dynamics are identical, that is, C_j, B_j, and Γ_j are identical for different j, then (2.26) degenerates to (2.20). That means that the homogeneous case is a special case of the heterogeneous case. Let $M = I_{mN} - CB\Gamma((L+D) \otimes I_m)$; of course, if $\rho(M) \le \rho_0 < 1$, $\forall t \in [0, T]$, then the consensus tracking erorr $\mathbf{e}_i(t)$ converges to zero along the iteration axis. The result is summarized in the following corollary.

Corollary 2.1 Assume that Assumptions 2.1–2.4 hold for time-varying nonlinear systems (2.1). Under the control law (2.24), if the learning gain matrix Γ satisfies the following condition:

$$\max_{t \in [0, T]} \rho \left(I_{mN} - CB\Gamma((L+D) \otimes I_m) \right) \le \rho_0 < 1,$$

for some $\rho_0 \in (0, 1)$, then we have that $\lim_{i \to \infty} \mathbf{y}_{i,j}(t) = \mathbf{y}_d(t)$ for all $t \in [0, \ T], j = 1, 2, \dots, N$.

However, such a convergence condition is specified at the group level, and one at agent level is preferred. Since $CB\Gamma$ and $(L+D) \otimes I_m$ do not commute in general, the matrix norm defined in the previous section cannot be directly applied. Let the learning gain $\Gamma_j = \gamma(C_j B_j)^T (C_j B_j (C_j B_j)^T)^{-1}$; then

$$M = I_{mN} - \gamma((L+D) \otimes I_m).$$

By using the matrix norm defined previously, we can get the following corollary.

Corollary 2.2 Let the learning gains $\Gamma_j = \gamma(C_jB_j)^T(C_jB_j(C_jB_j)^T)^{-1}$. Under the iterative learning rule (2.24), heterogeneous agent systems (2.1) can achieve perfect consensus tracking along the iteration axis, that is, $\lim_{i \to \infty} \mathbf{y}_{i,j}(t) = \mathbf{y}_d(t)$ for $t \in [0, T]$, if

$$\max_{j=1,2,\dots,N} |1 - \gamma\lambda_j| < 1, \tag{2.27}$$

where λ_j is an eigenvalue of $L + D$.

By setting the learning gains equal to $\Gamma_j = \gamma(C_jB_j)^T(C_jB_j(C_jB_j)^T)^{-1}$, the convergence condition in the heterogeneous case renders the same condition as in the homogeneous case. This can be seen from (2.23) and (2.27).

2.4 Optimal Learning Gain Design

Most ILC controllers converge asymptotically along the iteration axis. Fast and monotonic convergence is usually desired. In a continuous-time system, the convergence analysis relies on the λ-norm, which is essentially an exponentially time weighted norm. However, it is well known that the monotonic convergence of the λ-norm of error does not imply the monotonic convergence of actual error. In fact, the actual error may increase to a huge magnitude, and then gradually settle down to zero. Designing controllers that enable monotonic convergence of error is still an open topic for continuous-time systems.

In this section, we are not trying to solve the monotonic convergence problem. The optimal learning gain is designed in the sense that the λ-norm of error converges at the fastest rate, which indeed imposes a tighter bounding function for the actual tracking error. From the time weighted norm definition, it is not difficult to see that the λ-norm of error is equivalent to the norm of error, that is,

$$\|\mathbf{e}_i\|_\lambda \leq \|\mathbf{e}_i\| \leq e^{\lambda T}\|\mathbf{e}_i\|_\lambda,$$

where λ is a positive constant. From the above inequality, we notice that $\|\mathbf{e}_i\|$ is bounded by a constant times $\|\mathbf{e}_i\|_\lambda$. We still do not know exactly how the error behaves, but a faster decaying $\|\mathbf{e}_i\|_\lambda$ does impose a tighter bounding function for $\|\mathbf{e}_i\|$. This is the main rationale for the development here. The systems' transient responses depend on not only the controllers, but also the systems dynamics. However, the time-varying nonlinear terms $\mathbf{f}_j(t, \mathbf{x})$ are unknown. Thus ensuring the fastest convergence of $\|\mathbf{e}_i\|_\lambda$ is the best one can expect.

By setting $\Gamma_j = \gamma(C_jB_j)^T(C_jB_j(C_jB_j)^T)^{-1}$, it has been shown in the previous two sections that both homogeneous and heterogeneous systems render the same convergence condition. Thus, the optimal designs for the two different cases can be unified. The decay rate of $\|\mathbf{e}_i\|_\lambda$ is determined by

$$\max_\omega |1 - \omega\gamma|, \tag{2.28}$$

where $\omega \in \Lambda \triangleq \{\lambda_1, \lambda_2, \dots, \lambda_N\}$ is an eigenvalue of $L + D$. Depending on the communication graph, if the graph is undirected, the eigenvalues of $L + D$ are positive real numbers. Whereas, for directed graphs, the eigenvalues contain complex numbers in general, and all the eigenvalues have positive real parts. Hence, optimal learning gain designs for undirected and directed graphs have to be discussed separately.

To achieve the fastest convergence rate for $\|\mathbf{e}_i\|_\lambda$, (2.28) should be minimized by γ. Subsequently, we can formulate the optimal learning gain design problem as a *min-max* optimization problem, and the central task is to find an optimal γ^* such that the objective function given in (2.29) is minimized:

$$J = \min_\gamma \max_{\omega \in \Lambda} |1 - \omega\gamma|. \tag{2.29}$$

Theorem 2.2 gives the optimal solution to the optimization problem (2.29) when the underlying communication graph is undirected.

Theorem 2.2   When the communication graph among followers is undirected, that is, $L + D$ is symmetric, the solution to the optimization problem (2.29) is

$$\gamma^* = \frac{2}{\lambda_1 + \lambda_N}, \quad J_{min} = \frac{\lambda_N - \lambda_1}{\lambda_N + \lambda_1},$$

where $\lambda_1 = \min(\Lambda)$, $\lambda_N = \max(\Lambda)$.

Theorem 2.2 is a direct application of Proposition 2.3, therefore the proof is omitted here. In the theorem above, the minimum J can be written as

$$J_{min} = \frac{1 - \lambda_1/\lambda_N}{1 + \lambda_1/\lambda_N},$$

where $\frac{\lambda_1}{\lambda_N}$ is usually called the eigenratio (Altafini, 2013; Duan and Chen, 2012) in consensus (synchronization) literature. The smallest and largest eigenvalues of $L + D$ taken together determine the optimal convergence rate. If the eigenratio is close to 1, we get a faster convergence rate. This is a unique feature of the ILC based consensus algorithm.

When the communication topology is directed, the eigenvalues of $L + D$ can be complex. It is difficult to derive a general solution for (2.29) because Λ is a set of complex numbers. To find a suboptimal solution, define a compact region Ω, which contains all the eigenvalues of $L + D$,

$$\Omega = \{z \in \mathbb{C} \mid \mathfrak{R}(z) \geq \min_j(\mathfrak{R}(\lambda_j)) = \alpha_1, \ |z| \leq \max_j(|\lambda_j|) = \alpha_2\}.$$

Obviously, $\Lambda \subset \Omega$. Hence, the solution for

$$J = \min_\gamma \max_{\omega \in \Omega} |1 - \omega\gamma| \tag{2.30}$$

is a suboptimal solution for (2.29). Now it is straightforward to extend the result to a directed graph by using Lemma 2.1.

Theorem 2.3   When the communication graph among followers is directed, that is, $L + D$ is asymmetric, a suboptimal solution to the problem (2.29) is

$$\gamma^* = \frac{\alpha_1}{\alpha_2^2}, \quad J_{min} = \frac{\sqrt{\alpha_2^2 - \alpha_1^2}}{\alpha_2},$$

where $\alpha_1 = \min_{j=1,2,\ldots N}(\mathfrak{R}(\lambda_j))$, $\alpha_2 = \max_{j=1,2,\ldots N}(|\lambda_j|)$.

Proof: From above discussion, problem (2.30) is a suboptimal problem for (2.29), and problem (2.30) is the same problem as the one in Lemma 2.1. Therefore, Theorem 2.3 can be concluded by applying Lemma 2.1. ∎

Remark 2.9   To find the suboptimal solution, we purposely enlarge the search region from Λ to Ω, where $\Lambda \subset \Omega$. Hence, the suboptimal solution may be conservative. However, the suboptimal solution has two advantages. On the one hand, the method can be applied when the eigenvalues are not exactly known, but the region where they are located is known. On the other hand, the suboptimal solution is robust against small variations in the communication graph. When the exact eigenvalues are known, numerical methods should be adopted to minimize the objective function (2.29), which renders a global optimal solution.

Remark 2.10   For optimal learning gain designs, one needs complete information on the communication topology. This is a trade-off between systems information availability and performance. When the number of agents is huge, obtaining the eigenvalues of $L + D$ is computationally difficult. The analytical bounds of eigenvalues can be calculated from the results in Zhang (2011). Alternatively, Gershgorin disk theorem (Horn and Johnson, 1985) can be applied to estimate the eigenvalue region. When the detailed information of the graph is unknown, we can simply set $0 < \gamma < 1/\max_j(l_{jj} + d_j)$. If the leader agent has a path to any follower agent, this ensures that $\rho(M) < 1$ and consensus tracking can be achieved along the iteration axis.

2.5   Illustrative Example

To illustrate the efficacy of the proposed consensus schemes, consider a network consisting of four heterogeneous follower agents. The agent models are governed by

$$
\begin{cases}
\dot{\mathbf{x}}_{i,1} = \begin{bmatrix} 0.5\cos(x_{i,1_1}) + 0.2x_{i,1_2} \\ -0.1\sin(x_{i,1_2}) \end{bmatrix} + \begin{bmatrix} 1 & 1.5 \\ 1 & -1 \end{bmatrix} \mathbf{u}_{i,1}, \\
y_{i,1} = \begin{bmatrix} 1 & 1 \end{bmatrix} \mathbf{x}_{i,1};
\end{cases}
$$

$$
\begin{cases}
\dot{\mathbf{x}}_{i,2} = \begin{bmatrix} -0.1x_{i,2_1} + 0.2x_{i,2_2} \\ \sin(t)x_{i,2_1} \end{bmatrix} + \begin{bmatrix} 1 & 2 \\ 3 & 4 \end{bmatrix} \mathbf{u}_{i,2}, \\
y_{i,2} = \begin{bmatrix} 2 & 3 \end{bmatrix} \mathbf{x}_{i,2};
\end{cases}
$$

$$
\begin{cases}
\dot{\mathbf{x}}_{i,3} = \begin{bmatrix} 0.2\sin(x_{i,3_1}) \\ -0.2\sin(x_{i,3_2}) \\ -0.4\cos(x_{i,3_3}) \end{bmatrix} + \begin{bmatrix} 1 & 2 \\ 3 & 4 \\ 0.5 & 0.6 \end{bmatrix} \mathbf{u}_{i,3}, \\
y_{i,3} = \begin{bmatrix} 1 & 1 & 0.5 \end{bmatrix} \mathbf{x}_{i,3};
\end{cases}
$$

$$
\begin{cases}
\dot{\mathbf{x}}_{i,4} = \begin{bmatrix} -0.2x_{i,4_1} - 0.1x_{i,4_2} \\ \sin(x_{i,4_2}) \\ -0.3x_{i,4_3} \end{bmatrix} + \begin{bmatrix} 1 & -1 \\ 0.5 & 1 \\ 2 & 1 \end{bmatrix} \mathbf{u}_{i,4}, \\
y_{i,4} = \begin{bmatrix} 0.5 & 0.7 & 0.3 \end{bmatrix} \mathbf{x}_{i,4}.
\end{cases}
$$

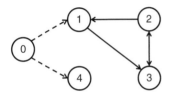

Figure 2.1 Communication topology among agents in the network.

The desired reference trajectory is

$$y_d = t + 2\sin(t), \ t \in [0, 5]. \tag{2.31}$$

The information exchange among followers is assumed to be fixed and directed.

Figure 2.1 shows the information flow among agents. The virtual leader is labeled by vertex 0 in the communication graph, and it has edges (dashed arrows) to agents 1 and 4. The communication among followers is depicted by solid arrows. Notice that the communication graph among followers is not connected. However, the communication graph including the leader contains a spanning tree with the leader being the root. We adopt 0–1 weighting; thus, the Laplacian for follower agents is

$$L = \begin{bmatrix} 1 & -1 & 0 & 0 \\ 0 & 1 & -1 & 0 \\ -1 & -1 & 2 & 0 \\ 0 & 0 & 0 & 0 \end{bmatrix},$$

and $D = \text{diag}(1, 0, 0, 1)$. The eigenvalues of $L + D$ are $\{0.16, 1.00, 2.42 \pm j0.61\}$ (j here denotes $\sqrt{-1}$). Since the eigenvalues are known exactly, we adopt a linear search algorithm to find out the optimal γ^* that minimizes the objective function (2.29). It turns out that $\gamma^* = 0.73$. Let the learning gains

$$\Gamma_j = \gamma^* (C_j B_j)^T (C_j B_j (C_j B_j)^T)^{-1},$$

where $j = 1, 2, 3, 4$. Check the convergence condition (2.27), we can obtain that

$$\max_{j=1,2,3,4} |1 - \lambda_j \gamma^*| = 0.88 < 1.$$

Therefore, the convergence condition in Corollary 2.2 is satisfied, and consensus tracking is achievable by the ILC rule (2.24).

In the simulation example, the identical initial condition (*iic*) is assumed, that is, $e_i(0) = 0$, and all the agents are reset to the same initial position after one iteration. The control signals at the zeroth iteration are set to zero, that is, $\mathbf{u}_{0,j} = 0$, for all agents. Figure 2.2 shows the agents' output tracking errors $y_d(t) - y_{i,j}(t)$ at the 1st, 15th, 30th, and 50th iterations. At the 1st iteration, the trajectories of followers have very large deviations from the desired one. As it can be seen from Figure 2.2, the tracking errors are gradually reduced by the learning controllers, and they are almost eliminated at the 50th iteration. Figure 2.3 shows the maximum error convergence profile versus iteration number. As the iteration number increases, all agents' outputs asymptotically converge to the desired trajectory (2.31).

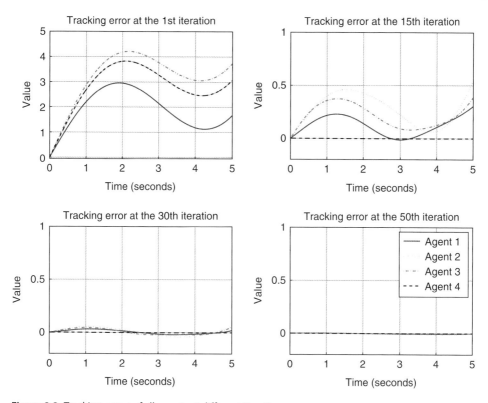

Figure 2.2 Tracking errors of all agents at different iterations.

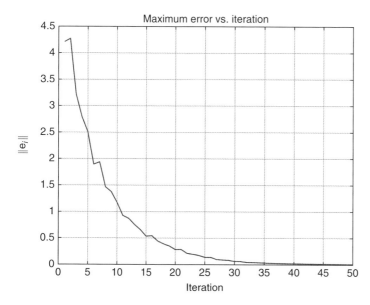

Figure 2.3 Maximum tracking error vs. iteration number.

2.6 Conclusion

In this chapter, a consensus tracking problem is formulated for a group of global Lipschitz nonlinear systems. A distributed D-type ILC control law is studied for the consensus tracking problem for both homogeneous and heterogeneous systems. By adoption of a graph dependent matrix norm, the convergence condition is specified at the agent level, which decouples the learning gain design from other agent dynamics. In addition, optimal learning gain design methods are developed for both undirected and directed graphs, in the sense that the λ-norm of the tracking error decays at the fastest rate, which imposes a tighter bounding function for the actual tracking error. A consensus tracking example for heterogeneous agent systems under a directed graph is given to demonstrate the effectiveness of the developed design methods.

3

Iterative Learning Control for Multi-agent Coordination Under Iteration-Varying Graph

3.1 Introduction

Multi-agent coordination and control problems are usually studied under a fixed communication topology. However, the fixed communication regime is restrictive and difficult to achieve for an MAS in reality. A switching communication regime is more general and has profound implications for implementation issues. If the developed controller works for a switching graph, this means that the controller is more robust to communication variations such as link failure and creation. Thus it is important to study switching communication topology and its impact on convergence results.

The consensus problem for a switching graph has been investigated by many researchers (Olfati-Saber and Murray, 2004; Hatano and Mesbahi, 2005; Moreau, 2005; Cao *et al.*, 2005; Wu, 2006; Hong *et al.*, 2006; Zhang and Tian, 2009). An excellent survey paper on communication assumptions and convergence results is Fang and Antsaklis (2006). Many interesting results are explored in the literature, for example, the consensus results under a uniformly connected union graph or a random graph. However, it is still unclear how switching communication affects the stability and convergence of ILC based controllers for multi-agent coordination. In the existing works in consensus tracking by ILC (Ahn and Chen, 2009; Xu *et al.*, 2011; Yang *et al.*, 2012; Yang and Xu, 2012; Meng *et al.*, 2012; Liu and Jia, 2012; Li and Li, 2013), the communication topology is assumed to be fixed, except for Liu and Jia (2012) in which the communication graph is time-varying but connected at every time instance. This chapter focuses on analyzing the stability and convergence properties of the learning controller under an iteration-varying communication topology. First, the convergence property is derived for a fixed strongly connected graph. The same analysis method will be used to derive more general results later. Next, the results are developed for an iteration-varying graph that is strongly connected in each iteration. Lastly, the results are generalized to the uniformly strongly connected graph along the iteration axis. There are some limited publications on application of ILC with switching graphs. The authors in Meng *et al.* (2013b,d, 2014) studied a similar problem and obtained some meaningful results. The fundamental difference is that they adopted a 2-dimensional (2D) system theory approach to analyze the controller convergence for discrete-time system models. As we discussed in the previous chapter, this approach is only applicable to linear systems, whereas, we are applying λ-norm methods for continuous-time nonlinear models.

Iterative Learning Control for Multi-agent Systems Coordination, First Edition.
Shiping Yang, Jian-Xin Xu, Xuefang Li, and Dong Shen.
© 2017 John Wiley & Sons Singapore Pte. Ltd. Published 2017 by John Wiley & Sons Singapore Pte. Ltd.

This chapter is organized as follows. The consensus tracking problem under switching communication is formulated in Section 3.2. Next, the convergence results under various communication assumptions are developed in Section 3.3. To verify the results, a numerical example is presented in Section 3.4. Lastly, Section 3.5 draws conclusions.

3.2 Problem Description

Consider a group of N dynamic agents, with the jth agent governed by the following nonlinear model:

$$\begin{cases} \dot{\mathbf{x}}_{i,j}(t) = \mathbf{f}(\mathbf{x}_{i,j}(t)) + B\mathbf{u}_{i,j}(t) \\ \mathbf{y}_{i,j}(t) = C\mathbf{x}_{i,j}(t) \end{cases} \forall j \in \mathcal{V}, \tag{3.1}$$

where i denotes the iteration number, $\mathbf{x}_{i,j} \in \mathbb{R}^n$ is the state vector, $\mathbf{y}_{i,j} \in \mathbb{R}^m$ is the output vector, $\mathbf{u}_{i,j} \in \mathbb{R}^p$ is the control input, $\mathbf{f}(\mathbf{x}_{i,j})$ is a global Lipschitz nonlinear function of $\mathbf{x}_{i,j}$, and B, C are constant matrices of compatible dimensions. In particular, CB is of full column rank, $\mathbf{f}(\cdot)$ is unknown and satisfies

$$|\mathbf{f}(\mathbf{z}_1) - \mathbf{f}(\mathbf{z}_2)| \le L_f|\mathbf{z}_1 - \mathbf{z}_2|, \text{ for any } \mathbf{z}_1, \mathbf{z}_2 \in \mathbb{R}^n,$$

where L_f is an unknown Lipschitz constant.

For simplicity, the time argument, t, is dropped when no confusion arises.

The desired trajectory $\mathbf{y}_d(t)$ is defined on a finite-time interval $[0, T]$, which is generated by the following dynamics:

$$\begin{cases} \dot{\mathbf{x}}_d = \mathbf{f}(\mathbf{x}_d) + B\mathbf{u}_d, \\ \mathbf{y}_d = C\mathbf{x}_d. \end{cases} \tag{3.2}$$

The communication topology among the agents is described by an iteration-varying graph $\mathcal{G}(i) = (\mathcal{V}, \mathcal{E}(i), \mathcal{A}(i))$. The desired trajectory can be treated as a virtual leader, and we denote it as vertex 0 in the graph representation. Assume that only a few of the agents in the network know the desired trajectory. Therefore vertex 0 has directed edges to those follower agents that have the knowledge of $\mathbf{y}_d$ in the graph representation. Together with $\mathcal{G}(i)$, the complete information flow among the followers and the leader can be depicted by a new graph $\overline{\mathcal{G}}(i) = (\mathcal{V} \cup \{0\}, \overline{\mathcal{E}}(i), \overline{\mathcal{A}}(i))$, where $\overline{\mathcal{E}}(i)$ and $\overline{\mathcal{A}}(i)$ are the corresponding edge set and adjacency matrix.

Based on the communication topology, let the extended tracking error for agent j at the ith iteration be

$$\xi_{i,j} = \sum_{k \in \mathcal{N}_j(i)} a_{j,k}(i)(\mathbf{y}_{i,k} - \mathbf{y}_{i,j}) + d_j(i)(\mathbf{y}_d - \mathbf{y}_{i,j}), \tag{3.3}$$

where $a_{j,k}(i)$ is the (j,k)th entry of the adjacency matrix $\mathcal{A}(i)$, $d_j(i) = 1$ if the jth agent can access the virtual leader's output at the ith iteration, and $d_j(i) = 0$ otherwise. The extended tracking error $\xi_{i,j}$ contains only local information. As such it can be used for distributed controller design.

The commonly applied D-type ILC rule is adopted in this chapter:

$$\mathbf{u}_{i+1,j} = \mathbf{u}_{i,j} + \Gamma_i \dot{\xi}_{i,j}, \tag{3.4}$$

where Γ_i is the learning gain to be designed.

Remark 3.1 The derivative term of $\xi_{i,j}$ is utilized in the learning rule (3.4). Note that $\xi_{i,j}$ is generated in the ith iteration, therefore it is already available at the $(i+1)$th iteration. As measured signals are usually contaminated by noise, some sophisticated numerical method should be applied to obtain $\dot{\xi}_{i,j}$ without generating large amount of noise. For example, the filtered differentiation method in Slotine and Li (Slotine and Li, 1991, p.202) is a simple but effective candidate. In fact, such kind of non-causalimplementation is one distinct feature of ILC.

Before presenting the convergence properties of control law (3.4) under various communication assumptions, the identical initialization condition (*iic*) is imposed.

Assumption 3.1 The initial condition of each agent is reset to the desired initial condition at every iteration, that is, $\mathbf{x}_{i,j}(0) = \mathbf{x}_d(0)$.

The *iic* is the most commonly used assumption in ILC literature. *Remark* 2.4 discusses the relaxation or removal of the *iic* at the cost of imperfect tracking performance. We will further investigate this issue in Chapter 4.

Initially we discuss the convergence properties under fixed communication topology, but eventually we will investigate the more general communication graph as below.

Definition 3.1 Consider an iteration-varying graph $\mathcal{G}(i)$. The graph is said to be uniformly strongly connected along the iteration axis, if there exists a constant integer K such that the union graph $\cup_{s=r}^{r+K}\mathcal{G}(s)$ is strongly connected for all r.

3.3 Main Results

In this section, convergence properties of control law (3.4) are investigated under three different communication assumptions. First, the convergence result is derived for a fixed strongly connected graph in Subsection 3.3.1. The proof is detailed here and it will be referred to by Subsections 3.3.2 and 3.3.3. Next the result is extended to an iteration-varying strongly connected graph in Subsection 3.3.2. Finally, generalization to a uniformly strongly connected graph is presented in Subsection 3.3.3.

3.3.1 Fixed Strongly Connected Graph

In this subsection, assume that the communication topology is a fixed strongly connected graph, and at least one of the follower agents can access the leader's trajectory. The following two lemmas are introduced here as they will be utilized in the main proof.

Lemma 3.1 For a given matrix M, if its spectral radius $\rho(M) < 1$, then there exist positive constants $c_1 > 0$, and $0 < \rho_0 < 1$, such that $\|M^k\| \leq c_1\rho_0^k$.

Proof: Since $\rho(M) < 1$, M is a stable matrix, that is, M^k converges to zero exponentially as k tends to infinity. Therefore, it is straightforward to conclude Lemma 3.1. ∎

In Lemma 3.1, c_1 is a positive constant. For convenience, choose $c_1 \geq 1$ in the following development, since when $c_1 < 1$, Lemma 3.1 still holds by setting $c_1 = 1$.

Lemma 3.2 Consider a positive sequence $\{a_i\}$ satisfying

$$a_{i+1} = c_1 \rho_0^i a_1 + \frac{c_2}{\lambda - L_f}(\rho_0^{i-1} a_1 + \rho_0^{i-2} a_2 + \cdots + a_i)$$

for $i \geq 1$, where L_f, c_1, and c_2 are positive constants, $0 < \rho_0 < 1$. If $\lambda > L_f + \frac{c_2}{1-\rho_0}$, then $a_i \to 0$.

Proof: By hypothesis,

$$a_{i+1} = c_1 \rho_0^i a_1 + \frac{c_2}{\lambda - L_f}(\rho_0^{i-1} a_1 + \rho_0^{i-2} a_2 + \cdots + a_i). \tag{3.5}$$

Then, for $i \geq 2$, we have

$$a_i = c_1 \rho_0^{i-1} a_1 + \frac{c_2}{\lambda - L_f}(\rho_0^{i-2} a_1 + \rho_0^{i-3} a_2 + \cdots + a_{i-1}), \tag{3.6}$$

From (3.5) and (3.6) we can obtain

$$a_{i+1} - \rho_0 a_i = \frac{c_2}{\lambda - L_f} a_i,$$

$$a_{i+1} = \left(\rho_0 + \frac{c_2}{\lambda - L_f} \right) a_i.$$

As $\lambda > L_f + \frac{c_2}{1-\rho_0}$, we have $\rho_0 + \frac{c_2}{\lambda - L_f} < 1$. Hence, $a_i \to 0$. ■

Let the learning gain be identical for all iterations since the communication graph is fixed,

$$\Gamma = \frac{1}{q}[(CB)^T CB]^{-1}(CB)^T, \tag{3.7}$$

where

$$q > \max_{j=1 \ldots N} \sum_{k=1}^{N} a_{j,k} + d_j.$$

To be more accurate, $a_{j,k}$ and d_j in the inequality above are supposed to be $a_{j,k}(i)$ are $d_j(i)$. As the graph is iteration-invariant, the iteration index i is omitted.

Now define the actual tracking error $\mathbf{e}_{i,j} = \mathbf{y}_d - \mathbf{y}_{i,j}$; together with (3.3), the control law (3.4) can be rewritten in a compact form,

$$\mathbf{u}_{i+1} = \mathbf{u}_i + (H \otimes \Gamma)\dot{\mathbf{e}}_i, \tag{3.8}$$

where $\mathbf{u}_i$ and $\mathbf{e}_i$ are the column stack vectors of $\mathbf{u}_{i,j}$ and $\mathbf{e}_{i,j}$, $H = L + D$, and L is the Laplacian matrix of $\mathcal{G}$, $D = \text{diag}(d_1, d_2, \ldots, d_N)$.

Theorem 3.1 Consider the MAS (3.1) under Assumption 3.1, the learning gain (3.7), and control law (3.8). If the communication topology is a fixed strongly connected graph, and at least one of the followers in the network has access to the virtual leader's trajectory, then the tracking error $\mathbf{e}_{i,j}$ converges to zero along the iteration axis, that is, $\lim_{i \to \infty} \mathbf{e}_{i,j} = 0$.

Proof: Define $\delta\mathbf{u}_{i,j} = \mathbf{u}_d - \mathbf{u}_{i,j}$, $\delta\mathbf{x}_{i,j} = \mathbf{x}_d - \mathbf{x}_{i,j}$, and $\delta\mathbf{f}(\mathbf{x}_{i,j}) = \mathbf{f}(\mathbf{x}_d) - \mathbf{f}(\mathbf{x}_{i,j})$. Subtracting (3.1) from (3.2) yields

$$\delta\dot{\mathbf{x}}_{i,j} = \delta\mathbf{f}(\mathbf{x}_{i,j}) + B\delta\mathbf{u}_{i,j}. \tag{3.9}$$

Let $\delta\mathbf{u}_i$, $\delta\mathbf{x}_i$, and $\delta\mathbf{f}(\mathbf{x}_i)$ be the column stack vectors of $\delta\mathbf{u}_{i,j}$, $\delta\mathbf{x}_{i,j}$, and $\delta\mathbf{f}(\mathbf{x}_{i,j})$, respectively. Then, (3.9) can be written as

$$\delta\dot{\mathbf{x}}_i = \delta\mathbf{f}(\mathbf{x}_i) + (I_N \otimes B)\delta\mathbf{u}_i, \tag{3.10}$$

where I is the identity matrix, and the subscript denotes its dimension.

Let us investigate the variation of $\delta\mathbf{u}_i$ between two consecutive iterations. From the learning rule (3.8), we have

$$\begin{aligned}\delta\mathbf{u}_{i+1} &= \delta\mathbf{u}_i - (H \otimes \Gamma)\dot{\mathbf{e}}_i, \\ &= \delta\mathbf{u}_i - (H \otimes \Gamma)(I_N \otimes C)\delta\dot{\mathbf{x}}_i. \end{aligned} \tag{3.11}$$

Substituting (3.10) in (3.11), and noticing the learning gain (3.7), we can obtain

$$\delta\mathbf{u}_{i+1} = \left(\left(I_N - \frac{1}{q}H\right) \otimes I_p\right)\delta\mathbf{u}_i - (H \otimes \Gamma C)\delta\mathbf{f}(\mathbf{x}_i). \tag{3.12}$$

For simplicity, denote $M = \left(I_N - \frac{1}{q}H\right) \otimes I_p$, and $F = H \otimes \Gamma C$. From (3.12), we can obtain the relation below between $\delta\mathbf{u}_{i+1}$ and $\delta\mathbf{u}_1$:

$$\delta\mathbf{u}_{i+1} = M^i\delta\mathbf{u}_1 - M^{i-1}F\delta\mathbf{f}(\mathbf{x}_1) - M^{i-2}F\delta\mathbf{f}(\mathbf{x}_2) - \cdots - F\delta\mathbf{f}(\mathbf{x}_i). \tag{3.13}$$

Since the communication graph is strongly connected, $\left(I_N - \frac{1}{q}H\right)$ must be an irreducible matrix. Notice that q is larger than the greatest diagonal entry of H, and D contains at least one positive entry, hence, at least one row sum of $\left(I_N - \frac{1}{q}H\right)$ is strictly less than one. Therefore, $\left(I_N - \frac{1}{q}H\right)$ is a substochastic matrix. By using *Corollary 6.2.28* of Horn and Johnson (Horn and Johnson, 1985, p.363), we can conclude that the spectral radius $\rho\left(I_N - \frac{1}{q}H\right) < 1$. Note that M and $\left(I_N - \frac{1}{q}H\right)$ have the same spectrum, except that each eigenvalue of M has multiplicity of p. Therefore, $\rho(M) < 1$. Taking the infinity norm of (3.13), and applying the global Lipschitz condition on $\mathbf{f}(\cdot)$ and Lemma 3.1, we have

$$\begin{aligned}|\delta\mathbf{u}_{i+1}| &\le c_1\rho_0^i|\delta\mathbf{u}_1| + c_1L_f\rho_0^{i-1}|F||\delta\mathbf{x}_1| + c_1L_f\rho_0^{i-2}|F||\delta\mathbf{x}_2| \\ &\quad + \cdots + L_f|F||\delta\mathbf{x}_i|, \end{aligned} \tag{3.14}$$

where $0 < \rho_0 < 1$ and satisfies the inequality in Lemma 3.1.

Integrating (3.10) together with Assumption 3.1 yields

$$\delta\mathbf{x}_i \le \int_0^t \left(\delta\mathbf{f}(\mathbf{x}_i) + (I_N \otimes B)\delta\mathbf{u}_i\right) d\tau. \tag{3.15}$$

Taking norm operations on both sides of (3.15), and applying Gronwall–Bellman's Lemma, we have

$$|\delta\mathbf{x}_i| \le \int_0^t e^{L_f(t-\tau)}|I_N \otimes B||\delta\mathbf{u}_i| d\tau. \tag{3.16}$$

Taking the λ-norm of (3.16) yields

$$\|\delta \mathbf{x}_i\|_\lambda \le \frac{1}{\lambda - L_f} \|I_N \otimes B\| \|\delta \mathbf{u}_i\|_\lambda. \tag{3.17}$$

Taking the λ-norm of (3.14), together with (3.17), yields

$$\|\delta \mathbf{u}_{i+1}\|_\lambda$$

$$\le c_1 \rho_0^i \|\delta \mathbf{u}_1\|_\lambda + c_1 \rho_0^{i-1} \frac{L_f}{\lambda - L_f} \|I_N \otimes B\| \|F\| \|\delta \mathbf{u}_1\|_\lambda \tag{3.18}$$

$$+ c_1 \rho_0^{i-2} \frac{L_f}{\lambda - L_f} \|I_N \otimes B\| \|F\| \|\delta \mathbf{u}_2\|_\lambda + \cdots + \frac{L_f}{\lambda - L_f} \|I_N \otimes B\| \|F\| \|\delta \mathbf{u}_i\|_\lambda.$$

We can purposely choose $c_1 \ge 1$, and denote $c_2 = c_1 L_f \|I_N \otimes B\| \|F\|$, thus, we have

$$\|\delta \mathbf{u}_{i+1}\|_\lambda$$

$$\le c_1 \rho_0^i \|\delta \mathbf{u}_1\|_\lambda + \frac{c_2}{\lambda - L_f} \rho_0^{i-1} \|\delta \mathbf{u}_1\|_\lambda$$

$$+ \frac{c_2}{\lambda - L_f} \rho_0^{i-2} \|\delta \mathbf{u}_2\|_\lambda + \cdots + \frac{c_2}{\lambda - L_f} \|\delta \mathbf{u}_i\|_\lambda. \tag{3.19}$$

Choose a $\lambda > L_f + \frac{c_2}{1 - \rho_0}$. Applying Lemma 3.2, it can be concluded that $\delta \mathbf{u}_{i,j} \to 0$ along the iteration axis, hence, $\lim_{i \to \infty} \mathbf{e}_{i,j} = 0$. ∎

Remark 3.2 From the proof to Lemma 3.2, the sequence $\{a_i\}$ converges to zero exponentially. As Theorem 3.1 is proved by using Lemma 3.2, the λ-norm of $\delta \mathbf{u}_i$, $\|\delta \mathbf{u}_i\|_\lambda$, should also converge to zero exponentially. From the definition of the λ-norm, it can be shown that $\|\delta \mathbf{u}_i\| \le e^{\lambda T} \|\delta \mathbf{u}_i\|_\lambda$. Note that $e^{\lambda T}$ is a constant. Therefore, $\|\delta \mathbf{u}_i\|$ and the tracking error $\|\mathbf{e}_i\|$ both converge to zero exponentially as well.

3.3.2 Iteration-Varying Strongly Connected Graph

In this subsection, we assume that the communication topology is iteration-varying, but that the graph is fixed and strongly connected in each iteration. Furthermore, at least one of the followers has access to the leader's trajectory in every iteration.

We adopt the following iteration-varying learning gain:

$$\Gamma(i) = \frac{1}{q(i)} \left((CB)^T CB \right)^{-1} (CB)^T, \tag{3.20}$$

where $\Gamma(i)$ depends on the iteration index i, and

$$q(i) > \max_{j=1...N} \sum_{k=1}^{N} a_{j,k}(i) + d_j(i).$$

ILC rule (3.8) becomes

$$\mathbf{u}_{i+1} = \mathbf{u}_i + (H(i) \otimes \Gamma(i)) \dot{\mathbf{e}}_i. \tag{3.21}$$

Since the communication topology is iteration-varying, the system matrix M in (3.12) is iteration-varying as well, and Lemma 3.1 is no longer applicable to show the convergence result. Therefore, we develop three preliminary results first before presenting the main convergence property.

Lemma 3.3   Let $\mathcal{M} \subset \mathbb{R}^{N \times N}$ denote the set of all irreducible substochastic matrices with positive diagonal entries, then we have

$$|M(N)M(N-1)\cdots M(1)| < 1,$$

where $M(k),\ k = 1, 2, \dots, N$, are N matrices arbitrarily selected from $\mathcal{M}$.

Proof: We show the result by induction. Let $\mathbf{1}_{(\cdot)}$ be a vector with all elements being 1. The subscript $(\cdot)$ denotes its dimension, and the subscript is omitted when the dimension is clear in the context.

Since $M(k)$ is a substochastic matrix, at least one element in the vector $M(1)\mathbf{1}$ must be strictly less than 1.

Next, assume that multiplication of k matrices from $\mathcal{M}$ has k row sums less than 1. Without loss of generality, assume the first k row sums less than 1, that is

$$M(k)M(k-1)\cdots M(1)\mathbf{1} = \begin{bmatrix} \boldsymbol{\alpha} \\ \mathbf{1}_{N-k} \end{bmatrix},$$

where $\boldsymbol{\alpha} \in \mathbb{R}^k$ with all elements less than 1.

Then, investigate the multiplication of $k + 1$ matrices from $\mathcal{M}$. Set

$$\begin{bmatrix} M_{11}(k+1) & M_{12}(k+1) \\ M_{21}(k+1) & M_{22}(k+1) \end{bmatrix} \begin{bmatrix} \boldsymbol{\alpha} \\ \mathbf{1}_{N-k} \end{bmatrix} = \begin{bmatrix} \boldsymbol{\beta} \\ \boldsymbol{\gamma} \end{bmatrix}.$$

Since the diagonal entries of $M_{11}(k+1)$ are positive, all the elements in $\boldsymbol{\beta}$ have to be less than 1. If $\boldsymbol{\gamma} = \mathbf{1}_{N-k}$, this implies that all elements of $M_{21}(k+1)$ are zero, which contradicts the fact $M(k+1)$ is irreducible. So multiplication of $k + 1$ matrices from $\mathcal{M}$ must have at least $k + 1$ row sums less than 1. This completes the proof. ∎

Lemma 3.4   Consider a positive sequence $\{a_i\}$ satisfying

$$a_{i+1} = \rho_0^{\lfloor \frac{i}{N} \rfloor} a_1 + \frac{c_2}{\lambda - L_f} \left(\rho_0^{\lfloor \frac{i-1}{N} \rfloor} a_1 + \rho_0^{\lfloor \frac{i-2}{N} \rfloor} a_2 + \cdots + \rho_0^{\lfloor \frac{i-i}{N} \rfloor} a_i \right), \tag{3.22}$$

for $i \geq 1$, where $\lfloor \cdot \rfloor$ stands for the floor function, L_f and c_2 are positive constants, $0 < \rho_0 < 1$. If $\rho_0 + \left(1 + \frac{c_2}{\lambda - L_f} \right)^N - 1 < 1$, then $a_i \to 0$.

Proof: The proof consists of three parts. In the first part, the relation between a_{N+1} and a_1 is investigated. In the second part, we present for any $k \geq 2$, $2 \leq j \leq N$, $a_{kN+j} = (1 + \frac{c_2}{\lambda - L_f})^{j-1} a_{kN+1}$. Lastly, the convergence of a_{kN+1} as k tends to infinity is given, which implies $\lim_{i \to \infty} a_i = 0$.

Part I. The relation between a_{N+1} and a_1: By the hypothesis (3.22), for $i = 1, 2, \dots, N - 1$, we have

$$a_{i+1} = a_1 + \frac{c_2}{\lambda - L_f}(a_1 + a_2 + \cdots + a_i)$$

$$= a_1 + \frac{c_2}{\lambda - L_f}(a_1 + a_2 + \cdots + a_{i-1}) + \frac{c_2}{\lambda - L_f} a_i$$

$$= a_i + \frac{c_2}{\lambda - L_f} a_i,$$

which implies

$$a_{i+1} = \left(1 + \frac{c_2}{\lambda - L_f}\right) a_i, \ i = 1, 2, \ldots, N - 1.$$

Subsequently, we can find a general formula for a_{i+1},

$$a_{i+1} = \left(1 + \frac{c_2}{\lambda - L_f}\right)^i a_1, \ i = 1, 2, \ldots, N - 1. \tag{3.23}$$

Next, let us study the relation between a_{N+1} and a_1. Since

$$a_{N+1} = \rho_0 a_1 + \frac{c_2}{\lambda - L_f}(a_1 + a_2 + \cdots + a_N)$$

$$= a_1 + \frac{c_2}{\lambda - L_f}(a_1 + a_2 + \cdots + a_{N-1}) + \rho_0 a_1 - a_1 + \frac{c_2}{\lambda - L_f}a_N$$

$$= a_N + \rho_0 a_1 - a_1 + \frac{c_2}{\lambda - L_f}a_N$$

$$= (\rho_0 - 1)a_1 + \left(1 + \frac{c_2}{\lambda - L_f}\right) a_N, \tag{3.24}$$

we can obtain

$$a_{N+1} = (\rho_0 - 1)a_1 + (1 + \frac{c_2}{\lambda - L_f})(1 + \frac{c_2}{\lambda - L_f})^{N-1}a_1$$

$$= \left(\rho_0 - 1 + \left(1 + \frac{c_2}{\lambda - L_f}\right)^N\right)a_1 \tag{3.25}$$

from (3.23).

Part II. The relation between a_{kN+j} and a_{kN+1}, $k \geq 2$: We show the result by induction. First of all, (3.22) gives

$$a_{kN+1} = \rho_0^k a_1 + \frac{c_2}{\lambda - L_f} \left(\rho_0^{k-1}(a_1 + \cdots + a_N) + \rho_0^{k-2}(a_{N+1}\right.$$

$$\left. + \cdots + a_{2N}) + \cdots + \rho_0^{k-k}(a_{(k-1)N+1} + \cdots + a_{kN})\right)$$

$$= \rho_0^k a_1 + \frac{c_2}{\lambda - L_f} \sum_{s=0}^{k-1} \rho_0^{k-1-s}(a_{sN+1} + \cdots + a_{sN+N}). \tag{3.26}$$

Similarly, by (3.26) we can obtain

$$a_{kN+2} = \rho_0^k a_1 + \frac{c_2}{\lambda - L_f} \sum_{s=0}^{k-1} \rho_0^{k-1-s}(\rho_0 a_{sN+1} + \cdots + a_{sN+N}) + \frac{c_2}{\lambda - L_f}a_{kN+1}$$

$$= a_{kN+1} + \frac{c_2}{\lambda - L_f}a_{kN+1}$$

$$= \left(1 + \frac{c_2}{\lambda - L_f}\right) a_{kN+1}. \tag{3.27}$$

Next, assume that for any $3 \leq j \leq N - 1$,

$$a_{kN+j} = \left(1 + \frac{c_2}{\lambda - L_f}\right)^{j-1} a_{kN+1}. \tag{3.28}$$

Then, for a_{kN+j+1}, we have

$$
a_{kN+j+1} = \rho_0^k a_1 + \frac{c_2}{\lambda - L_f} \sum_{s=0}^{k-1} \rho_0^{k-1-s} (\rho_0 a_{sN+1} + \cdots + \rho_0 a_{sN+j}
$$

$$
+ a_{sN+j+1} + \cdots + a_{sN+N}) + \frac{c_2}{\lambda - L_f} \sum_{l=1}^{j} a_{kN+l}
$$

$$
= a_{kN+1} + \frac{c_2}{\lambda - L_f} \sum_{l=1}^{j} a_{kN+l}
$$

$$
= \left(1 + \frac{c_2}{\lambda - L_f} + \frac{c_2}{\lambda - L_f} \left(1 + \frac{c_2}{\lambda - L_f} \right) + \cdots \right.
$$

$$
\left. + \frac{c_2}{\lambda - L_f} \left(1 + \frac{c_2}{\lambda - L_f} \right)^{j-1} \right) a_{kN+1}
$$

$$
= \left(1 + \frac{c_2}{\lambda - L_f} \right)^{j} a_{kN+1}. \tag{3.29}
$$

Hence, we have

$$
a_{kN+j} = \left(1 + \frac{c_2}{\lambda - L_f} \right)^{j-1} a_{kN+1}, \ 1 \le j \le N. \tag{3.30}
$$

Part III. The convergence of a_{kN+1}:
Similar to (3.26), we have

$$
a_{(k+1)N+1} = \rho_0^{k+1} a_1 + \frac{c_2}{\lambda - L_f} \sum_{s=0}^{k} \rho_0^{k-s} (a_{sN+1} + \cdots + a_{sN+N})
$$

$$
= \rho_0 \left(\rho_0^k + \frac{c_2}{\lambda - L_f} \sum_{s=0}^{k-1} \rho_0^{k-1-s} (a_{sN+1} + \cdots + a_{sN+N}) \right)
$$

$$
+ \frac{c_2}{\lambda - L_f} \sum_{j=1}^{N} a_{kN+j}
$$

$$
= \rho_0 a_{kN+1} + \frac{c_2}{\lambda - L_f} \sum_{j=1}^{N} a_{kN+j}. \tag{3.31}
$$

Applying (3.30) and (3.31) leads to

$$
a_{(k+1)N+1} = \rho_0 a_{kN+1} + \frac{c_2}{\lambda - L_f} \left(\sum_{j=1}^{N} (1 + \frac{c_2}{\lambda - L_f})^{(j-1)} \right) a_{kN+1}
$$

$$
= \left(\rho_0 - 1 + 1 + \frac{c_2}{\lambda - L_f} \sum_{j=1}^{N} (1 + \frac{c_2}{\lambda - L_f})^{(j-1)} \right) a_{kN+1}
$$

$$= \left(\rho_0 - 1 + (1 + \frac{c_2}{\lambda - L_f})^N \right) a_{kN+1}$$

$$= \left(\rho_0 - 1 + (1 + \frac{c_2}{\lambda - L_f})^N \right)^{k+1} a_1. \tag{3.32}$$

Therefore, if $\rho_0 - 1 + \left(1 + \frac{c_2}{\lambda - L_f}\right)^N < 1$ holds, we have $\lim_{k \to \infty} a_{kN+1} = 0$, which implies $\lim_{k \to \infty} a_{kN+j} = 0, j = 1, 2, \dots, N$ from (3.30). Finally, we can obtain that $\lim_{i \to \infty} a_i = 0$. ∎

Lemma 3.5 If the sequence $\{a_i\}$ in Lemma 3.4 converges to zero and the sequence $\{b_i\}$ with $b_1 = a_1$ satisfies

$$b_{i+1} \leq \eta^{\lfloor \frac{i}{N} \rfloor} b_1 + \beta \left(\eta^{\lfloor \frac{i-1}{N} \rfloor} b_1 + \eta^{\lfloor \frac{i-2}{N} \rfloor} b_2 + \dots + \eta^{\lfloor \frac{i-i}{N} \rfloor} b_i \right), \tag{3.33}$$

then we have $b_i \to 0$ as well.

Proof: By induction, it is easy to obtain that $b_i \leq a_i$ for $i \geq 1$. As such, the convergence of $\{a_i\}$ implies that the sequence $\{b_i\}$ converges to zero immediately. ∎

The next theorem shows the convergence result for iteration-varying communication.

Theorem 3.2 Consider the MAS (3.1) under Assumption 3.1, the learning gain (3.20), and control law (3.21). If the communication graph is iteration-varying, but in each iteration, the graph is fixed and strongly connected, and at least one of the followers in the network has access to the virtual leader's trajectory, then the tracking error $\mathbf{e}_{i,j}$ converges to zero along the iteration axis, that is, $\lim_{i \to \infty} \mathbf{e}_{i,j} = 0$.

Proof: Denote $M(i) = \left(I_N - \frac{1}{q(i)} H(i)\right) \otimes I_p$, and $F(i) = H(i) \otimes \Gamma(i) C$. For notational simplicity, define $M(i,k) = M(i)M(i-1) \cdots M(k)$ for $i \geq k$.

Following the analysis framework in the proof of Theorem 3.1, we can obtain

$$\delta \mathbf{u}_{i+1} = M(i,1)\delta \mathbf{u}_1 - M(i-1,1)F(1)\delta \mathbf{f}(\mathbf{x}_1)$$
$$- M(i-2,2)F(2)\delta \mathbf{f}(\mathbf{x}_2) - \dots - F(i)\delta \mathbf{f}(\mathbf{x}_i). \tag{3.34}$$

From the definition of $\Gamma(i)$ in (3.20), we can see that $M(i)$ is a substochastic matrix with positive diagonal entries. Taking the infinity norm on both sides of Equation (3.34), we can obtain

$$|\delta \mathbf{u}_{i+1}| \leq |M(i,1)||\delta \mathbf{u}_1| + L_f |M(i-1,1)||F(1)||\delta \mathbf{x}_1|$$
$$+ L_f |M(i-2,2)||F(2)||\delta \mathbf{x}_2| + \dots + L_f |F(i)||\delta \mathbf{x}_i|. \tag{3.35}$$

Divide each $M(i,k)$ in (3.35) into groups of size N, and calculate the product of each group. Then, applying Lemma 3.3, we have

$$|\delta \mathbf{u}_{i+1}| \leq \rho_0^{\lfloor \frac{i}{N} \rfloor} |\delta \mathbf{u}_1| + L_f \rho_0^{\lfloor \frac{i-1}{N} \rfloor} |F(1)||\delta \mathbf{x}_1|$$
$$+ L_f \rho_0^{\lfloor \frac{i-2}{N} \rfloor} |F(2)||\delta \mathbf{x}_2| + \dots + L_f |F(i)||\delta \mathbf{x}_i|. \tag{3.36}$$

Taking the λ-norm of (3.36), and substituting in Equation (3.17) yields

$$\|\delta \mathbf{u}_{i+1}\|_\lambda \le \rho_0^{\lfloor \frac{i}{N} \rfloor} \|\delta \mathbf{u}_1\|_\lambda + \frac{L_f}{\lambda - L_f} \|I_N \otimes B\| \rho_0^{\lfloor \frac{i-1}{N} \rfloor} \|F(1)\| \|\delta \mathbf{u}_1\|_\lambda$$

$$+ \frac{L_f}{\lambda - L_f} \|I_N \otimes B\| \rho_0^{\lfloor \frac{i-2}{N} \rfloor} \|F(2)\| \|\delta \mathbf{u}_2\|_\lambda$$

$$+ \cdots + \frac{L_f}{\lambda - L_f} \|I_N \otimes B\| \|F(i)\| \|\delta \mathbf{u}_i\|_\lambda. \tag{3.37}$$

Let $c_2 = \max_{k=1}^{i} L_f \|F(k)\| \|I_N \otimes B\|$; then (3.37) becomes

$$\|\delta \mathbf{u}_{i+1}\|_\lambda \le \rho_0^{\lfloor \frac{i}{N} \rfloor} \|\delta \mathbf{u}_1\|_\lambda + \frac{c_2}{\lambda - L_f} \left(\rho_0^{\lfloor \frac{i-1}{N} \rfloor} \|\delta \mathbf{u}_1\|_\lambda + \rho_0^{\lfloor \frac{i-2}{N} \rfloor} \|\delta \mathbf{u}_2\|_\lambda + \cdots \right.$$

$$\left. + \rho_0^{\lfloor \frac{i-i}{N} \rfloor} \|\delta \mathbf{u}_i\|_\lambda \right).$$

Based on the result in Lemma 3.6, we can conclude that $\delta \mathbf{u}_{i,j} \to 0$ along the iteration axis, hence, $\lim_{i \to \infty} \mathbf{e}_{i,j} = 0$ as well. ∎

3.3.3 Uniformly Strongly Connected Graph

In this subsection, we further generalize the results in the previous two subsections. Let the communication topology be a uniformly strongly connected graph as defined in Definition 3.1. Such a condition is much more flexible than the communication assumptions in Theorems 3.1 and 3.2. The following lemma is required to generalize our results obtained previously.

Lemma 3.6 Let $i \ge 2$ be an integer and let $Q_1, Q_2, \ldots, Q_i \in \mathbb{R}^{N \times N}$ be nonnegative matrices. Suppose that the diagonal entries of Q_k ($k = 1, 2, \ldots, i$) are positive, then there exists a $\gamma > 0$ such that

$$Q_1 Q_2 \cdots Q_i \ge \gamma \left(Q_1 + Q_2 + \cdots + Q_i \right),$$

where $\ge$ is defined entrywise.

The proof of Lemma 3.5 can be found in Jadbabaie *et al.* (2003).

The theorem below presents the most general convergence result in this chapter.

Theorem 3.3 Consider the MAS (3.1) under Assumption 3.1, the learning gain (3.20), and control law (3.21). If the communication graph is uniformly strongly connected along the iteration axis within a duration that is upper bounded by a constant number K, and in each of K consecutive iterations there is at least one follower in the network having access to the virtual leader's trajectory, then the tracking error $\mathbf{e}_{i,j}$ converges to zero along the iteration axis, that is, $\lim_{i \to \infty} \mathbf{e}_{i,j} = 0$.

Proof: Let $\tilde{M}(i) = I_N - \frac{1}{q(i)} H(i)$. $\tilde{M}(i)$ is a nonnegative matrix with positive diagonal entries. Since the communication graph is uniformly strongly connected along the iteration axis with upper bound K, and applying Lemma 3.5, we can derive that $\tilde{M}(i + K, i)$ is an irreducible nonnegative matrix for any i. Furthermore, if there is at least one follower in the network having access to the leader's trajectory within the iteration interval $[i, i + K]$, then $\tilde{M}(i + K, i)$ is a substochastic matrix.

Now, following the same process as in the proof of Theorem 3.2, we can, eventually, obtain that

$$\|\delta \mathbf{u}_{i+1}\|_\lambda \leq \rho_0^{\lfloor \frac{i}{NK} \rfloor} \|\delta \mathbf{u}_1\|_\lambda + \frac{c_2}{\lambda - L_f} \left(\rho_0^{\lfloor \frac{i-1}{NK} \rfloor} \|\delta \mathbf{u}_1\|_\lambda + \rho_0^{\lfloor \frac{i-2}{NK} \rfloor} \|\delta \mathbf{u}_2\|_\lambda + \cdots \right.$$

$$\left. + \rho_0^{\lfloor \frac{i-i}{NK} \rfloor} \|\delta \mathbf{u}_i\|_\lambda \right).$$

Therefore, if $\rho_0 + (1 + \frac{c_2}{\lambda - L_f})^{NK} - 1 < 1$, according to Lemma 3.4, we can conclude that the tracking error $\mathbf{e}_{i,j}$ converges to zero as the iteration index tends to infinity. ∎

Remark 3.3 Intuitively, the convergence rate will become slow when the communication is weak, for instance the uniformly strongly connected graph assumption. This point can be seen in the proof of Theorem 3.3, because the convergence rate of $\|\delta \mathbf{u}_i\|_\lambda$ is proportional to the factor $\rho_0^{\lfloor \frac{i}{NK} \rfloor}$, hence the larger the K, the slower the convergence.

3.4 Illustrative Example

The results in Theorems 3.1 and 3.2 are special cases of Theorem 3.3. In order to test all the theorems, it is therefore sufficient to verify the result in Theorem 3.3. Hence, a tracking example under the uniformly strongly connected topology is provided here.

Consider four dynamic agents, with their system parameters given by

$$\mathbf{f}(\mathbf{x}) = \begin{bmatrix} x_2 \\ \cos(x_1) - x_2 \end{bmatrix}, \quad B = \begin{bmatrix} 0 \\ 1 \end{bmatrix}, \quad C = \begin{bmatrix} 1 & 0 \\ 2 & 1 \end{bmatrix}.$$

The leader's input $u_d = t + 4\sin(2t)$, $t \in [0, 5]$. u_d is unknown to any of the followers.

The complete communication topology is depicted in Figure 3.1, in which there are four graphs. In each iteration, only one of the communication graphs is activated, chosen by the selection function $i = 4k + j$, $j = 1, 2, 3, 4$, and k is a nonnegative integer; for example, if $i = 1$ then the graph in the left top corner is selected. In each of the individual graphs, the topology is not even connected. However, the union of all the four graphs is indeed strongly connected. Therefore, the communication topology is uniformly strongly connected along the iteration axis with upper bound $K = 4$. Thus, the communication requirement in Theorem 3.3 is satisfied.

To design the learning gain, we need to investigate the matrices $H(i) = L(i) + D(i)$, and the first four of them are listed below:

$$H(1) = \begin{bmatrix} 1 & 0 & 0 & 0 \\ 0 & 0 & 0 & 0 \\ 0 & -1 & 1 & 0 \\ 0 & 0 & -1 & 1 \end{bmatrix}, H(2) = \begin{bmatrix} 0 & 0 & 0 & 0 \\ -1 & 1 & 0 & 0 \\ 0 & 0 & 1 & 0 \\ -1 & 0 & 0 & 1 \end{bmatrix}$$

$$H(3) = \begin{bmatrix} 0 & 0 & 0 & 0 \\ 0 & 0 & 0 & 0 \\ 0 & -1 & 1 & 0 \\ 0 & 0 & 0 & 0 \end{bmatrix}, H(4) = \begin{bmatrix} 1 & 0 & 0 & -1 \\ 0 & 0 & 0 & 0 \\ 0 & 0 & 0 & 0 \\ 0 & 0 & 0 & 0 \end{bmatrix}.$$

The maximal diagonal entry of $H(i)$ is 1, hence, we can choose the learning gain $q_s = 2$ for switching topology.

Denote the union graph of the four subgraphs in Figure 3.1 by $\mathcal{G}_u$. The corresponding graph Laplacian L_u and matrix D_u are

$$
L_u = \begin{bmatrix}
1 & 0 & 0 & -1 \\
-1 & 1 & 0 & 0 \\
0 & -1 & 1 & 0 \\
-1 & 0 & -1 & 2
\end{bmatrix}
$$

and $D_u = \text{diag}(1, 0, 1, 0)$ respectively. As a comparative study, we also investigate the convergence performance of the four agents under the fixed graph $\mathcal{G}_u$. In this case, let the learning gain $q_f = 3$. Notice that q_s and q_f are slightly different. This is because we choose the value such that the control performances are the best for the two cases respectively.

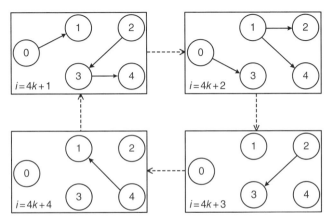

Figure 3.1 Communication topology among agents in the network.

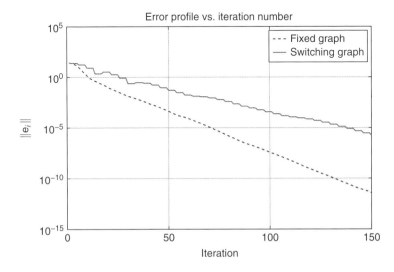

Figure 3.2 Maximum norm of error vs. iteration number.

Figure 3.2 describes how the maximum error $\|\mathbf{e}_i\| = \sup_{t \in [0,T]} |\mathbf{e}_i(t)|$ evolves along the iteration axis for both switching and fixed graphs. The y-axis is plotted on a log-scale. This means that the convergence rates in both cases are exponentially fast. It can be seen from Figure 3.2 that the dashed line is below the solid line. This observation indicates that the intermittent communication slows down the convergence rate. This also shows that the simulation results match perfectly with our theoretical predictions.

3.5   Conclusion

In this chapter, a typical D-type ILC rule for consensus tracking is investigated under an iteration-varying graph. It is shown that if the iteration-varying graph is uniformly strongly connected along the iteration axis, and at least one of the followers has access to the leader's trajectory, then the proposed ILC rule can perfectly synchronize the output trajectories of all follower agents. The results complement those in the previous chapter by generalizing from a fixed communication model to an iteration-varying graph. A simulation study demonstrates that the learning controller is very robust to communication variations.

4

Iterative Learning Control for Multi-agent Coordination with Initial State Error

4.1 Introduction

In the discussions of Chapters 2 and 3, the perfect identical initial condition (*iic*) is assumed. Similarly, most existing ILC for MAS coordination assume this restrictive assumption, for example Ahn and Chen (2009), Xu *et al.* (2011), Yang *et al.* (2012), Yang and Xu (2012), Liu and Jia (2012), Meng *et al.* (2012), and Li and Li (2013). Notice that the *iic* is one of the fundamental problems pertaining to the ILC literature and its applicability. It is required that the initial state is the same as the desired initial state for perfect tracking to be achievable. However, the controllers in a MAS are distributed in nature, and all the agents are independent entities. The communication among agents may not be complete so that many agents are not aware of the desired initial state. It is difficult to ensure perfect *iic* for all agents in general. Even if all the agents are aware of the desired initial state, it may not be possible for all the agents to adjust their initial state. Thus, the initial condition problem in MAS requires further investigation.

There are many excellent ideas in the ILC literature to deal with imperfect initial conditions. Many researchers have analyzed and developed algorithms that do not require *iic*, at the cost of imperfect tracking. The initial state learning method, developed in Chen *et al.* (1999), is applicable when the initial state is measurable and manipulatable. Some extra system knowledge is required by the learning rule. When the learning gain is chosen correctly, the initial state converges to the desired one, and the tracking error due to initial state error is bounded. Initial rectifying action control is introduced by Sun and Wang (2002), which modifies the desired trajectory such that the initial output of the new reference coincides with the actual initial output. Because of the nature of this control method, the actual output converges to a modified desired trajectory. These two approaches are adopted for multi-agent control with initial state error, for example initial state learning is applied in Yang *et al.* (2012), and initial rectifying action is utilized in Meng *et al.* (2012). The variable initial state for discrete-time system is discussed in Fang and Chow (2003) via 2-D analysis. Park *et al.* (1999) study the robustness of a PID-type updating rule. In Park (2005), an average operator based PD-type rule is developed to improve the control performance against variable initial state error. Furthermore, Xu and Yan (2005) fully investigate five different initial conditions and their convergence properties.

In this chapter, we investigate the initial state error and its impact on the control performance in the multi-agent consensus tracking problem. At the beginning of a task execution, the initial state of each agent is reset to a fixed position which is different

Iterative Learning Control for Multi-agent Systems Coordination, First Edition.
Shiping Yang, Jian-Xin Xu, Xuefang Li, and Dong Shen.
© 2017 John Wiley & Sons Singapore Pte. Ltd. Published 2017 by John Wiley & Sons Singapore Pte. Ltd.

from the desired one. We show that under the sparse communication assumption, the D-type (Xu and Tan, 2003) ILC rule is still applicable and convergence can be guaranteed when the communication graph contains a spanning tree with the leader being its root. However, in terms of performance, the final output of each agent has large deviations from the desired reference. In order to improve the performance, a PD-type (Park *et al.*, 1999) rule is indicated, and it turns out that the new updating rule gives the designer more freedom to tune the final performance. The analysis method itself in this chapter is of independent interest, and it contributes to the ILC literature on the initial state problem.

This chapter is organized as follows. The consensus tracking problem with initial state error is formulated in Section 4.2. Next, the D-type, PD-type updating rules, and convergence results are developed in Section 4.3. To demonstrate the effectiveness of the results, two numerical examples are presented in Section 4.4. Lastly, Section 4.5 draws conclusions.

4.2 Problem Description

Consider a group of N homogeneous dynamic agents, with the jth agent governed by the following linear time-invariant model:

$$\begin{cases} \dot{\mathbf{x}}_{i,j}(t) = A\mathbf{x}_{i,j}(t) + B\mathbf{u}_{i,j}(t) \\ \mathbf{y}_{i,j}(t) = C\mathbf{x}_{i,j}(t) \end{cases} \forall j \in \mathcal{V}, \tag{4.1}$$

where i denotes the iteration number, $\mathbf{x}_{i,j} \in \mathbb{R}^n$, $\mathbf{y}_{i,j} \in \mathbb{R}^p$, and $\mathbf{u}_{i,j} \in \mathbb{R}^m$ are the state vector, output vector, and control input respectively, and A, B, C are constant matrices of compatible dimensions. For simplicity, the time argument, t, is dropped when no confusion arises.

The leader's trajectory, or the desired consensus trajectory $\mathbf{y}_d(t)$ is defined on a finite-time interval $[0, T]$, and it is generated by the following dynamics:

$$\begin{cases} \dot{\mathbf{x}}_d = A\mathbf{x}_d + B\mathbf{u}_d, \\ \mathbf{y}_d = C\mathbf{x}_d, \end{cases} \tag{4.2}$$

where $\mathbf{u}_d$ is the continuous and unique desired control input.

Due to communication or sensor limitations, the leader's trajectory is only accessible to a small portion of the followers. Let the communication among followers be described by the graph $\mathcal{G}(\mathcal{V}, \mathcal{E})$. If the leader is labeled by vertex 0, then the complete information flow among all the agents can be characterized by a new graph $\overline{\mathcal{G}} = (0 \cup \mathcal{V}, \overline{\mathcal{E}})$, where $\overline{\mathcal{E}}$ is the new edge set. The major task is to design a set of distributed ILC rules such that each individual agent in the network is able to track the leader's trajectory under the sparse communication graph $\overline{\mathcal{G}}$.

To simplify the controller design and convergence analysis, the following two assumptions are imposed.

Assumption 4.1 CB is of full column rank.

Remark 4.1   The full column rank assumption is a necessary condition to find out a suitable learning gain such that ILC D-type learning rule satisfies the contraction-mapping criterion. Assumption 4.1 implies that the relative degree of system (4.1) is well defined and it is exactly 1. When CB is not of full rank, a high-order derivative of the tracking error can be utilized in the controller design.

Assumption 4.2   The initial state of an agent is reset to the same position at every iteration, which is not equal to the desired state, that is, $\mathbf{x}_{i,j}(0) = \mathbf{x}_{1,j}(0) \neq \mathbf{x}_d(0)$ for all $i \geq 1$.

Remark 4.2   Assumption 4.2 is referred as the resetting condition in the ILC literature. Since the initial state is different from the desired state, that is $\mathbf{y}_{i,j}(0) \neq \mathbf{y}_d(0)$, it is impossible to achieve perfect tracking. The ILC rule should force the output trajectory of each agent to be as close as possible to the leader's trajectory.

4.3 Main Results

The main results contain two subsections. In Subsection 4.3.1, the distributed D-type updating rule and its convergence properties are fully analyzed. To improve the tracking performance of learning rules, a PD-type updating rule is proposed and analyzed in Subsection 4.3.2.

4.3.1 Distributed D-type Updating Rule

Let $\xi_{i,j}$ be the distributed measurement by agent j at the ith iteration over the graph $\overline{\mathcal{G}}$, and it is defined as

$$\xi_{i,j} = \sum_{k \in \mathcal{N}_j} a_{j,k}(\mathbf{y}_{i,k} - \mathbf{y}_{i,j}) + d_j(\mathbf{y}_d - \mathbf{y}_{i,j}), \tag{4.3}$$

where $d_j = 1$ if $(0, j) \in \overline{\mathcal{E}}$, otherwise $d_j = 0$.

Note that the actual tracking error $\mathbf{e}_{i,j} = \mathbf{y}_d - \mathbf{y}_{i,j}$ cannot be utilized in the controller design as only a small number of followers have access to the leader's trajectory. Therefore, $\mathbf{e}_{i,j}$ is not available to many of the followers. It is natural to incorporate the distributed measurement $\xi_{i,j}$ in the ILC design. Hence, the following D-type ILC updating rule is adopted in this chapter:

$$\mathbf{u}_{i+1,j} = \mathbf{u}_{i,j} + Q\dot{\xi}_{i,j}, \quad \mathbf{u}_{0,j} = 0, \ \forall j \in \mathcal{V}, \tag{4.4}$$

where Q is the learning gain to be designed. For simplicity, the initial control input $\mathbf{u}_{0,j}$ is set to zero. However, in practical implementation, the initial control input can be generated by certain feedback mechanisms such that the system is stable. This may improve the transient performance of the learning controller. Note that $\xi_{i,j}$ is already available at the $(i + 1)$th iteration. Therefore, the derivative of $\xi_{i,j}$ can be obtained by any advanced numerical differentiation that does not generate a large amount of noise.

Let $\hat{\mathbf{x}}_{v,j}$, $\hat{\mathbf{u}}_{v,j}$, and $\hat{\mathbf{y}}_{v,j} = C\hat{\mathbf{x}}_{v,j}$ satisfy the following virtual dynamics:

$$\hat{\mathbf{x}}_{v,j}(t) = \mathbf{x}_{i,j}(0) + \int_0^t \left[A\hat{\mathbf{x}}_{v,j}(\tau) + B\hat{\mathbf{u}}_{v,j}(\tau) \right] d\tau, \tag{4.5}$$

and

$$Q(\dot{\mathbf{y}}_d - \dot{\hat{\mathbf{y}}}_{v,j}) = 0. \tag{4.6}$$

Before investigating the existence and uniqueness of $\hat{\mathbf{u}}_{v,j}$, $\hat{\mathbf{x}}_{v,j}$, and $\hat{\mathbf{y}}_{v,j}$ satisfying Equations (4.5) and (4.6), we have the following convergence results.

Theorem 4.1   Consider the MAS (4.1), under Assumptions 4.1 and 4.2, the communication graph $\bar{\mathcal{G}}$, and distributed D-type updating rule (4.4). If the learning gain Q is chosen such that

$$|I_{mN} - H \otimes QCB| \le \rho_0 < 1,$$

where $I_{(\cdot)}$ is the identity matrix with the subscript denoting its dimension, ρ_0 is a constant, $H = L + D$, L is the Laplacian matrix of $\mathcal{G}$, and $D = \mathrm{diag}(d_1, d_2, \dots, d_N)$, then the control input $\mathbf{u}_{i,j}$ and output $\mathbf{y}_{i,j}$ converge to $\hat{\mathbf{u}}_{v,j}$ and $\hat{\mathbf{y}}_{v,j}$ respectively as the iteration number tends to infinity.

Proof: From the definition of the distributed measurement $\xi_{i,j}$ in (4.3) and the equality constraint (4.6), one obtains that

$$
\begin{aligned}
Q\dot{\xi}_{i,j} &= Q\left(\sum_{k \in \mathcal{N}_j} a_{j,k}(\dot{\mathbf{y}}_{i,k} - \dot{\mathbf{y}}_{i,j}) + d_j(\dot{\mathbf{y}}_d - \dot{\mathbf{y}}_{i,j}) \right) \\
&= \sum_{k \in \mathcal{N}_j} a_{j,k}(Q\dot{\mathbf{y}}_{i,k} - Q\dot{\mathbf{y}}_{i,j}) + d_j(Q\dot{\mathbf{y}}_d - Q\dot{\mathbf{y}}_{i,j}) \\
&= \sum_{k \in \mathcal{N}_j} a_{j,k}(Q\dot{\mathbf{y}}_d - Q\dot{\mathbf{y}}_{i,j} - Q\dot{\mathbf{y}}_d + Q\dot{\mathbf{y}}_{i,k}) + d_j(Q\dot{\mathbf{y}}_d - Q\dot{\mathbf{y}}_{i,j}) \\
&= \sum_{k \in \mathcal{N}_j} a_{j,k}(Q\dot{\hat{\mathbf{y}}}_{v,j} - Q\dot{\mathbf{y}}_{i,j} - Q\dot{\hat{\mathbf{y}}}_{v,k} + Q\dot{\mathbf{y}}_{i,k}) + d_j(Q\dot{\hat{\mathbf{y}}}_{v,j} - Q\dot{\mathbf{y}}_{i,j}).
\end{aligned}
\tag{4.7}
$$

Define the virtual tracking error $\epsilon_{i,j} = \hat{\mathbf{y}}_{v,j} - \mathbf{y}_{i,j}$; then (4.7) can be simplified as

$$Q\dot{\xi}_{i,j} = Q\left(\sum_{k \in \mathcal{N}_j} a_{j,k}(\dot{\epsilon}_{i,j} - \dot{\epsilon}_{i,k}) + d_j \dot{\epsilon}_{i,j} \right). \tag{4.8}$$

Define the following notations: $\delta\mathbf{u}_{i,j} = \hat{\mathbf{u}}_{v,j} - \mathbf{u}_{i,j}$ and $\delta\mathbf{x}_{i,j} = \hat{\mathbf{x}}_{v,j} - \mathbf{x}_{i,j}$. From (4.4) and (4.8) we have

$$\delta\mathbf{u}_{i+1,j} = \delta\mathbf{u}_{i,j} - Q\left(\sum_{k \in \mathcal{N}_j} a_{j,k}(\dot{\epsilon}_{i,j} - \dot{\epsilon}_{i,k}) + d_j \dot{\epsilon}_{i,j} \right). \tag{4.9}$$

Let $\delta\mathbf{u}_i$ and ϵ_i be the column stack vectors of $\delta\mathbf{u}_{i,j}$ and $\epsilon_{i,j}$, thus, (4.9) can be written in the following compact form,

$$\delta\mathbf{u}_{i+1} = \delta\mathbf{u}_i - (H \otimes Q)\dot{\epsilon}_i. \tag{4.10}$$

Taking the derivative from both sides of (4.5), and subtracting (4.1) yields

$$\delta\dot{\mathbf{x}}_{i,j} = A\delta\mathbf{x}_{i,j} + B\delta\mathbf{u}_{i,j}. \tag{4.11}$$

Rewriting (4.11) in the compact form yields

$$\delta\dot{\mathbf{x}}_i = (I_N \otimes A)\delta\mathbf{x}_i + (I_N \otimes B)\delta\mathbf{u}_i, \tag{4.12}$$

where $\delta\mathbf{x}_i$ is the column stack vector of $\delta\mathbf{x}_{i,j}$.

Note that $\epsilon_i = (I_N \otimes C)\delta\mathbf{x}_i$. Substituting (4.12) in (4.10) yields

$$\delta\mathbf{u}_{i+1} = \delta\mathbf{u}_i - (H \otimes Q)(I_N \otimes C)((I_N \otimes A)\delta\mathbf{x}_i + (I_N \otimes B)\delta\mathbf{u}_i)$$
$$= (I_{mN} - H \otimes QCB)\delta\mathbf{u}_i - (H \otimes QCA)\delta\mathbf{x}_i. \tag{4.13}$$

Taking the λ-norm operation on (4.13) yields

$$\|\delta\mathbf{u}_{i+1}\|_\lambda \le \rho_0 \|\delta\mathbf{u}_i\|_\lambda + b_1 \|\delta\mathbf{x}_i\|_\lambda, \tag{4.14}$$

where $b_1 = |H \otimes QCA|$.

It can be shown that $\hat{\mathbf{x}}_{v,j}(0) = \mathbf{x}_{i,j}(0)$ from (4.5), that is $\delta\mathbf{x}_{i,j}(0) = 0$. Therefore, solving $\delta\mathbf{x}_{i,j}$ from (4.12) we have

$$\delta\mathbf{x}_i = \int_0^t e^{(I_N \otimes A)(t-\tau)} B\delta\mathbf{u}_i(\tau)\, d\tau. \tag{4.15}$$

Taking any generic norm on both sides of (4.15), we have

$$|\delta\mathbf{x}_i| \le |B| \int_0^t e^{|I_N \otimes A|(t-\tau)} |\delta\mathbf{u}_i(\tau)|\, d\tau,$$

$$e^{-\lambda t}|\delta\mathbf{x}_i| \le e^{-\lambda t}|B| \int_0^t e^{|I_N \otimes A|(t-\tau)} |\delta\mathbf{u}_i(\tau)|\, d\tau,$$

$$e^{-\lambda t}|\delta\mathbf{x}_i| \le |B| \int_0^t e^{-(\lambda - |I_N \otimes A|)(t-\tau)} e^{-\lambda\tau} |\delta\mathbf{u}_i(\tau)|\, d\tau. \tag{4.16}$$

Again taking the λ-norm on (4.16) yields

$$\|\delta\mathbf{x}_i\|_\lambda \le \frac{|B|}{\lambda - a}|\delta\mathbf{u}_i|_\lambda, \tag{4.17}$$

where $a = |I_N \otimes A|$. Substituting (4.17) in (4.14), we have

$$\|\delta\mathbf{u}_{i+1}\|_\lambda \le \left(\rho_0 + \frac{b_1|B|}{\lambda - a}\right)\|\delta\mathbf{u}_i\|_\lambda = \rho_1\|\delta\mathbf{u}_i\|_\lambda, \tag{4.18}$$

where $\rho_1 = \rho_0 + \frac{b_1|B|}{\lambda - a}$.

If $\lambda > \frac{b_1|B|}{\rho_0} + a$, then $\rho_1 < 1$. Therefore, $\|\delta\mathbf{u}_i\|_\lambda$ converges to zero as the iteration number increases, that is, $\mathbf{u}_{i,j} \to \hat{\mathbf{u}}_{v,j}$ and $\mathbf{y}_{i,j} \to \hat{\mathbf{y}}_{v,j}$. ∎

By using the graph dependent matrix norm methods in Chapter 2, we can convert the norm inequality convergence condition in Theorem 4.1 to the spectral radius condition. Therefore, we have the following corollary.

Corollary 4.1 Consider the MAS (4.1), under Assumptions 4.1, 4.2, the communication graph $\overline{\mathcal{G}}$, and distributed D-type updating rule (4.4). If the learning gain Q is chosen such that

$$\rho(I_{mN} - H \otimes QCB) < 1,$$

where $\rho(\cdot)$ denotes the spectral radius of a matrix, then the control input $\mathbf{u}_{i,j}$ and output $\mathbf{y}_{i,j}$ converge to $\hat{\mathbf{u}}_{v,j}$ and $\hat{\mathbf{y}}_{v,j}$ respectively as iteration tends to infinity.

As the spectral radius of any given matrix is the infimum of any matrix norm (Horn and Johnson, 1985), the spectral radius inequality in Corollary 4.1 is indeed more general than the norm inequality condition in Theorem 4.1.

To satisfy either the norm inequality condition or the spectral radius condition, the communication topology plays an important role for multi-agent coordination to be realizable. If the communication assumption is not strong enough, the coordination goal may not be achievable. For example, if there is an isolated agent or cluster of agents, to which the leader's information cannot be relayed, it is impossible for the multi-agent system to achieve consensus tracking in general. The following lemma reveals a very useful algebraic property of the communication graph.

Lemma 4.1 (Ren and Beard, 2008) If the communication graph $\bar{\mathcal{G}}$ contains a spanning tree with the leader being the root, then all the eigenvalues of matrix H have positive real parts.

With the help of Lemma 4.1, we have the following sufficient communication requirement for the coordination problem.

Lemma 4.2 If the communication graph $\bar{\mathcal{G}}$ contains a spanning tree with the leader being the root, there always exists a learning gain Q such that the convergence condition in Corollary 4.1 holds, that is,

$$\rho(I_{mN} - H \otimes QCB) < 1.$$

Proof: Due to Assumption 4.1 that CB is of full column rank, let Q be the pseudoinverse of CB times a positive scalar gain $q > 0$, that is, $Q = q\left((CB^T)CB\right)^{-1}(CB)^T$. Denote $M = I_{mN} - H \otimes QCB$; hence,

$$M = I_{mN} - qH \otimes I_m.$$

Let $\sigma(H) = \{\lambda_1, \lambda_2, \ldots, \lambda_N\}$ be the spectrum of H. Based on Lemma 4.1, all the eigenvalues have positive real parts. The spectrum of $I_N - qH$ is $\sigma(I_N - qH) = \{1 - q\lambda_1, 1 - q\lambda_2, \ldots, 1 - q\lambda_N\}$. Therefore, we can choose q sufficiently small such that all the magnitudes of $\sigma(I_N - qH)$ are strictly less than 1. Note the property of the Kronecker product, the spectrum of M is the same as $\sigma(I_N - qH)$ except that each eigenvalue has multiplicity of m. Hence, $\rho(M) < 1$. ∎

The following result will be used to derive the existence and uniqueness of $\hat{\mathbf{u}}_{v,j}$ and $\hat{\mathbf{y}}_{v,j}$.

Lemma 4.3 If $\rho(I_{mN} - H \otimes QCB) < 1$ holds, then QCB is nonsingular.

Proof: This can be shown by contradiction. Assume QCB is singular, that is, QCB has at least one eigenvalue equal to zero. Therefore, $I_{mN} - H \otimes QCB$ has at least one eigenvalue equal to one. Subsequently, we can obtain that $\rho(I_{mN} - H \otimes QCB) \geq 1$, which contradicts the hypothesis. This completes the proof. ∎

The results in Theorem 4.1 and Corollary 4.1 guarantee that the input and output trajectories of each individual agent converge to the corresponding virtual dynamics. We are now in a position to investigate the properties of the virtual dynamics.

Theorem 4.2 If the D-type rule (4.4) converges, there exist unique $\hat{\mathbf{u}}_{v,j}$ and $\hat{\mathbf{y}}_{v,j}$ satisfying the virtual dynamics (4.5) and (4.6), specifically,

$$\hat{\mathbf{u}}_{v,j} = \mathbf{u}_d + (QCB)^{-1}QCAe^{F_D t}(\mathbf{x}_d(0) - \mathbf{x}_{i,j}(0)),$$

and

$$\hat{\mathbf{y}}_{v,j} = \mathbf{y}_d - Ce^{F_D t}(\mathbf{x}_d(0) - \mathbf{x}_{i,j}(0)),$$

where $F_D = (I - B(QCB)^{-1}QC)A$.

Proof: From (4.6), and the dynamics (4.5) and (4.2), we have

$$Q\dot{\mathbf{y}}_d = Q\dot{\hat{\mathbf{y}}}_{v,j}$$
$$QCA\mathbf{x}_d + QCB\mathbf{u}_d = QCA\hat{\mathbf{x}}_{v,j} + QCB\hat{\mathbf{u}}_{v,j}$$
$$QCB(\mathbf{u}_d - \hat{\mathbf{u}}_{v,j}) = -QCA(\mathbf{x}_d - \hat{\mathbf{x}}_{v,j}) \tag{4.19}$$

Define $\delta\hat{\mathbf{u}}_{v,j} = \mathbf{u}_d - \hat{\mathbf{u}}_{v,j}$, and $\delta\hat{\mathbf{x}}_{v,j} = \mathbf{x}_d - \hat{\mathbf{x}}_{v,j}$. Note that QCB is nonsingular, hence (4.19) can be written as

$$\delta\hat{\mathbf{u}}_{v,j} = -(QCB)^{-1}QCA\delta\hat{\mathbf{x}}_{v,j}. \tag{4.20}$$

Taking the derivative on both sides of (4.5) and subtracting it from (4.2) yields

$$\delta\dot{\hat{\mathbf{x}}}_{v,j} = A\delta\hat{\mathbf{x}}_{v,j} + B\delta\hat{\mathbf{u}}_{v,j}. \tag{4.21}$$

With initial condition $\delta\hat{\mathbf{x}}_{v,j}(0) = \mathbf{x}_d(0) - \mathbf{x}_{i,j}(0)$, substituting (4.20) in (4.21) yields

$$\delta\dot{\hat{\mathbf{x}}}_{v,j} = (I_n - B(QCB)^{-1}QC)A\delta\hat{\mathbf{x}}_{v,j}. \tag{4.22}$$

As (4.22) is a linear differential equation, its solution exists and is unique. Therefore, both $\hat{\mathbf{u}}_{v,j}$ and $\hat{\mathbf{y}}_{v,j}$ exist and are unique.

Denote $F_D = (I_n - B(QCB)^{-1}QC)A$. Solving $\delta\hat{\mathbf{x}}_{v,j}$ from (4.22), we have

$$\delta\hat{\mathbf{x}}_{v,j} = e^{F_D t}(\mathbf{x}_d(0) - \mathbf{x}_{i,j}(0)).$$

Therefore, from (4.20) we have

$$\hat{\mathbf{u}}_{v,j} = \mathbf{u}_d + (QCB)^{-1}QCAe^{F_D t}(\mathbf{x}_d(0) - \mathbf{x}_{i,j}(0)),$$

and

$$\hat{\mathbf{y}}_{v,j} = \mathbf{y}_d - Ce^{F_D t}(\mathbf{x}_d(0) - \mathbf{x}_{i,j}(0)).$$

∎

The learning gain Q is designed independent of A. It is interesting to note that the system matrix A does not affect the convergence property of the ILC algorithm. However, Theorem 4.2 says that when there is a discrepancy between the initial condition and the desired one, the final output trajectory is decided by all the system parameters. In the D-type ILC rule, we only have the freedom to tune one parameter, that is the learning gain Q. It is hard to ensure that the ILC convergence condition is met while also minimizing the effect of $e^{F_D t}$. This motivates us to consider a PD-type updating rule in the next section, which gives us two degrees of freedom to ensure both convergence and final performance.

Corollary 4.2 If Q is of full column rank, then

$$\hat{\mathbf{y}}_{v,j} = \mathbf{y}_d - C(\mathbf{x}_d(0) - \mathbf{x}_{i,j}(0)).$$

Proof: If Q is of full column rank, it can be shown that $\dot{\mathbf{y}}_d - \dot{\hat{\mathbf{y}}}_{v,j}$ has to be zero in order to satisfy the equality constraint (4.6). Hence, by integration we have

$$\mathbf{y}_d(t) - \mathbf{y}_d(0) = \hat{\mathbf{y}}_{v,j}(t) - \hat{\mathbf{y}}_{v,j}(0),$$
$$\hat{\mathbf{y}}_{v,j}(t) = \mathbf{y}_d(t) - \mathbf{y}_d(0) + \hat{\mathbf{y}}_{v,j}(0),$$
$$\hat{\mathbf{y}}_{v,j}(t) = \mathbf{y}_d(t) - C(\mathbf{x}_d(0) - \mathbf{x}_{i,j}(0)).$$

■

Corollary 4.2 implies that when Q is of full column rank or just nonsingular, the final output trajectory of each follower is identical to the leader trajectory with a constant shift, and the discrepancy is simply the initial output difference $C(\mathbf{x}_d(0) - \mathbf{x}_{i,j}(0))$.

4.3.2 Distributed PD-type Updating Rule

It was noted in the previous section that the D-type ILC rule only has one degree of freedom. To improve the final performance as the iteration number gets large, we can apply a PD-type updating rule. Let the distributed measurement be the same as the one in the D-type case. The proposed PD-type updating rule for the jth agent is

$$\mathbf{u}_{i+1,j} = \mathbf{u}_{i,j} + Q(\dot{\boldsymbol{\xi}}_{i,j} + R\boldsymbol{\xi}_{i,j}), \tag{4.23}$$

where R is another learning gain to be designed.

Comparing (4.23) with (4.4), the -type rule has one extra term $R\boldsymbol{\xi}_{i,j}$ compared with the D-type rule. It will be shown that it is this extra term that enables us to tune the final performance.

Theorem 4.3 Consider the MAS (4.1), under Assumptions 4.1 and 4.2, the communication graph $\bar{\mathcal{G}}$, and the distributed PD-type updating rule (4.23). If the learning gain Q is chosen such that

$$\rho(I_{mN} - H \otimes QCB) < 1,$$

then the ILC rule is stable and the output trajectory of any follower converges, in particular as the iteration number tends to infinity

$$\mathbf{u}_{i,j} = \mathbf{u}_d + (QCB)^{-1}Q(CA + RC)e^{F_{PD} t}(\mathbf{x}_d(0) - \mathbf{x}_{i,j}(0)),$$

and

$$\mathbf{y}_{i,j} = \mathbf{y}_d - Ce^{F_{PD} t}(\mathbf{x}_d(0) - \mathbf{x}_{i,j}(0)),$$

where $F_{PD} = A - B(QCB)^{-1}Q(CA + RC)$.

The proof of Theorem 4.3 is omitted here as it can be derived analogously to Theorem 4.1. In Theorem 4.3, Q can be tuned to make the ILC rule stable, and R is used to modify the final performance as the iteration number gets large. Hence, R should be designed such that F_{PD} is Hurwitz. Then, the final output trajectory of each individual follower will converge exponentially to the leader's trajectory.

Rewrite F_{PD} as

$$F_{PD} = (I_n - B(QCB)^{-1}QC)A + B(QCB)^{-1}Q(-R)C,$$

which can be interpreted as a static output feedback stabilization problem (Syrmos *et al.*, 1997) with system matrix $(I_n - B(QCB)^{-1}QC)A$, input gain $B(QCB)^{-1}Q$, and output matrix C. R is the output feedback gain. R should be chosen such that the output feedback system is stable.

4.4 Illustrative Examples

To verify the theoretical results in the previous section, two illustrative examples are presented here. Consider a group of four followers with their dynamics governed by the following model:

$$A = \begin{bmatrix} 0 & 1 \\ -6 & -5 \end{bmatrix}, B = \begin{bmatrix} 0 \\ 1 \end{bmatrix}, C = \begin{bmatrix} 1 & 0 \\ 2 & 1 \end{bmatrix}.$$

The communication among all the agents in the network is depicted in Figure 4.1. Vertex 0 represents the leader agent. It has two direct edges (dashed lines) to agents 1 and 3, which means that agents 1 and 3 are able to access the leader's information. The communications among followers are represented by the solid lines. It is easy to verify that the complete information flow graph $\bar{G}$ contains a spanning tree with the leader being its root. From Figure 4.1, we can write down the graph Laplacian for the followers below:

$$L = \begin{bmatrix} 1 & 0 & -1 & 0 \\ -1 & 1 & 0 & 0 \\ 0 & -1 & 2 & -1 \\ 0 & -1 & 0 & 1 \end{bmatrix},$$

and $D = \text{diag}(1, 0, 1, 0)$ which represents the information flow from leader to followers.

The leader's input is chosen as $u_d = t + 4\sin(2t)$, $t \in [0, 5]$, and initial condition $x_d(0) = [0, 0]^T$. The initial conditions for followers are $x_{i,1}(0) = [1, 0.8]^T$, $x_{i,2}(0) = [-0.7, 1]^T$, $x_{i,3}(0) = [0.5, 0.6]^T$, and $x_{i,4}(0) = [-1, -1]^T$. Obviously, initial

Figure 4.1 Communication topology among agents in the network.

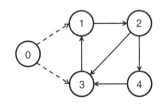

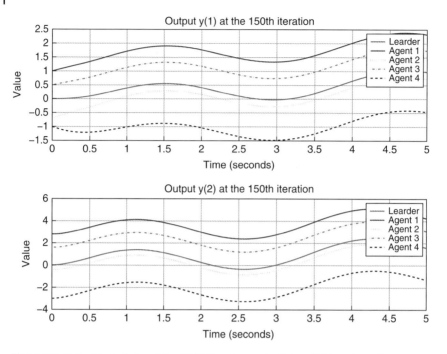

Figure 4.2 Output trajectories at the 150th iteration under D-type ILC learning rule.

state errors are nonzero. Note that the output is only accessible to agents 1 and 3. We apply the updating rules (4.4) and (4.23) with learning gains $Q = [0.0667, 0.333]$ and $R = \text{diag}(5, 5)$. We check the convergence condition,

$$\rho(I_4 - H \otimes QCB) = 0.7715 < 1,$$

which satisfies the convergence requirement in Corollary 4.1. Figure 4.2 shows the output profiles of all agents at the 150th iteration under a D-type updating rule. The followers' output is able to track the general trend of the leader, but large deviations exist. Simple calculation shows that the spectrum of F_D is $\sigma(F_D) = \{0, -2.2\}$. This means that one eigenmode converges to zero exponentially; the other one is a constant. This implies that the discrepancies between the leader's and followers' trajectories approach a constant and will remain thus. The theoretical prediction perfectly matches the observation in simulation. Figures 4.3 and 4.4 describe the trajectory and error profiles under the PD-type rule. It can be seen that the tracking error converges exponentially to zero. Calculating the spectrum of F_{PD}, we have $\sigma(F_{PD}) = \{-2.2, -5\}$, which are all stable. This also matches the actual simulation.

4.5 Conclusion

This chapter investigates the initial state error problem in the multi-agent setup where each agent is able to be reset to a fixed initial position. Such an assumption is less restrictive than the perfect initial condition. It has been shown that under the imperfect initial

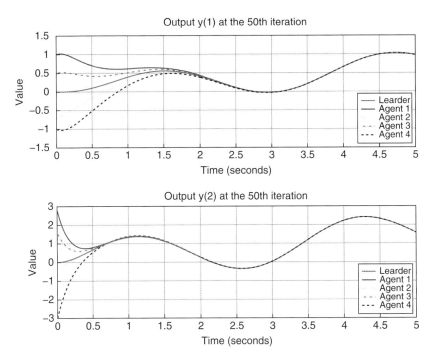

Figure 4.3 Output trajectories at the 50th iteration under PD-type ILC learning rule.

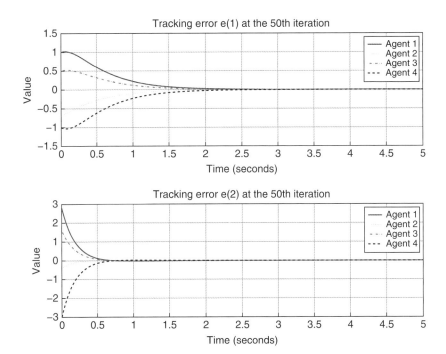

Figure 4.4 Tracking error profiles at the 50th iteration under PD-type ILC learning rule.

condition, the D-type learning rule is still convergent. However, the final trajectories have deviations from that of the leader. To improve the performance, a PD-type updating rule is proposed, which gives the designer more freedom to tune the final control performance. Numerical examples verify the obtained results. The obtained results can be easily generalized to PID-type learning rules.

5

Multi-agent Consensus Tracking with Input Sharing by Iterative Learning Control

5.1 Introduction

A number of ILC works for formation control and consensus tracking problems have been reported in the literature (Ahn and Chen, 2009; Liu and Jia, 2012; Meng and Jia, 2014; Xu *et al.*, 2011; Yang *et al.*, 2012; Yang and Xu, 2012; Li and Li, 2013). The distributed ILC laws in these works have a common structure, that is, each individual agent maintains its own learning process, and the correction term is synthesized by local measurement (extended tracking error). Furthermore, they have no communication over their learned information. However, if the control inputs for some agents are already close to the desired control signals, these agents may help other agents by sharing their learned information. Therefore, all the agents in the system will be better off toward the global objective, that is, reaching consensus. Based on this idea, a new type of learning controller is developed in this chapter. The new controller has two types of learning mechanisms. On the one hand, each agent observes the behavior of the agents in their neighborhood, and constructs a correction term through its local measurement. This is the typical learning method in the majority of consensus literature. On the other hand, each agent shares its learned control input signal with its neighbors. As such, the two learning mechanisms are combined in the hope of enhancing the learning process. The main contribution of this chapter is the incorporation of input sharing into the learning controller. The convergence condition of the proposed controller is rigorously derived and analyzed. To demonstrate the robustness of the proposed controller to communication variations, the ILC with input sharing is extended to a MAS with iteration-varying graph. As the new learning controller combines two learning mechanisms, the traditional ILC renders a special case. This point is verified by the convergence condition. To demonstrate the performance of the new learning controller, two numerical examples are provided at the end. These show that the new learning controller not only improves the convergence rate, but also smooths the transient performance.

This chapter is organized as follows. In Section 5.2, the consensus tracking problem is formulated. Then, the controller design and convergence analysis are developed in Sections 5.3 and 5.4. To demonstrate the effectiveness of the results, two numerical examples are presented in Section 5.5. Lastly, Section 5.6 draws conclusions.

Iterative Learning Control for Multi-agent Systems Coordination, First Edition.
Shiping Yang, Jian-Xin Xu, Xuefang Li, and Dong Shen.
© 2017 John Wiley & Sons Singapore Pte. Ltd. Published 2017 by John Wiley & Sons Singapore Pte. Ltd.

5.2 Problem Formulation

Consider a group of N homogeneous dynamic agents, with the jth agent governed by the following linear time-invariant model:

$$\begin{cases} \dot{\mathbf{x}}_{i,j}(t) = A\mathbf{x}_{i,j}(t) + B\mathbf{u}_{i,j}(t) \\ \mathbf{y}_{i,j}(t) = C\mathbf{x}_{i,j}(t) \end{cases} \forall j \in \mathcal{V}, \tag{5.1}$$

where i denotes the iteration number, $\mathbf{x}_{i,j} \in \mathbb{R}^n$ is the state vector, $\mathbf{y}_{i,j} \in \mathbb{R}^p$ is the output vector, $\mathbf{u}_{i,j} \in \mathbb{R}^m$ is the control input, and A, B, C are constant matrices of compatible dimensions. For simplicity, the time argument, t, is dropped when no confusion arises.

The desired consensus trajectory, or the (virtual) leader's trajectory, is $\mathbf{y}_d(t)$ defined on a finite-time interval $[0, T]$, which is generated by the following dynamics:

$$\begin{cases} \dot{\mathbf{x}}_d = A\mathbf{x}_d + B\mathbf{u}_d, \\ \mathbf{y}_d = C\mathbf{x}_d, \end{cases} \tag{5.2}$$

where $\mathbf{u}_d$ is the continuous and unique desired control input.

Due to communication or sensor limitations, the leader's trajectory is only accessible to a small portion of the followers. Let the communication among followers be described by the graph $\mathcal{G}$. If the leader is labeled by vertex 0, then the complete information flow among all the agents can be characterized by a new graph $\overline{\mathcal{G}} = \{0 \cup \mathcal{V}, \overline{\mathcal{E}}\}$, where $\overline{\mathcal{E}}$ is the new edge set.

Let the tracking error for the jth agent at the ith iteration be $\mathbf{e}_{i,j} = \mathbf{y}_d - \mathbf{y}_{i,j}$. Assuming the tracking task is repeatable, the control objective is to design a set of distributed learning controllers such that the tracking error converges to zero along the iteration axis, that is, $\lim_{i \to \infty} \|\mathbf{e}_{i,j}\| = 0$ for $j \in \mathcal{V}$.

To simplify the controller design and convergence analysis, the following two assumptions are imposed.

Assumption 5.1 CB is of full column rank.

Remark 5.1   Assumption 5.1 implies that the relative degree of system (5.1) is well defined and it is exactly 1. When CB is not of full rank, a high-order derivative of the tracking error can be utilized in the controller design, and perfect consensus tracking can still be achieved.

Assumption 5.2   The initial state of all agents are reset to the desired initial state at every iteration, that is, $\mathbf{x}_{i,j}(0) = \mathbf{x}_d(0)$.

Remark 5.2   Assumption 5.2 is referred as the identical initialization condition (*iic*), which is widely used in the ILC literature Xu and Tan (2003). Please see Remark 2.4 for detailed discussions and how to relax this assumption.

5.3 Controller Design and Convergence Analysis

In this section, the ILC algorithm with input sharing is developed and the convergence condition is rigorously derived.

5.3.1 Controller Design Without Leader's Input Sharing

The learning controller consists of two learning mechanisms. The first one is the correction term by distributed measurement over the communication topology $\bar{\mathcal{G}}$. Define the extended tracking error $\xi_{i,j}$ as

$$\xi_{i,j} = \sum_{k \in \mathcal{N}_j} a_{j,k}(\mathbf{y}_{i,k} - \mathbf{y}_{i,j}) + d_j(\mathbf{y}_d - \mathbf{y}_{i,j}), \tag{5.3}$$

where $d_j = 1$ if $(0, j) \in \bar{\mathcal{E}}$, otherwise $d_j = 0$.

The second mechanism is the input sharing. For simplicity, the leader's input is not included in this section. However, it can be easily incorporated in the controller, which will be discussed in the next section. Before presenting the input sharing mechanism, define the following sparse matrix set:

$$S = \{S \in \mathbb{R}^{N \times N} \mid S_{k,j} = 0 \text{ if } (j, k) \notin \mathcal{E} \text{ with } j \neq k\},$$

where $S_{k,j}$ is the (k, j)th entry of the matrix S. Let the input sharing of agent j be the weighted average of the control inputs from its neighbors. As such the input sharing can be defined as

$$\mathbf{u}_{i,j}^s = \sum_{k \in \mathcal{N}_j} W_{j,k} \mathbf{u}_{i,k}, \tag{5.4}$$

where $W = (W_{j,k}) \in S$, $W\mathbf{1} = \mathbf{1}$, and $\mathbf{1}$ is a $N \times 1$ vector with all its elements being ones.

Now combining the two learning resources in (5.3) and (5.4), the new learning structure is

$$\mathbf{u}_{i+1,j} = \mathbf{u}_{i,j}^s + \Gamma\dot{\xi}_{i,j}, \quad \mathbf{u}_{0,j}^s = 0, \tag{5.5}$$

where Γ is the learning gain to be designed.

Remark 5.3   The initial input sharing $\mathbf{u}_{0,j}^s$ is set to identically zero for all agents in the learning controller (5.5). However, in the actual implementation, $\mathbf{u}_{0,j}^s$ can be initialized by some feedback control method such that the control systems at the first iteration are stable. This can help improve the transient performance.

To analyze the convergence property of the learning rule, we rewrite (5.5) in the compact matrix form and define $H \triangleq L + D$. By using the definition of the tracking error $\mathbf{e}_{i,j}$, the extended tracking error $\xi_{i,j}$ can be written as

$$\begin{aligned}
\xi_i &= -(L \otimes I_p)(\mathbf{y}_i - \mathbf{y}_d) + (D \otimes I_p)\mathbf{e}_i \\
&= (L \otimes I_p)\mathbf{e}_i + (D \otimes I_p)\mathbf{e}_i \\
&= ((L + D) \otimes I_p)\mathbf{e}_i \\
&= (H \otimes I_p)\mathbf{e}_i, \tag{5.6}
\end{aligned}$$

where ξ_i, $\mathbf{y}_i$ and $\mathbf{e}_i$ are the column stack vectors of $\xi_{i,j}$, $\mathbf{y}_{i,j}$, and $\mathbf{e}_{i,j}$ respectively, $I_{(\cdot)}$ denotes the identity matrix with the subscript indicating its dimension, L is the graph Laplacian of $\mathcal{G}$, $D = \text{diag}(d_1, \ldots, d_N)$, and $H \triangleq L + D$. By using (5.6), (5.5) becomes

$$\mathbf{u}_{i+1} = (W \otimes I_m)\mathbf{u}_i + (H \otimes \Gamma)\dot{\mathbf{e}}_i, \tag{5.7}$$

where $\mathbf{u}_i$ is the column stack vector of $\mathbf{u}_{i,j}$.

The main result is summarized in the following theorem.

Theorem 5.1 For the system in (5.1), under Assumptions 5.1 and 5.2, and the learning rule (5.5), if

$$|W \otimes I_m - H \otimes \Gamma CB| < 1, \tag{5.8}$$

the individual tracking error $\mathbf{e}_{i,j}$ for $j \in \mathcal{V}$ converges to zero along the iteration domain, that is, $\lim_{i \to \infty} \|\mathbf{e}_{i,j}\| = 0$.

Proof: Define $\delta \mathbf{u}_{i,j} \triangleq \mathbf{u}_d - \mathbf{u}_{i,j}$, and $\delta \mathbf{x}_{i,j} \triangleq \mathbf{x}_d - \mathbf{x}_{i,j}$. $\delta \mathbf{u}_i$ and $\delta \mathbf{x}_i$ are the column stack vectors of $\delta \mathbf{u}_{i,j}$ and $\delta \mathbf{x}_{i,j}$, respectively. Noticing that $W\mathbf{1} = \mathbf{1}$, Equation (5.7) leads to

$$\delta \mathbf{u}_{i+1} = (W \otimes I_m)\delta \mathbf{u}_i - (H \otimes \Gamma)\dot{\mathbf{e}}_i. \tag{5.9}$$

From the systems dynamics (5.1) and (5.2), $\dot{\mathbf{e}}_{i,j}$ can be calculated as

$$\begin{aligned}
\dot{\mathbf{e}}_{i,j} &= \dot{\mathbf{y}}_d - \dot{\mathbf{y}}_{i,j} \\
&= CA\mathbf{x}_d + CB\mathbf{u}_d - CA\mathbf{x}_{i,j} + CB\mathbf{u}_{i,j} \\
&= CA\delta\mathbf{x}_{i,j} + CB\delta\mathbf{u}_{i,j}.
\end{aligned} \tag{5.10}$$

Rewriting (5.10) in the compact matrix form yields

$$\dot{\mathbf{e}}_i = (I_N \otimes CA)\delta\mathbf{x}_i + (I_N \otimes CB)\delta\mathbf{u}_i. \tag{5.11}$$

Substituting (5.11) in (5.9) yields

$$\delta\mathbf{u}_{i+1} = (W \otimes I_m - H \otimes \Gamma CB)\delta\mathbf{u}_i - (H \otimes \Gamma CA)\delta\mathbf{x}_i. \tag{5.12}$$

For simplicity, define the following constants:

$$\begin{aligned}
a &\triangleq |I_N \otimes A|, \\
M &\triangleq W \otimes I_m - H \otimes \Gamma CB, \\
b_1 &\triangleq |H \otimes \Gamma CA|, \\
b_2 &\triangleq |I_N \otimes B|.
\end{aligned}$$

Taking the λ-norm on both sides of (5.12), we have

$$\|\delta\mathbf{u}_{i+1}\|_\lambda \leq |M| \|\delta\mathbf{u}_i\|_\lambda + b_1 \|\delta\mathbf{x}_i\|_\lambda. \tag{5.13}$$

To derive the convergence condition, the relation between $\|\delta\mathbf{u}_i\|_\lambda$ and $\|\delta\mathbf{x}_i\|_\lambda$ should be investigated. From the dynamics of (5.1) and (5.2), $\delta\mathbf{x}_i$ satisfies

$$\delta\dot{\mathbf{x}}_i = (I_N \otimes A)\delta\mathbf{x}_i + (I_N \otimes B)\delta\mathbf{u}_i.$$

Together with Assumption 5.2 ($\delta\mathbf{x}_i = 0$), the above differential equation renders the following solution:

$$\delta\mathbf{x}_i = \int_0^t e^{I_N \otimes A(t-\tau)}(I_N \otimes B)\delta\mathbf{u}_i(\tau)\, d\tau. \tag{5.14}$$

Therefore, based on (5.14) we have

$$|\delta\mathbf{x}_i| \leq \int_0^t \left| e^{I_N \otimes A(t-\tau)} \right| |I_N \otimes B| |\delta\mathbf{u}_i(\tau)|\, d\tau, \tag{5.15}$$

which implies that

$$|\delta \mathbf{x}_i| \le \int_0^t e^{a(t-\tau)} b_2 |\delta \mathbf{u}_i(\tau)| \, d\tau. \tag{5.16}$$

Taking the λ-norm on both sides of (5.15) yields,

$$\|\delta \mathbf{x}_i\|_\lambda \le \frac{b_2}{\lambda - a} \|\delta \mathbf{u}_i\|_\lambda. \tag{5.17}$$

Substituting (5.17) in (5.13), we can obtain

$$\|\delta \mathbf{u}_{i+1}\|_\lambda \le \left(|M| + \frac{b_1 b_2}{\lambda - a} \right) \|\delta \mathbf{u}_i\|_\lambda. \tag{5.18}$$

If $|M| < 1$, it is possible to choose a sufficiently large λ such that $|M| + \frac{b_1 b_2}{\lambda - a} \triangleq \eta < 1$, which implies that $\|\delta \mathbf{u}_{i+1}\|_\lambda \le \eta \|\delta \mathbf{u}_i\|_\lambda$, where $\eta < 1$ is a positive constant. Therefore, by contraction-mapping, $\|\delta \mathbf{u}_i\|_\lambda$ converges to zero as the iteration number tends to infinity. As a result, we have $\delta \mathbf{u}_i \to 0$. According to the convergence of $\delta \mathbf{u}_i$ and the inequality (17), it is obvious that $\lim_{i \to \infty} \delta \mathbf{x}_i = 0$. Since $\mathbf{e}_i = (I_N \otimes C) \delta \mathbf{x}_i$, it follows that $\lim_{i \to \infty} \mathbf{e}_i \to 0$, that is, $\lim_{i \to \infty} \mathbf{e}_{i,j} \to 0$. ∎

Remark 5.4   Although the agent model (5.1) is linear time-invariant system, the analysis here can be easily extended to time-varying linear system as well as a global Lipschitz nonlinear system. The controller structure and convergence results remain the same. Detailed analysis of ILC for global Lipschitz nonlinear systems can be found in Xu and Tan (2003).

Based on Lemma 5.6.10 in Horn and Johnson (1985), the norm inequality condition (5.8) can be simplified to the spectral radius condition, which makes the convergence condition more general; see detailed discussion and development on this topic in Chapter 2.

Corollary 5.1   For the system in (5.1), under Assumptions 5.1 and 5.2, and the learning rule (5.5), if

$$\rho \left(W \otimes I_m - H \otimes \Gamma CB \right) < 1, \tag{5.19}$$

the individual tracking error $\mathbf{e}_{i,j}$ for $j \in \mathcal{V}$ converges to zero along the iteration domain, that is, $\lim_{i \to \infty} \|\mathbf{e}_{i,j}\| = 0$.

Although Theorem 5.1 and Corollary 5.1 do not reveal any connection between the communication topology and the convergence condition, communication among agents is one of the indispensable requirements for the consensus tracking problem. Thus, it is very important to analyze under what kind of communication requirement the convergence condition (5.8) can be satisfied. It turns out that the communication requirement is quite weak. Theorem 5.2 presents the communication requirement.

Theorem 5.2   There exists a learning gain Γ satisfying the convergence condition (5.19) if and only if the communication graph $\overline{\mathcal{G}}$ contains a spanning tree with the leader being its root.

Proof: Necessity: If the communication requirement is not satisfied, there must exist one or a group of agents that are isolated from the leader. The leader's information cannot reach that group of isolated agents. Therefore, the consensus tracking cannot be achieved.

Sufficiency: As CB is of full column rank, choose the learning gain

$$\Gamma = \gamma((CB)^T CB)^{-1}(CB)^T.$$

Furthermore, choose $W = I_N$ ($I_N \in S$ and $I_N \mathbf{1} = \mathbf{1}$). Therefore, the convergence condition becomes

$$\rho\left(I_{mN} - \gamma H \otimes I_m\right) < 1.$$

Since the communication requirement is satisfied, then $L + D$ is nonsingular and all the eigenvalues have positive real parts (Ren and Cao, 2011). Therefore, we can find a sufficiently small γ such that the spectral radius condition holds. ∎

Remark 5.5 In most ILC work for multi-agent coordination, the learned inputs are not shared among agents (Ahn and Chen, 2009; Yang *et al.*, 2012). The typical controller structure has the following form:

$$\mathbf{u}_{i+1,j} = \mathbf{u}_{i,j} + \Gamma \dot{\xi}_{i,j}. \tag{5.20}$$

Comparing this with (5.5), the only difference is that (5.5) has the input sharing component. The convergence condition for (5.20) is

$$|I_{mN} - H \otimes \Gamma CB| < 1.$$

The convergence condition is identical to (5.8) if we select $W = I_N$. Therefore, it can be seen that (5.20) is a special case of (5.5).

5.3.2 Optimal Design Without Leader's Input Sharing

For practical systems, in order to improve efficiency and also reduce operational costs, it is desired to achieve the fastest convergence rate. However, in the ILC system, it seems that it is not possible to directly achieve the fastest convergence rate for tracking error, as the actual convergence profile depends on both controller and system parameters. Also, the system parameters are usually not precisely known, and some of them are not even required for the learning controller, for example, the matrix A in (5.1). The best thing we can do is to achieve the fastest convergence of $\|\delta \mathbf{u}_{i,j}\|_\lambda$. Therefore, we can try to minimize the left hand sides of (5.8) or (5.19). Minimizing the spectral radius is extremely difficult in general, as the spectral radius is not a convex function (Xiao and Boyd, 2004). Some of the minimization problems related to spectral radius are NP-hard. As such, we will focus on the minimization of (5.8). Set the learning gain as follows:

$$\Gamma = \gamma((CB)^T CB)^{-1}(CB)^T, \tag{5.21}$$

where γ is a positive constant. Thus, the minimization problem can be formulated as below:

$$\underset{W, \gamma}{\text{minimize}} \quad |W - \gamma H|$$

$$\text{subject to} \quad W \in S, W\mathbf{1} = \mathbf{1},$$

$$\gamma > 0.$$

By introducing a new variable r and applying the Schur complement formula, the minimization problem can be converted to an equivalent linear matrix inequality (LMI) problem:

$$\underset{W,\gamma}{\text{minimize}} \quad r$$

$$\text{subject to} \quad \begin{bmatrix} rI_N & (W - \gamma H)^T \\ W - \gamma H & rI_N \end{bmatrix} \geq 0,$$

$$W \in S, W\mathbf{1} = \mathbf{1},$$

$$\gamma > 0,$$

which can by efficiently solved by MATLAB LMI toolbox.

Remark 5.6 It is worthwhile to note that (5.21) requires accurate information about CB. However, the learning gain (5.21) is only one illustrative example to simplify the optimization problem. To design learning gain when CB is unknown is another topic and is dependent on what information about CB is available. In an analogous way to robust control design, by specifying the uncertainty types and prior knowledge on CB or other systems parameters, the robustness could be formulated and incorporated into ILC designs.

5.3.3 Controller Design with Leader's Input Sharing

When the leader's input is available to some of the agents in the network, the shared input in (5.4) should be incorporated with the leader's input $\mathbf{u}_d$. For instance, we can adopt the following sharing mechanism:

$$\mathbf{u}^s_{i,j} = \mathbf{u}_{i,j} + \epsilon \left(\sum_{k \in \mathcal{N}_j} a_{j,k}(\mathbf{u}_{i,k} - \mathbf{u}_{i,j}) + d_j(\mathbf{u}_d - \mathbf{u}_{i,j}) \right), \tag{5.22}$$

where ϵ is a positive learning gain. Substituting (5.22) in the controller equation (5.5), the learning controller becomes

$$\mathbf{u}_{i+1,j} = \mathbf{u}_{i,j} + \epsilon \left(\sum_{k \in \mathcal{N}_j} a_{j,k}(\mathbf{u}_{i,k} - \mathbf{u}_{i,j}) + d_j(\mathbf{u}_d - \mathbf{u}_{i,j}) \right) + \Gamma \dot{\xi}_{i,j}. \tag{5.23}$$

Rewriting the new learning rule with leader's input (5.23) in matrix form, we have

$$\mathbf{u}_{i+1} = \mathbf{u}_i + \epsilon(H \otimes I_m)\delta\mathbf{u}_i + (H \otimes \Gamma)\dot{\mathbf{e}}_i. \tag{5.24}$$

By applying a similar convergence analysis to that in the proof of Theorem 5.1, the following corollary can be derived.

Corollary 5.2 For the system in (5.1), under Assumptions 5.1 and 5.2, and the learning rule (5.23), if

$$\rho\left(I_{mN} - \epsilon H \otimes I_m - H \otimes \Gamma CB\right) < 1, \tag{5.25}$$

the individual tracking error $\mathbf{e}_{i,j}$ for $j \in \mathcal{V}$ converges to zero along the iteration domain, that is, $\lim_{i \to \infty} \|\mathbf{e}_{i,j}\| = 0$.

If we adopt the learning gain in (5.21), the convergence condition in (5.25) becomes

$$\rho\left(I_N - (\epsilon + \gamma)H\right) < 1, \tag{5.26}$$

which is an interesting result. Examine the controller without input sharing (5.20), and set the learning gain to be

$$\Gamma = (\epsilon + \gamma)((CB)^T CB)^{-1}(CB)^T.$$

In this case, the convergence condition for (5.20) is exactly the same as (5.26). As the convergence condition is the same, but the controllers are different, we may expect different control performance. Furthermore, as controller (5.23) contains more learning resources than (5.20), it is possible that (5.23) may outperform (5.20). This result will be verified in the simulation section.

5.4 Extension to Iteration-Varying Graph

In this section, the proposed ILC scheme with input sharing will be extended to an iteration-varying graph.

5.4.1 Iteration-Varying Graph with Spanning Trees

Assume that the communication topology is iteration-varying but the graph is fixed and contains a spanning tree at each iteration. If the controller (5.5) is applied, the convergence condition becomes

$$|W(i) \otimes I_m - H(i) \otimes \Gamma(i)CB| < 1, \tag{5.27}$$

where $\Gamma(i)$ and $W(i)$ depend on the iteration index i. Let $N_g > 0$ be the number of graphs with a spanning tree among all the agents and define $\mathcal{H} \triangleq \{\text{All possible } H(i)\}$. It is obvious that $\mathcal{H}$ has N_g elements. Fix an order over $\mathcal{H}$ and denote H_k the kth element of $\mathcal{H}$, namely, $\mathcal{H} = \{H_k, k = 1, 2, \ldots, N_g\}$. Then the convergence condition can be rewritten as

$$\max_{k=1,2,\ldots,N_g} |W_k \otimes I_m - H_k \otimes \Gamma_k CB| < 1, \tag{5.28}$$

where W_k and Γ_k denote the design gains corresponding to the matrix H_k. Similar to Subsection 5.3.2, the optimal design can be reformulated as N_g minimization problems and converted to N_g equivalent LMI problems. In the simulation section, we will investigate the performance by a numerical example.

5.4.2 Iteration-Varying Strongly Connected Graph

In this subsection, we assume the communication topology is iteration-varying, but the graph is fixed and strongly connected in each iteration. Furthermore, at least one of the followers has access to the leader's trajectory in every iteration.

Consider the most general control law (5.24). Since the communication topology is iteration-varying, we adopt the following iteration-varying learning gains:

$$\Gamma(i) = \frac{1}{2q(i)}[(CB)^T(CB)]^{-1}(CB)^T, \tag{5.29}$$

$$\epsilon(i) = \frac{1}{2q(i)} \tag{5.30}$$

where $\Gamma(i)$ and $\epsilon(i)$ depend on the iteration index i, and

$$q(i) > \max_{j=1,\ldots,N} \sum_{k=1}^{N} a_{j,k}(i) + d_j(i).$$

Then ILC law (5.24) becomes

$$\delta \mathbf{u}_{i+1} = \delta \mathbf{u}_i + \epsilon(i)(H(i) \otimes I_m)\delta \mathbf{u}_i + (H(i) \otimes \Gamma(i))\dot{\mathbf{e}}_i. \tag{5.31}$$

Adopting the learning gains (5.29)–(5.30) and the error dynamics (5.11) gives

$$\delta \mathbf{u}_{i+1} = M(i)\delta \mathbf{u}_i - F(i)\delta \mathbf{x}_i, \tag{5.32}$$

where

$$M(i) \triangleq (I_N - \frac{1}{q(i)}H(i)) \otimes I_m,$$
$$F(i) \triangleq H(i) \otimes \Gamma(i)CA.$$

Since the communication graph is strongly connected in each iteration, $M(i)$ must be an irreducible matrix. Notice that $q(i)$ is larger than the greatest diagonal entry of $H(i)$, and D contains at least one positive diagonal entry, hence, at least one row sum of $M(i)$ is strictly less than one. Therefore, $M(i)$ is an irreducible substochastic matrix with positive diagonal entries.

The convergence of the proposed ILC scheme under iteration-varying strongly connected graph is summarized in the following theorem.

Theorem 5.3 Consider the systems (5.1), under Assumptions 5.1 and 5.2, the learning rule (5.23), and the learning gains (5.29) and (5.30). If the communication graph is iteration-varying, but in each iteration the graph is fixed and strongly connected, and at least one of the followers in the network has access to the virtual leader's trajectory, then the individual tracking error $\mathbf{e}_{i,j}$ for $j \in \mathcal{V}$ converges to zero along the iteration domain, that is, $\lim_{i \to \infty} \|\mathbf{e}_{i,j}\| = 0$.

Proof: From (5.32), the relation between $\delta \mathbf{u}_{i+1}$ and $\delta \mathbf{u}_1$ can be given as follows:

$$\begin{aligned} \delta \mathbf{u}_{i+1} = M(i,1)\delta \mathbf{u}_1 &- M(i,2)F(1)\delta \mathbf{x}_1 \\ &-M(i,3)F(2)\delta \mathbf{x}_2 - \cdots - F(i)\delta \mathbf{x}_i, \end{aligned} \tag{5.33}$$

where $M(i,k) \triangleq M(i)M(i-1)\cdots M(k)$ for $i \geq k$. Taking the infinity norm on both sides of Equation (5.33), we can obtain

$$\begin{aligned} |\delta \mathbf{u}_{i+1}| \leq |M(i,1)||\delta \mathbf{u}_1| &+ |M(i,2)||F(1)||\delta \mathbf{x}_1| \\ &+|M(i,3)||F(2)||\delta \mathbf{x}_2| + \cdots + |F(i)||\delta \mathbf{x}_i|. \end{aligned} \tag{5.34}$$

Group every N matrix products together in (5.34), and apply Lemma 3.3, we have

$$\begin{aligned} |\delta \mathbf{u}_{i+1}| \leq \eta^{\lfloor \frac{i}{N} \rfloor}|\delta \mathbf{u}_1| &+ \eta^{\lfloor \frac{i-1}{N} \rfloor}|F(1)||\delta \mathbf{x}_1| \\ &+\eta^{\lfloor \frac{i-2}{N} \rfloor}|F(2)||\delta \mathbf{x}_2| + \cdots + |F(i)||\delta \mathbf{x}_i|, \end{aligned} \tag{5.35}$$

where $0 < \eta < 1$. Taking the λ-norm on both sides of (5.35) and substituting in Equation (5.17) yields

$$|\delta\mathbf{u}_{i+1}|_\lambda \leq \eta^{\lfloor\frac{i}{N}\rfloor}|\delta\mathbf{u}_1|_\lambda + \frac{b_2}{\lambda - a}\eta^{\lfloor\frac{i-1}{N}\rfloor}|F(1)||\delta\mathbf{u}_1|_\lambda$$

$$+\frac{b_2}{\lambda - a}\eta^{\lfloor\frac{i-2}{N}\rfloor}|F(2)||\delta\mathbf{u}_2|_\lambda$$

$$+\cdots+\frac{b_2}{\lambda - a}|F(i)||\delta\mathbf{u}_i|_\lambda. \tag{5.36}$$

Let $c_2 \triangleq \max_{k=1}^{i} b_2|F(k)|$. Then (5.36) becomes

$$|\delta\mathbf{u}_{i+1}|_\lambda \leq \eta^{\lfloor\frac{i}{N}\rfloor}|\delta\mathbf{u}_1|_\lambda + \frac{c_2}{\lambda - a}\left(\eta^{\lfloor\frac{i-1}{N}\rfloor}|\delta\mathbf{u}_1|_\lambda\right.$$

$$\left.+\eta^{\lfloor\frac{i-2}{N}\rfloor}|\delta\mathbf{u}_2|_\lambda + \cdots + |\delta\mathbf{u}_i|_\lambda\right). \tag{5.37}$$

According to the result in Lemma 3.4, we can conclude that $\delta\mathbf{u}_{i,j} \to 0$ along the iteration axis, hence, $\lim_{i\to\infty}\mathbf{e}_{i,j} = 0$ as well. ∎

5.4.3 Uniformly Strongly Connected Graph

In this subsection, we further generalize the proposed controller to multi-agent systems with uniformly strongly connected graph. The theorem below presents the most general convergence result in this chapter.

Theorem 5.4 Consider the systems (5.1), under Assumptions 5.1 and 5.2, the learning rule (5.23), and the learning gains (5.29) and (5.30). If the communication graph is uniformly strongly connected along the iteration axis with upper bound K, and in every K consecutive iterations, at least one of the followers in the network has access to the virtual leader's trajectory, then the individual tracking error $\mathbf{e}_{i,j}$ for $j \in \mathcal{V}$ converges to zero along the iteration domain, that is, $\lim_{i\to\infty}\|\mathbf{e}_{i,j}\| = 0$.

Proof: First of all, from the definition of $q(i)$, we have $M(i) = (I_N - \frac{1}{q(i)}H(i)) \otimes I_m$ is a nonnegative matrix with positive diagonal entries. Thus, by applying Lemma 3.6, we can obtain that $M(i + K, i)$ is a nonnegative matrix with positive diagonal entries for any i. In addition, since the union graph $\bigcup_{s=r}^{r+K}\mathcal{G}(s)$ is strongly connected, $M(i + K, i)$ is irreducible as well. Furthermore, because at least one of the followers in the network has access to the virtual leader's trajectory within $[i, i + K]$, then $M(i + K, i)$ is a substochastic matrix.

Now, following a similar approach to the proof of Theorem 5.3 leads to

$$|\delta\mathbf{u}_{i+1}|_\lambda \leq \eta^{\lfloor\frac{i}{NK}\rfloor}|\delta\mathbf{u}_1|_\lambda + \frac{c_2}{\lambda - a}\left(\eta^{\lfloor\frac{i-1}{NK}\rfloor}|\delta\mathbf{u}_1|_\lambda\right.$$

$$\left.+\eta^{\lfloor\frac{i-2}{NK}\rfloor}|\delta\mathbf{u}_2|_\lambda + \cdots + |\delta\mathbf{u}_i|_\lambda\right). \tag{5.38}$$

Therefore, if $\eta - 1 + (1 + \beta)^{NK} < 1$, Lemma 3.4 implies that the tracking error $\mathbf{e}_{i,j}$ converges to zero as $i \to \infty$. ∎

5.5   Illustrative Examples

In order to demonstrate the efficiency of the proposed controller, three numerical examples are considered.

5.5.1   Example 1: Iteration-Invariant Communication Graph

In this subsection, we investigate the tracking performances of three different types of controllers under an iteration-invariant communication graph. Consider four linear time-invariant dynamic agents, with their system matrices given by:

$$A = \begin{bmatrix} 0 & 1 \\ 6 & -5 \end{bmatrix}, B = \begin{bmatrix} 0 \\ 1 \end{bmatrix}, C = \begin{bmatrix} 1 & 0 \\ 2 & 1 \end{bmatrix}.$$

The leader's input is $u_d = t + 4\sin(2t)$, $t \in [0, 5]$. The communication among all the agents in the network is depicted in Figure 5.1. Vertex 0 represents the leader agent. It has two direct edges (dashed lines) to agents 1 and 3, which means that the agents 1 and 3 are able to access to the leader's information. The communication among followers are represented by the solid lines. From Figure 5.1, we can write down the matrix H below:

$$H = \begin{bmatrix} 2 & 0 & -1 & 0 \\ -1 & 1 & 0 & 0 \\ 0 & -1 & 3 & -1 \\ 0 & -1 & 0 & 1 \end{bmatrix}.$$

In the actual simulation, the learning structure (5.23) is adopted as it is the most general case. Meanwhile, the learning gain Γ in the correction term is chosen in the form of (5.21). Thus,

$$\Gamma = \gamma((CB)^T CB)^{-1}(CB)^T = \gamma \begin{bmatrix} 0 & 1 \end{bmatrix}.$$

Three types of controllers are implemented: (1) learning rule without input sharing; (2) learning rule without correction term; and (3) learning rule with both input sharing and correction term. As such the learning gains for the three different cases are given below:

1) $\epsilon = 0, \gamma = \frac{1}{3}$,

2) $\epsilon = \frac{1}{3}, \gamma = 0$,

3) $\epsilon = \frac{1}{3}\left(1 - e^{-0.1*\text{iteration}}\right), \gamma = \frac{1}{3}e^{-0.1*\text{iteration}}$.

Figure 5.1 Iteration-invariant communication topology among agents in the network.

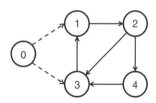

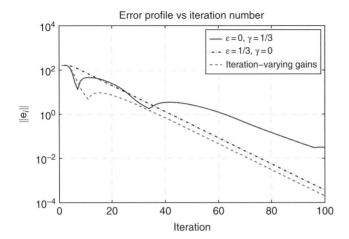

Figure 5.2 Supremum norm of error in log scale vs. iteration number.

From learning rule (5.23), it can be seen that if $\epsilon = 0$, the input sharing is shut down. Similarly, if $\gamma = 0$, the correction term is set to zero. Based on Corollary 5.2, and the learning gain selections above, the three cases render the same convergence condition as $\epsilon + \gamma = \frac{1}{3}$, which is a constant. The iteration-varying gain in the third case is intended; the reason will be made clear later. Checking the spectral radius condition in (5.25),

$$\rho\left(I_4 - \frac{1}{3}H\right) = 0.8727 < 1,$$

confirms that the convergence condition is fulfilled. All the three cases should converge in the iteration domain.

Figure 5.2 shows the supremum errors in three cases versus iteration number. It can be seen from the plots that case 1 has very fast convergence rates at the very first eight iterations, that is when the error magnitudes are large. However, the transient performance is fluctuating. In contrast, case 2 has very smooth transient performance, but the convergence rate is relatively slow. Case 3 contains two updating components, namely, both the input sharing and correction terms, which inherits the advantages of both algorithms that have only one updating component each. Thus, it is easy to interpret why the case 3 controller can outperform both case 1 and case 2. In the first several learning iterations, the error magnitudes are large, the correction term dominates the learning resource in case 3, and the controller tends to reduce the error magnitudes. After some iterations, the error magnitudes are small, and the input sharing becomes the dominant learning resource, which makes the transient profile smooth. That is the reason why the developed learning controller can enjoy both fast convergence and smooth transient performance.

5.5.2 Example 2: Iteration-Varying Communication Graph

In this subsection, we will illustrate that the proposed controller works well under iteration-varying communication graphs. Without loss of generality, we apply the proposed controller (5.5).

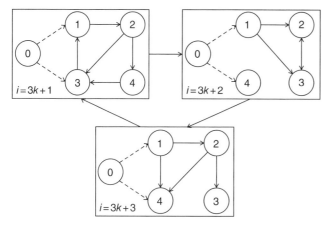

Figure 5.3 Iteration-varying communication topology among agents in the network.

Consider the same MAS as Example 1, and the communication topology depicted in Figure 5.3, in which there are three graphs. In each iteration, only one of the communication graphs is activated, chosen by the selection function $i = 3k + j, j = 1, 2, 3$, and k is a nonnegative integer. For example, if $i = 1$, the graph on the top left corner is selected. In each of the graphs, there exists a spanning tree with the leader being its root. Thus, the communication requirement in Theorem 5.2 is satisfied. To design the learning gain, we need to investigate the matrices $H(i)$, and the first three of them are listed below:

$$H(1) = \begin{bmatrix} 2 & 0 & -1 & 0 \\ -1 & 1 & 0 & 0 \\ 0 & -1 & 3 & -1 \\ 0 & -1 & 0 & 1 \end{bmatrix},$$

$$H(2) = \begin{bmatrix} 2 & -1 & 0 & 0 \\ 0 & 1 & -1 & 0 \\ -1 & -1 & 2 & 0 \\ 0 & 0 & 0 & 1 \end{bmatrix},$$

$$H(3) = \begin{bmatrix} 1 & 0 & 0 & 0 \\ -1 & 1 & 0 & 0 \\ 0 & -1 & 1 & 0 \\ -1 & -1 & 0 & 3 \end{bmatrix}.$$

Adopting the learning gains $\Gamma(i) = \gamma_i((CB)^T CB)^{-1}(CB)^T = \gamma_i[0, 1], i = 1, 2, 3$ and solving three LMI problems with respect to $H(1), H(2)$ and $H(3)$, we can obtain that $\gamma_1 = 0.27$, $\gamma_2 = 0.4824$, $\gamma_3 = 0.2309$, and

$$W(1) = \begin{bmatrix} 0.4457 & 0 & 0.5543 & 0 \\ 0.5263 & 0.4737 & 0 & 0 \\ 0 & 0 & 1 & 0 \\ 0 & 0.1312 & 0 & 0.8688 \end{bmatrix},$$

$$W(2) = \begin{bmatrix} 1 & 0 & 0 & 0 \\ 0 & 0.6218 & 0.3782 & 0 \\ 0.1507 & 0 & 0.8493 & 0 \\ 0 & 0 & 0 & 1 \end{bmatrix},$$

$$W(3) = \begin{bmatrix} 0.6204 & 0.3796 & 0 & 0 \\ 0 & 0.3298 & 0.6702 & 0 \\ 0.2938 & 0.2026 & 0.5036 & 0 \\ 0 & 0 & 0 & 1 \end{bmatrix}.$$

The corresponding convergence conditions are

$$|W(1) - \gamma_1 H(1)| = 0.8778, \quad |W(2) - \gamma_2 H(2)| = 0.8766, \quad |W(3) - \gamma_3 H(3)| = 0.9710,$$

respectively. The convergence of supremum tracking error is shown in Figure 5.4. In addition, to demonstrate the superiority of the proposed controller, the convergence of tracking error for the controller (5.20) under iteration-varying communication graphs is also presented, where the optimal learning gains for the controller (5.20) with respect to $H(1)$, $H(2)$ and $H(3)$ are $\gamma_1 = 0.21$, $\gamma_2 = 0.4238$ and $\gamma_3 = 0.152$, and the corresponding convergence conditions are $|I_{mN} - \gamma_1 H(1)| = 0.9768$, $|I_{mN} - \gamma_2 H(2)| = 0.9623$ and $|I_{mN} - \gamma_3 H(3)| = 0.9866$, respectively. We can see that the controller (5.5) with input sharing has a very fast convergence rate, namely, the controller (5.5) with input sharing outperforms the typical controller (5.20).

5.5.3   Example 3: Uniformly Strongly Connected Graph

Since the result in Theorem 5.3 is a special case of Theorem 5.4, it is sufficient to verify the result in Theorem 5.4 to show the effectiveness of the proposed controller.

Consider the same multi-agent systems as Example 1, with the complete communication topology depicted in Figure 5.5, in which there are four graphs. In each iteration, only one of the communication graphs is activated, chosen by the selection function

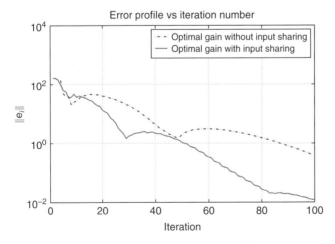

Figure 5.4 Supremum norm of error under iteration-varying graph vs. iteration number.

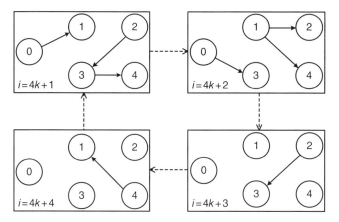

Figure 5.5 Uniformly strongly connected communication graph for agents in the network.

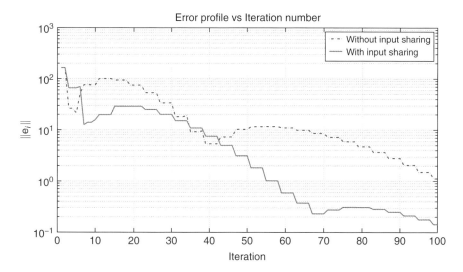

Figure 5.6 Supremum norm of error under uniformly strongly connected graph vs. iteration number.

$i = 4k + j, j = 1, 2, 3, 4$, and k is a nonnegative integer. In each of the graphs, the topology is not even connected. However, the union of all the four graphs is indeed strongly connected. Therefore, the communication topology is uniformly strongly connected along the iteration axis with upper bound $K = 4$. Thus, the communication requirement in Theorem 5.4 is satisfied. To design the learning gain, we need to investigate the matrices $H(i)$, and the first four of them are listed below:

$$H(1) = \begin{bmatrix} 1 & 0 & 0 & 0 \\ 0 & 0 & 0 & 0 \\ 0 & -1 & 1 & 0 \\ 0 & 0 & -1 & 1 \end{bmatrix}, H(2) = \begin{bmatrix} 0 & 0 & 0 & 0 \\ -1 & 1 & 0 & 0 \\ 0 & 0 & 1 & 0 \\ -1 & 0 & 0 & 1 \end{bmatrix}$$

$$H(3) = \begin{bmatrix} 0 & 0 & 0 & 0 \\ 0 & 0 & 0 & 0 \\ 0 & -1 & 1 & 0 \\ 0 & 0 & 0 & 0 \end{bmatrix}, \ H(4) = \begin{bmatrix} 1 & 0 & 0 & -1 \\ 0 & 0 & 0 & 0 \\ 0 & 0 & 0 & 0 \\ 0 & 0 & 0 & 0 \end{bmatrix}.$$

The maximal diagonal entry of $H(i)$ is 1, hence, we can choose the learning gain as $q(i) = 2$ for switching topology. The convergence of the tracking error profile for the controller (5.23), with input sharing, is presented in Figure 5.6. To show the advantage of the proposed controller, the convergence of the tracking error for the controller (5.20) under a uniformly strongly connected graph without input sharing, is also presented, where the learning gain is $\Gamma = \frac{1}{q(i)}[(CB)^T(CB)]^{-1}(CB)^T$. Although with the given learning gains the controllers (5.20) and (5.23) have the same convergence condition $\rho(I_4 - \frac{1}{q(i)}H(i)) < 1$, the one with input sharing demonstrates a faster convergence rate, as shown in Figure 5.6.

5.6 Conclusion

In this chapter, a new iterative learning control structure is developed for multi-agent consensus tracking, consisting of two types of learning updates: the correction term and input sharing. The first of these is commonly used in multi-agent coordination control, whereas the second has not been investigated until now. As more learning resources are available, the learning process is improved. The convergence condition for the proposed learning rule is rigorously studied. It shows that the traditional ILC rule renders as a special case. The proposed controller is also extended to an MAS under an iteration-varying graph. Numerical examples demonstrate that the learning rule with input sharing can simultaneously both improve the convergence rate and smooth the transient performance.

6

A HOIM-Based Iterative Learning Control Scheme for Multi-agent Formation

6.1 Introduction

In Chapters 2–5 we have investigated the consensus tracking problem from different perspectives. Specifically, in Chapters 3 and 4 we relax some assumptions in communication requirement or initial condition. In Chapter 5 we develop a new type of iterative learning method by enabling each agent in the network to share their learned information. All these problems have a common basis, that is, the coordination task is fixed. In other words, the target trajectory is invariant over iterations. Indeed, task invariance is one of the fundamental assumptions in ILC. There are few works in the literature that extend ILC to iteration-varying control tasks. Noticeably, Liu *et al.* (2010) models the desired tracking trajectories over consecutive iterations by a high-order internal model (HOIM). Subsequently, the HOIM is incorporated into the learning algorithm, which shows that learning convergence can be achieved. Inspired by this work, we consider an iteratively switching formation problem in this chapter.

To simplify the problem, we assume a tree-like communication topology, and the formation tasks are based on the communication topology structure. Specifically, if two agents can communicate with each other or one agent can detect the other's position, the agents have to keep a certain relative distance between them to fulfill the formation task. The agent models are described by a kinematic model with nonholonomic constraints. Specifically, the agent is modeled by a point mass with a forward moving speed and an angular velocity. The mathematical formulation for the model will be presented in later sections. Many realistic systems, like ground moving vehicles and fixed wing unmanned aerial vehicles, can be abstracted with this model. The formation tasks alternate between two related geometric formations, and the relationship is determined by an HOIM. This HOIM is known to all followers. The control task is to devise an appropriate local coordination rule for each agent such that the iteration switching formation control can be realized.

The rest of this chapter is organized as follows. In Section 6.2, the kinematic model with nonholonomic constraints and its properties are presented. In Section 6.3, the formation problem is formulated, and we propose the learning controller for each agent and verify that by using the proposed control algorithm, the formation error of the multi-agent system converges to zero. In Section 6.4, a formation control example is presented to demonstrate the effectiveness of the control algorithm. Lastly, we conclude this chapter in Section 6.5.

Iterative Learning Control for Multi-agent Systems Coordination, First Edition.
Shiping Yang, Jian-Xin Xu, Xuefang Li, and Dong Shen.
© 2017 John Wiley & Sons Singapore Pte. Ltd. Published 2017 by John Wiley & Sons Singapore Pte. Ltd.

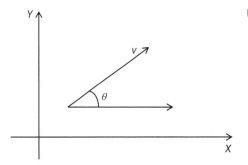

Figure 6.1 Kinematics model.

6.2 Kinematic Model Formulation

The kinematic motion of the agents is illustrated by Figure 6.1. It has two steering inputs, the forward moving speed and angular velocity. Mathematically the motion of the agent can be described below (Kang *et al.*, 2005):

$$\begin{bmatrix} \dot{x} \\ \dot{y} \\ \dot{\theta} \end{bmatrix} = \begin{bmatrix} \cos\theta & 0 \\ \sin\theta & 0 \\ 0 & 1 \end{bmatrix} \begin{bmatrix} v \\ w \end{bmatrix}, \tag{6.1}$$

where x, y are the position coordinates, θ is the angle between the heading direction and x-axis, v is the moving speed, and w is the angular velocity. In terms of the standard *arctan* function which is defined in the domain $(-\frac{\pi}{2}, \frac{\pi}{2})$, θ can be expressed as follows:

$$\theta = atan2(\dot{y}, \dot{x})$$

$$= \begin{cases} \arctan(\frac{\dot{y}}{\dot{x}}), & \dot{x} > 0, \\ \pi + \arctan(\frac{\dot{y}}{\dot{x}}), & \dot{y} \geq 0, \dot{x} < 0, \\ -\pi + \arctan(\frac{\dot{y}}{\dot{x}}), & \dot{y} < 0, \dot{x} < 0, \\ \frac{\pi}{2}, & \dot{y} > 0, \dot{x} = 0, \\ -\frac{\pi}{2}, & \dot{y} < 0, \dot{x} = 0, \\ \text{undefined}, & \dot{y} = 0, \dot{x} = 0. \end{cases}$$

Using the agent state $q = [x, y, \theta]^T$ and augmented velocity $u = [v, w]^T$, the kinematic motion can be expressed as

$$\dot{q}(t) = B(q, t)u(t),$$
$$z(t) = q(t), \tag{6.2}$$

where $B(q, t)$ is the first matrix on the right hand side of (6.1), and $z(t)$ is the formation output. The above kinematic model (6.2) has the following two properties (Kang *et al.*, 2005).

Property 6.1   The matrix function $B(q, t)$ is Lipschitz in q, that is

$$\|B(q_1, t) - B(q_2, t)\| \leq c_B \|q_1 - q_2\| \tag{6.3}$$

for some positive constant c_B.

Property 6.2   The matrix $B(q, t)$ is bounded, that is $\|B(q, t)\| \leq b_B$, where b_B is a finite positive constant. Furthermore, $B(q, t)$ is full rank for any (q, t).

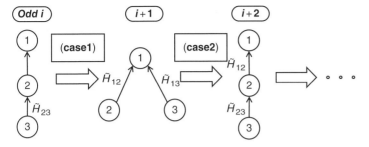

Figure 6.2 Switching process between two different structure formations.

6.3 HOIM-Based ILC for Multi-agent Formation

For simplicity, we consider three agents since extension to an arbitrary number of agents can be derived analogously. Let Agent 1 be the leader with its own desired trajectory. Agent 2 tracks the leader with certain desired relative distance. Agent 3 tracks Agent 2 at odd iterations, and tracks Agent 1 at even iterations. Furthermore, the relative distance between agents alternates over consecutive iterations. Figure 6.2 illustrates the switching process between two different structure formations. For such kinds of switching tasks, the algorithms previously developed in Chapters 2–5 cannot be applied. We will model the switching process with a high-order internal model (HOIM), and derive the learning controller for each agent by incorporating the HOIM.

Denote the desired output trajectory of the leader by $z_{i,1}^d$. The desired relative distances between different agents are defined as $z_{i,jk}^d$, $j, k = 1, 2, 3$ and $j < k$. Here, i denotes the iteration number and j, k denote the agent indexes. For multi-agent system formation control, we impose the following assumptions.

Assumption 6.1 Assume for the desired output trajectory $z_{i,1}^d$, there exits a unique control signal $u_{i,1}^d$ that satisfies the following desired dynamics:

$$\dot{q}_{i,1}^d(t) = B(q_{i,1}^d, t)u_{i,1}^d(t),$$
$$z_{i,1}^d(t) = q_{i,1}^d(t).$$

Assumption 6.2 Assume that for the desired relative distance between Agent j and Agent k, there exits a unique control signal $u_{i,k}^d$ for Agent k that satisfies the following desired dynamics:

$$\dot{q}_{i,jk}^d = B(q_{i,k}^d, t)u_{i,k}^d - B(q_{i,j}, t)u_{i,j}$$
$$z_{i,jk}^d = q_{i,jk}^d,$$

where $u_{i,j}$ is the input signal for Agent j at the ith iteration.

Assumption 6.3 The desired control input $u_{i,k}^d$ is bounded, that is $\|u_{i,k}^d\| \leq b_{u_{dk}}$, $k = 1, 2, 3$.

Assumption 6.4 The perfect resetting condition is assumed.

Assumptions 6.1 and 6.2 imply that the predefined formation can theoretically be achieved by all the agents. If these two assumptions are not satisfied, no learning controller will achieve the control task. Assumption 6.3 says the desired control signal is bounded. This is reasonable as the control input is always bounded in the physical world. The perfect resetting condition or *iic* in Assumption 6.4 is very stringent. It is imposed here to simplify the derivation. the issue of imperfect initial conditions in MAS is discussed in Chapter 4. The focus of this chapter is the learning controller design for iteratively switching tasks.

Control law designs for the three agents are illustrated in the following subsections together with the convergence proof.

6.3.1 Control Law for Agent 1

The control law for the leader can be considered as a tracking problem since it has its own independent desired trajectory. The iteration-varying references are generated by the following HOIM:

$$z_{i+1,1}^d = \tilde{H}_1 z_{i,1}^d, \tag{6.4}$$

where

$$\tilde{H}_1 = \begin{bmatrix} h_{11} & 0 & 0 \\ 0 & h_{12} & 0 \\ 0 & 0 & h_{13} \end{bmatrix}$$

is a constant diagonal matrix. Such a HOIM will be adopted to describe the switching geometric relations for the formation problem.

Taking the derivative of the HOIM in (6.4) yields

$$\dot{z}_{i+1,1}^d = \tilde{H}_1 \dot{z}_{i,1}^d.$$

From Assumption 6.1, we can obtain

$$B(q_{i+1,1}^d, t)u_{i+1,1}^d = \tilde{H}_1 B(q_{i,1}^d, t)u_{i,1}^d. \tag{6.5}$$

Multiplying $B^T(q_{i+1,1}^d, t)$ on both sides of (6.5), and noticing the property $B^T B = I$, we obtain the relation between desired control input over two consecutive iterations,

$$u_{i+1,1}^d = B^T(q_{i+1,1}^d, t)\tilde{H}_1 B(q_{i,1}^d, t)u_{i,1}^d.$$

Denote $H_1 \triangleq B^T(q_{i+1,1}^d, t)\tilde{H}_1 B(q_{i,1}^d, t)$, the desired iteration-varying input now is

$$u_{i+1,1}^d = H_1 u_{i,1}^d.$$

Then the iteration-varying output problem can be converted into iteration-varying input problem with the following constraint:

$$\tan(h_{13}\theta_i^d) = \frac{h_{12}}{h_{11}} \tan(\theta_i^d). \tag{6.6}$$

Detailed derivation is given in Appendix B.1.

We propose an iterative learning rule for steering the agents as:

$$u_{i+1,1} = H_1 u_{i,1} + \Gamma_1 \dot{e}_{i,1}, \tag{6.7}$$

where Γ_1 is an appropriate learning gain matrix with a bound $\|\Gamma_1\| \le b_{\Gamma_1}$, b_{Γ_1} is a positive constant, and $e_{i,1} = z_{i,1}^d - z_{i,1}$ is the output tracking error for the leader. The constraints on selecting Γ_1 will be discussed later.

For the ith iteration, taking the derivative of the tracking error yields $\dot{e}_{i,1} = \dot{z}_{i,1}^d - \dot{z}_{i,1}$. The control input error is defined to be $\Delta u_{i,1} = u_{i,1}^d - u_{i,1}$.

The control input error at $(i + 1)$th iteration becomes

$$
\begin{aligned}
&\Delta u_{i+1,1} \\
&= u_{i+1,1}^d - u_{i+1,1} \\
&= H_1 u_{i,1}^d - H_1 u_{i,1} - \Gamma_1 \dot{e}_{i,1} \\
&= H_1 \Delta u_{i,1} - \Gamma_1 (\dot{z}_{i,1}^d - \dot{z}_{i,1}) \\
&= H_1 \Delta u_{i,1} - \Gamma_1 (B(q_{i,1}^d, t) u_{i,1}^d(t) - B(q_{i,1}, t) u_{i,1}) \\
&= H_1 \Delta u_{i,1} - \Gamma_1 [(B(q_{i,1}^d, t) - B(q_{i,1}, t)) u_{i,1}^d(t) - B(q_{i,1}, t)(u_{i,1} - u_{i,1}^d(t))] \\
&= (H_1 - \Gamma_1 B(q_{i,1}, t)) \Delta u_{i,1} - \Gamma_1 (B(q_{i,1}^d, t) - B(q_{i,1})) u_{i,1}^d(t).
\end{aligned}
\tag{6.8}
$$

Taking any matrix norm on both sides of (6.8) and applying Property 6.2 and Assumption 6.1, we obtain

$$
\begin{aligned}
\|\Delta u_{i+1,1}\| &\le \|H_1 - \Gamma_1 B(q_{i,1}, t)\| \|\Delta u_{i,1}\| + \|\Gamma_1\| \|B(q_{i,1}^d, t) - B(q_{i,1})\| \|u_{i,1}^d(t)\| \\
&\le \|H_1 - \Gamma_1 B(q_{i,1}, t)\| \|\Delta u_{i,1}\| + b_{u_{d1}} c_{B_1} b_{\Gamma_1} \|q_{i,1}^d - q_{i,1}\|.
\end{aligned}
\tag{6.9}
$$

Consider $\|q_{i,1}^d - q_{i,1}\|$:

$$
\begin{aligned}
\|q_{i,1}^d - q_{i,1}\| &= \|q^d(0) - q_i(0)\| + \int_0^t \|B(q_{i,1}^d, \tau) u_{i,1}^d(\tau) - B(q_{i,1}, \tau) u_{i,1}(\tau)\| d\tau \\
&\le \int_0^t \|(B(q_{i,1}^d, \tau) - B(q_{i,1}, \tau)) u_{i,1}^d(\tau)\| d\tau \\
&\quad + \int_0^t \|B(q_{i,1}, \tau) u_{i,1}^d(\tau) - B(q_{i,1}, \tau) u_{i,1}(\tau)\| d\tau.
\end{aligned}
$$

Applying Properties 6.1, 6.2, and Assumption 6.1, we obtain

$$
\|q_{i,1}^d - q_{i,1}\| \le b_{u_{d1}} c_{B_1} \int_0^t \|q_{i,1}^d - q_{i,1}\| d\tau + b_{B_1} \int_0^t \|\Delta u_{i,1}\| d\tau.
\tag{6.10}
$$

Applying Gronwall–Bellman's Lemma (Khalil, 2002) to (6.10), we have

$$
\begin{aligned}
\|q_{i,1}^d - q_{i,1}\| &\le b_{B_1} \int_0^t \|\Delta u_{i,1}\| e^{\int_\tau^t \beta_1 ds} d\tau \\
&\le b_{B_1} \int_0^t \|\Delta u_{i,1}\| e^{\beta_1(t-\tau)} d\tau,
\end{aligned}
\tag{6.11}
$$

where $\beta_1 = b_{u_{d_1}} c_{B_1}$. Applying the λ-norm to (6.11) yields

$$\|q_{i,1}^d - q_{i,1}\|_\lambda = \sup_{t\in[0,T]} e^{-\lambda t}\|q_{i,1}^d - q_{i,1}\|$$

$$\leq \sup_{t\in[0,T]} e^{-\lambda t} b_{B_1} \int_0^t \|\Delta u_{i,1}\| e^{\beta_1(t-\tau)} d\tau$$

$$\leq \frac{1 - e^{(\beta_1-\lambda)T}}{\lambda - \beta_1} \|\Delta u_{i,1}\|_\lambda$$

Applying the λ-norm to (6.9), we have

$$\|\Delta u_{i+1,1}\|_\lambda \leq \|H_1 - \Gamma_1 B(q_{i,1}, t)\| \|\Delta u_{i,1}\|_\lambda + \delta_1 \|\Delta u_{i,1}\|_\lambda, \tag{6.12}$$

where $\delta_1 = \frac{1-e^{(\beta_1-\lambda)T}}{\lambda-\beta_1}$ is a function of λ and can be made arbitrarily small with a sufficiently large λ. Consequently, by choosing a proper learning gain Γ_1 such that $\|H_1 - \Gamma_1 B(q_{i,1}, t)\| < 1$, $u_{i,1}$ converges to $u_{i,1}^d$ as i tends to infinity since $\|\Delta u_{i+1,1}\|_\lambda$ converges to zero.

6.3.2 Control Law for Agent 2

The relative distance between leader Agent 1 and follower Agent 2 is given as $z_{i,12} = z_{i,2} - z_{i,1}$; then we have

$$\dot{z}_{i,12} = \dot{z}_{i,2} - \dot{z}_{i,1} \tag{6.13}$$
$$= B(q_{i,2}, t)u_{i,2} - B(q_{i,1}, t)u_{i,1}.$$

The desired relative distance between Agent 1 and Agent 2 is iteration-varying, which is

$$z_{i+1,12}^d = \tilde{H}_{12} z_{i,12}^d, \tag{6.14}$$

and

$$\tilde{H}_{12} = \begin{bmatrix} h_{121} & 0 & 0 \\ 0 & h_{122} & 0 \\ 0 & 0 & h_{123} \end{bmatrix}. \tag{6.15}$$

$\tilde{H}_{12}$ must satisfy the following constraint (see Appendix B.1):

$$\tan(h_{123}\theta_{i,2}^d + (h_{13} - h_{123})\theta_{i,1}^d) = \frac{h_{122}\dot{y}_{i,2}^d + (h_{12} - h_{122})\dot{y}_{i,1}^d}{h_{121}\dot{x}_{i,2}^d + (h_{11} - h_{121})\dot{x}_{i,1}^d}. \tag{6.16}$$

Noticing that

$$\dot{z}_{i+1,2}^d - \dot{z}_{i+1,1}^d = \tilde{H}_{12}(\dot{z}_{i,2}^d - \dot{z}_{i,1}^d),$$

substituting in $\dot{z}_{i+1,1}^d = \tilde{H}_1 \dot{z}_{i,1}^d$ and pre-multiplying both sides with $B^T(q_{i+1,2}^d, t)$ yields

$$u_{i+1,2}^d = B^T(q_{i+1,2}^d, t)\tilde{H}_{12}B(q_{i,2}^d, t)u_{i,2}^d + B^T(q_{i+1,2}^d, t)(\tilde{H}_1 - \tilde{H}_{12})B(q_{i,1}^d, t)u_{i,1}^d.$$

Denote

$$H_2 \triangleq B^T(q_{i+1,2}^d, t)\tilde{H}_{12}B(q_{i,2}^d, t)$$
$$H_{12} \triangleq B^T(q_{i+1,2}^d, t)(\tilde{H}_1 - \tilde{H}_{12})B(q_{i,1}^d, t).$$

The desired iteration-varying input is

$$u_{i+1,2}^d = H_2 u_{i,2}^d + H_{12} u_{i,1}^d.$$

We propose the iterative learning rule for steering Agent 2 as:

$$u_{i+1,2} = H_2 u_{i,2} + H_{12} u_{i,1} + \Gamma_{12} \dot{e}_{i,12}, \tag{6.17}$$

where Γ_{12} is an appropriate learning gain with a bound $\|\Gamma_{12}\| \leq b_{\Gamma_{12}}$. Since

$$\begin{aligned}
\dot{e}_{i,12} &= \dot{z}_{i,12}^d - \dot{z}_{i,12} \\
&= [B(q_{i,2}^d, t)u_{i,2}^d - B(q_{i,1}, t)u_{i,1}] - [B(q_{i,2}, t)u_{i,2} - B(q_{i,1}, t)u_{i,1}] \\
&= B(q_{i,2}^d, t)u_{i,2}^d - B(q_{i,2}, t)u_{i,2}, \tag{6.18}
\end{aligned}$$

from (6.17) and (6.18) we obtain

$$\begin{aligned}
\Delta u_{i+1,2} &= u_{i+1,2}^d - u_{i+1,2} \\
&= H_2 \Delta u_{i,2} - H_{12} \Delta u_{i,1} - \Gamma_{12}[B(q_{i,2}^d, t)u_{i,2}^d - B(q_{i,2}, t)u_{i,2}].
\end{aligned}$$

Applying Gronwall–Bellman's Lemma and the λ-norm , we have

$$\begin{aligned}
\|\Delta u_{i+1,2}\|_\lambda &\leq \|H_2 - \Gamma_{12}B(q_{i,2}, t)\|\|\Delta u_{i,2}\|_\lambda + \|H_{12}\|\|\Delta u_{i,1}\|_\lambda + \delta_2 \|\Delta u_{i,2}\|_\lambda \\
&= \|H_2 - \Gamma_{12}B(q_{i,2}, t)\|\|\Delta u_{i,2}\|_\lambda + \|H_{12}\|\|\Delta u_{i,1}\|_\lambda. \tag{6.19}
\end{aligned}$$

The last equality in (6.19) is based on the condition that δ_2 can be made arbitrarily small by choosing a sufficiently large λ. The derivation for (6.19) is very similar to the previous subsection, hence is not detailed here.

6.3.3 Control Law for Agent 3

Since the formation structures are dynamically changing for Agent 3, we have to discuss the two structures separately. For the odd ith iteration, the relative distance between Agent 2 and Agent 3 is given by $z_{i,23} = z_{i,3} - z_{i,2}$.

Applying Assumption 6.2, there exits a unique control signal $u_{i,3}^d$ that satisfies

$$\begin{aligned}
\dot{q}_{i,23}^d &= B(q_{i,3}^d, t)u_{i,3}^d - B(q_{i,2}, t)u_{i,2} \\
z_{i,23}^d &= q_{i,23}^d,
\end{aligned}$$

and the desired control input $u_{i,3}^d$ is bounded, that is $\|u_{i,3}^d\| \leq b_{u_{d3}}$, where $b_{u_{d3}}$ is some finite positive constant.

The desired relative distance between Agent 1 and Agent 3 is iteration-varying, which can be modeled by the following HOIM:

$$z_{i+1,13}^d = \tilde{H}_{13} z_{i,13}^d,$$

and

$$\tilde{H}_{13} = \begin{bmatrix} h_{131} & 0 & 0 \\ 0 & h_{132} & 0 \\ 0 & 0 & h_{133} \end{bmatrix} \tag{6.20}$$

with the constraint that

$$\tan(h_{133}\theta_{i,3}^d + (h_{13} - h_{133})\theta_{i,1}^d) = \frac{h_{132}\dot{y}_{i,3}^d + (h_{12} - h_{132})\dot{y}_{i,1}^d}{h_{131}\dot{x}_{i,3}^d + (h_{11} - h_{131})\dot{x}_{i,1}^d}. \tag{6.21}$$

The corresponding iteration-varying desired input is

$$u_{i+1,3}^d = H_3 u_{i,3}^d + H_{13} u_{i,1}^d,$$

where

$$H_3 \triangleq B^T(q_{i+1,3}^d, t)\tilde{H}_{13} B(q_{i,3}^d, t)$$
$$H_{13} \triangleq B^T(q_{i+1,3}^d, t)(\tilde{H}_1 - \tilde{H}_{13}) B(q_{i,1}^d, t).$$

We propose the iterative learning rule for Agent 3 at the odd ith iteration as:

$$u_{i+1,3} = H_3 u_{i,3} + H_{13} u_{i,1} + \Gamma_{13} \dot{e}_{i,13}, \tag{6.22}$$

where Γ_{13} is the desired learning gain satisfying $\|\Gamma_{13}\| \leq b_{\Gamma_{13}}$ and $b_{\Gamma_{13}}$ is a finite positive constant. Since $e_{i,13} = e_{i,12} + e_{i,23}$, we have

$$\dot{e}_{i,23} = \dot{z}_{i,23}^d - \dot{z}_{i,23}$$
$$= [B(q_{i,3}^d, t)u_{i,3}^d - B(q_{i,2}, t)u_{i,2}] - [B(q_{i,3}, t)u_{i,3} - B(q_{i,2}, t)u_{i,2}]$$
$$= B(q_{i,3}^d, t)u_{i,3}^d - B(q_{i,3}, t)u_{i,3},$$

and

$$\dot{e}_{i,13} = \dot{e}_{i,12} + \dot{e}_{i,23}$$
$$= [B(q_{i,2}^d, t)u_{i,2}^d - B(q_{i,2}, t)u_{i,2}] + [B(q_{i,3}^d, t)u_{i,3}^d - B(q_{i,3}, t)u_{i,3}]. \tag{6.23}$$

With the help of (6.22) and (6.23), the control input error can be derived below:

$$\Delta u_{i+1,3} = u_{i+1,3}^d - u_{i+1,3}$$
$$= H_3 \Delta u_{i,3} - H_{13} \Delta u_{i,1} - \Gamma_{13}[B(q_{i,2}^d, t)u_{i,2}^d - B(q_{i,2}, t)u_{i,2}]$$
$$+ (B(q_{i,3}^d, t)u_{i,3}^d - B(q_{i,3}, t)u_{i,3}).$$

Applying Gronwall–Bellman's Lemma and the λ-norm, we have

$$\|\Delta u_{i+1,3}\|_\lambda \leq \|H_3 - \Gamma_{13}B(q_{i,3}, t)\| \|\Delta u_{i,3}\|_\lambda + \delta_{13}\|\Delta u_{i,3}\|_\lambda \tag{6.24}$$
$$+ \delta_2'\|\Delta u_{i,2}\|_\lambda + b_{\Gamma_{13}} b_{B2}\|\Delta u_{i,2}\|_\lambda + \|H_{13}\| \|\Delta u_{i,1}\|$$
$$\leq \|H_3 - \Gamma_{13}B(q_{i,3}, t)\| \|\Delta u_{i,3}\|_\lambda + b_{\Gamma_{13}} b_{B_2}\|\Delta u_{i,2}\|_\lambda + \|H_{13}\| \|\Delta u_{i,1}\|,$$

where δ_{13} and δ_2' can be made arbitrarily small.

Similarly for even i, the relative distance between Agent 1 and Agent 3 is given as $z_{i,13} = z_{i,3} - z_{i,1}$.

Applying Assumption 6.2, there exits a desired control input $u_{i,3}^d$ that satisfies

$$\dot{q}_{i,13}^d = B(q_{i,3}^d, t)u_{i,3}^d - B(q_{i,1}, t)u_{i,1}$$
$$z_{i,13}^d = q_{i,13}^d,$$

and the desired control input $u_{i,3}^d$ is bounded, that is, $\|u_{i,3}^d\| \leq b_{u_{d3}}$.

The desired relative distance between Agent 2 and Agent 3 is iteration-varying and modeled by the following HOIM:

$$z_{i+1,23}^d = \tilde{H}_{23} z_{i,23}^d,$$

and

$$\tilde{H}_{23} = \begin{bmatrix} h_{231} & 0 & 0 \\ 0 & h_{232} & 0 \\ 0 & 0 & h_{233} \end{bmatrix},$$

with the constraint

$$\tan(h_{233}\theta_{i,3}^d + (h_{123} - h_{233})\theta_{i,2}^d) + (h_{13} - h_{123})\theta_{i,1}^d)$$
$$= \frac{h_{232}\dot{y}_{i,3}^d + (h_{122} - h_{232})\dot{y}_{i,2}^d + (h_{12} - h_{122})\dot{y}_{i,1}^d}{h_{231}\dot{x}_{i,3}^d + (h_{121} - h_{231})\dot{x}_{i,2}^d + (h_{11} - h_{121})\dot{x}_{i,1}^d}. \tag{6.25}$$

The corresponding desired input is iteration-varying and given by the following equation:

$$u_{i+1,3}^d = H_3' u_{i,3}^d + H_{23} u_{i,2}^d,$$

where

$$H_3' \triangleq B^T(q_{i+1,3}^d, t)\tilde{H}_{23}B(q_{i,3}^d, t)$$
$$H_{23} \triangleq B^T(q_{i+1,3}^d, t)(\tilde{H}_{12} - \tilde{H}_{23})B(q_{i,2}^d, t).$$

We propose the iterative learning rule below for steering Agent 3 as:

$$u_{i+1,3} = H_3' u_{i,3} + H_{23} u_{i,2} + \Gamma_{23}\dot{e}_{i,23}, \tag{6.26}$$

where Γ_{23} is learning gain satisfyig $\|\Gamma_{23}\| \le b_{\Gamma_{23}}$. Since $e_{i,23} = e_{i,13} - e_{i,12}$, we have

$$\begin{aligned} \dot{e}_{i,13} &= \dot{z}_{i,13}^d - \dot{z}_{i,13} \\ &= [B(q_{i,3}^d, t)u_{i,3}^d - B(q_{i,1}, t)u_{i,1}] - [B(q_{i,3}, t)u_{i,3} - B(q_{i,1}, t)u_{i,1}] \\ &= B(q_{i,3}^d, t)u_{i,3}^d - B(q_{i,3}, t)u_{i,3}, \end{aligned}$$

and

$$\begin{aligned} \dot{e}_{i,23} &= \dot{e}_{i,13} - \dot{e}_{i,12} \\ &= (B(q_{i,3}^d, t)u_{i,3}^d - B(q_{i,3}, t)u_{i,3}) - (B(q_{i,2}^d, t)u_{i,2}^d - B(q_{i,2}, t)u_{i,2}). \end{aligned} \tag{6.27}$$

With the help of (6.26) and (6.27), the control input error can be derived as

$$\begin{aligned} \Delta u_{i+1,3} &= u_{i+1,3}^d - u_{i+1,3} \\ &= H_3'\Delta u_{i,3} - H_{23}\Delta u_{i,2} - \Gamma_{23}[(B(q_{i,3}^d, t)u_{i,3}^d \\ &\quad - B(q_{i,3}, t)u_{i,3}] - [B(q_{i,2}^d, t)u_{i,2}^d - B(q_{i,2}, t)u_{i,2})]. \end{aligned} \tag{6.28}$$

Applying Gronwall–Bellman's Lemma and the λ-norm, we have

$$\begin{aligned} \|\Delta u_{i+1,3}\|_\lambda &\le \|H_3' - \Gamma_{23}B(q_{i,3}, t)\|\|\Delta u_{i,3}\|_\lambda + \delta_{23}\|\Delta u_{i,3}\|_\lambda \\ &\quad + \delta_2''\|\Delta u_{i,2}\|_\lambda + \|H_{23} - \Gamma_{23}B(q_{i,2}, t)\|\|\Delta u_{i,2}\|_\lambda \\ &\le \|H_3 - \Gamma_{23}B(q_{i,3}, t)\|\|\Delta u_{i,3}\|_\lambda + \|H_{23} - \Gamma_{23}B(q_{i,2}, t)\|\|\Delta u_{i,2}\|_\lambda. \end{aligned} \tag{6.29}$$

where δ_{23} and δ_2'' can be made arbitrarily small by choosing a sufficiently large λ.

Remark 6.1 $\delta_2, \delta_2', \delta_2'', \delta_{13}, \delta_{23}$ can be derived in a similar way to δ_1.

6.3.4 Switching Between Two Structures

Finally we discuss the learning rule convergence property for the varying formations associated with the control input error Equations (6.12), (6.19), (6.24), and (6.30).

For the odd ith iteration, we have an assembled formation matrix W_1 as shown below:

$$\begin{bmatrix} H_1 - \Gamma_1 B(q_{i,1}, t) & 0 & 0 \\ \|H_{12}\| & H_2 - \Gamma_{12} B(q_{i,2}, t) & 0 \\ \|H_{13}\| & b_{\Gamma_{13}} b_{B2} & H_3 - \Gamma_{13} B(q_{i,3}, t) \end{bmatrix}.$$

For the even ith iteration, we have another assembled formation matrix W_2 as shown below:

$$\begin{bmatrix} H_1 - \Gamma_1 B(q_{i,1}, t) & 0 & 0 \\ \|H_{12}\| & H_2 - \Gamma_{12} B(q_{i,2}, t) & 0 \\ 0 & H_{23} - \Gamma_{23} B(q_{i,2}, t) & H_3' - \Gamma_{23} B(q_{i,3}, t) \end{bmatrix}.$$

We can choose an appropriate learning gain matrix such that the spectral radius of W_1 and W_2 are strictly less than 1. Then control input $u_{i,j}$ converges to the desired $u_{i,j}^d$ as i tends to infinity, which implies that the formation error eventually converges to zero.

6.4 Illustrative Example

For numerical verification, we consider three agents with kinematic models as shown in (6.2). Agent 1 is the leader that has its own independent desired trajectory. Agent 2 is connected to Agent 1 and they keep a fixed distance. In the odd iterations, Agent 3 is connected to Agent 2. In the even iterations, Agent 3 is connected to Agent 1. The desired trajectory and desired relative distances are given below:

$$z_{1,1}^d = [10 \cos t, 10 \sin t, t + \tfrac{\pi}{2}]^T, \quad z_{1,12}^d = [0, -10, 0]^T,$$
$$z_{1,23}^d = [10, 0, 0]^T, \quad z_{2,13}^d = [10, -20, 0]^T.$$

The HOIM parameters for iteration-varying references $\tilde{H}_1, \tilde{H}_{12}, \tilde{H}_{23}, \tilde{H}_{13}$ should satisfy constraints (6.6), (6.16), (6.21), (6.25), respectively. In this example, we choose

$$\tilde{H}_1 = \begin{bmatrix} 1 & 0 & 0 \\ 0 & 1 & 0 \\ 0 & 0 & 1 \end{bmatrix}, \quad \tilde{H}_{12} = \begin{bmatrix} 1 & 0 & 0 \\ 0 & 1 & 0 \\ 0 & 0 & 1 \end{bmatrix},$$

$$\tilde{H}_{23} = \begin{bmatrix} 1 & 0 & 0 \\ 0 & 0 & 0 \\ 0 & 0 & 1 \end{bmatrix}, \quad \tilde{H}_{13} = \begin{bmatrix} 1 & 0 & 0 \\ 0 & 2 & 0 \\ 0 & 0 & 1 \end{bmatrix}.$$

The learning gain matrix should be chosen such that the spectral radius of W_1 and W_2 are strictly less than 1. Noticing that $B^T(q_{i,j}) B(q_{i,j}) = I$, the learning gains are set as

$$\Gamma_1 = 0.5 B^T(q_{i,1}, t), \quad \Gamma_{12} = 0.5 B^T(q_{i,2}, t),$$
$$\Gamma_{13} = 1.25 B^T(q_{i,3}, t), \quad \Gamma_{23} = 0.75 B^T(q_{i,3}, t).$$

The ILC control laws are given in (6.7), (6.17), (6.22) and (6.26). Figure 6.3 shows the odd number iteration formation in which Agent 3 is connected to Agent 2. Figure 6.4 shows

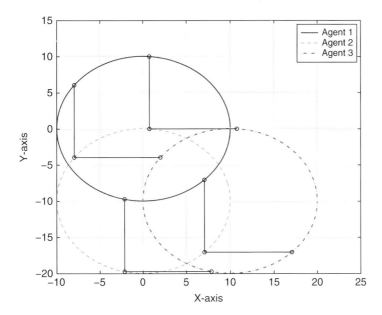

Figure 6.3 Multi-Agent formation at 29th (odd) iteration.

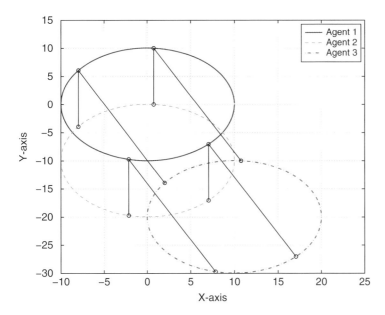

Figure 6.4 Multi-Agent formation at 30th (even) iteration.

the even number iteration formation in which Agent 3 is connected to Agent 1. The multi-agent formation structure is switching between these two different structures. It can be seen from Figure 6.5 that the formation tracking error converges to zero along the iteration axis by using the proposed ILC updating rules.

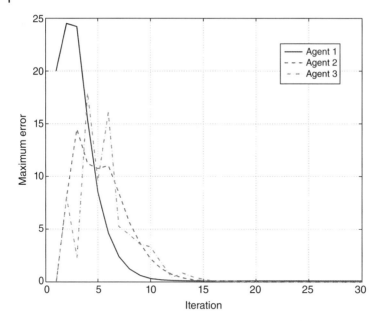

Figure 6.5 Error convergence along the iteration axis.

6.5 Conclusion

In this chapter, a HOIM-based ILC algorithm is proposed for multi-agent formation in terms of the kinematic model. The learning convergence property and associated condition are derived and analyzed. The proposed HOIM-based ILC algorithm could be employed for formation control of a MAS when the connections between agents are dynamically changing, that is, the formation configuration is different between consecutive iterations but can be modeled with a HOIM. By employing the proposed ILC algorithm, the desired formation trajectory has been achieved. Both theoretical analysis and numerical example results have demonstrated the effectiveness of the proposed HOIM-based ILC algorithm.

7

P-type Iterative Learning for Non-parameterized Systems with Uncertain Local Lipschitz Terms

7.1 Introduction

Traditional ILC has two fundamental postulates, namely system repeatability and global Lipschitz condition (GLC). The system repeatability consists of three conditions over the iteration axis: the identical initialization condition (*iic*), identical system dynamics, and identical control task. Relaxation or removal of each repeatability condition has led to great advances in recent ILC research. Much effort has been devoted to relaxing or even removing the *iic* (Xu and Qu, 1998; Chen *et al.*, 1999; Sun and Wang, 2002; Fang and Chow, 2003; Park, 2005; Xu and Yan, 2005). For a detailed discussion on the initial condition problem, please refer to Chapter 4. Relaxation or removal of the identical system dynamics and identical control task conditions have relatively few results available. The results in Saab (1994) show that if the fluctuations in system dynamics can be modeled by bounded term, uniform convergence can be obtained. Chen *et al.* (1997) shows that if the variation of reference trajectories between two consecutive iterations is bounded, the tracking error will converge to some bounded ball. Liu *et al.* (2010) and Yin *et al.* (2010) successfully apply the ILC learning rule to iteration-varying control tasks and iteration-varying dynamic systems by introducing a high-order internal model to the controller design. A similar idea is applied in Chapter 6 for MAS to perform the switching formation control problem.

Unlike in the rich literature on system repeatability issues, it seems impossible to remove the GLC in contraction-mapping (CM)-based ILC. To date, it is not clear whether or not CM-based ILC is applicable to local Lipschitz systems. The composite energy function (CEF)-based ILC (Xu and Tan, 2002a) is able to handle local Lipschitz systems. However, the system dynamics must be in linear in parameter form, and full state information must be used for feedback or nonlinear compensation. When dealing with real-life control problems we always consider CM-based ILC first of all because of the following advantages: extremely simple algorithms, little knowledge requirement on the state dynamics, exponential convergence rate, and being applicable to non-affine-in-input processes. Thus, it is very suitable and highly desired for practical control problems. The major disadvantage is due to the requirement of GLC. Traditional CM-based ILC is essentially a feedforward control system, and only output information is used. There is no feedback control for state dynamics. GLC is required for the drift term in the unforced system, such that in the worst case when the unforced system is unstable, the finite escape time phenomenon will not occur. From the practical point of view, most control engineering systems are inherently stable, for

Iterative Learning Control for Multi-agent Systems Coordination, First Edition.
Shiping Yang, Jian-Xin Xu, Xuefang Li, and Dong Shen.
© 2017 John Wiley & Sons Singapore Pte. Ltd. Published 2017 by John Wiley & Sons Singapore Pte. Ltd.

instance mechanical systems with friction, thermal systems with heat dissipation, and electrical systems with inner resistance. In traditional CM-based ILC analysis, the drift term is always treated as an unstable term by taking a norm operation, even though the drift term could originally be a stabilizing term. This limitation is innate in CM-based analysis. To overcome this limitation, in this chapter we introduce the Lyapunov method into CM-based ILC analysis, aiming at fully making use of the inherent stability of the unforced system. Following this, ILC can be applied to more generic systems with local Lipschitz nonlinearities. Our objective is to retain the CM-based ILC, whilst incorporating the Lyapunov method to facilitate ILC convergence analysis, and hence to widen the applicability of CM-based ILC to more generic dynamic systems with nonlinearities beyond GLC.

The extension of the ILC approach to local Lipschitz systems has a great positive impact on multi-agent coordination problems. In the previous several chapters, all the agent models are assumed to be global Lipschitz continuous, and D-type rules are adopted to achieve the coordination tasks. Although it is straightforward to generalize the learning rule design to P-type rules, the agent dynamics have to be global Lipschitz continuous. By using the results in this chapter, it is possible to design distributed iterative learning rules for certain classes of local Lipschitz MAS. Concrete examples will be provided to illustrate the generalization.

The rest of this chapter is organized as follows. The motivation and problem formulation are described in Section 7.2. Sections 7.3 and 7.4 present the main results. Specifically, Subsection 7.3.1 introduces some preliminary results that will be used to prove the main theorems. Subsection 7.3.2 develops the Lyapunov function based sufficient convergence conditions. Subsection 7.3.3 studies the local Lipschitz system whose unstable part is global Lipschitz continuous. Subsection 7.4.1 studies the system with a bounded drift term. Subsection 7.4.2 investigates the class of bounded energy bounded state systems subject to control saturation. Application of a P-type rule for MAS coordination under local Lipschitz uncertainties are provided in Section 7.5. Finally, conclusions are drawn in Section 7.6.

7.2 Motivation and Problem Description

7.2.1 Motivation

ILC is a partially model-free control as the controller design does not require detailed information on the system model. To illustrate the motivation of the problem in this chapter, consider the following first-order dynamics:

$$\dot{x} = f(x) + u,$$

where $f(x)$ is the drift term that determines the stability of the unforced system. $f(x)$ may contain both stable and unstable terms. In traditional ILC analysis, norm operations are taken on the system dynamics in order to find out the relation between the magnitudes of system state and control input. As a result, both the stable and unstable terms in $f(x)$ are treated as if they were unstable. As such, the bounding function for system state is too conservative, which makes the analysis method only applicable to the system whose $f(x)$ satisfies GLC. This restricts the application of ILC to a very limited class of systems.

To date, it is not clear how the local Lipschitz terms in $f(x)$ affect the convergence property of ILC rule. However, it has been observed that many local Lipschitz systems demonstrate convergence under ILC. This motivates us to investigate to what kind of local Lipschitz system the ILC rule is applicable. Three classes of local Lipschitz systems are studied in this chapter. First, we distinguish the stable and unstable terms in the system dynamics, and develop several sufficient conditions in the form of Lyapunov function based criteria. Next, we study bounded local Lipschitz terms. Finally, the convergence property of the uniformly bounded energy bounded state system under control saturation is analyzed.

7.2.2 Problem Description

Consider the system model in (7.1):

$$\dot{\mathbf{x}}(t) = \mathbf{f}(\mathbf{x}(t)) + \mathbf{g}(\mathbf{x}(t))\mathbf{u}(t), \tag{7.1a}$$

$$\mathbf{y}(t) = \mathbf{h}(\mathbf{x}(t), \mathbf{u}(t)), \tag{7.1b}$$

where $\mathbf{x} \in \mathbb{R}^n$, $\mathbf{u} \in \mathbb{R}^p$, and $\mathbf{y} \in \mathbb{R}^m$ are the system state, input and output vectors, $\mathbf{f}(\mathbf{x})$ belongs to a class of local Lipschitz functions, $\mathbf{g}(\mathbf{x})$ is a continuously differentiable and bounded function for all $\mathbf{x} \in \mathbb{R}^n$, and $\mathbf{h}(\mathbf{x}, \mathbf{u})$ is a differentiable function. Furthermore, $\frac{\partial \mathbf{h}}{\partial \mathbf{u}}(\mathbf{x}, \mathbf{u})$ is of full rank, and both $\frac{\partial \mathbf{h}}{\partial \mathbf{u}}(\mathbf{x}, \mathbf{u})$ and $\frac{\partial \mathbf{h}}{\partial \mathbf{x}}(\mathbf{x}, \mathbf{u})$ are bounded for all $\mathbf{x} \in \mathbb{R}^n$ and $\mathbf{u} \in \mathbb{R}^p$. For notional simplicity, the following three functional norms are defined:

$$\|\mathbf{g}(\mathbf{x})\| = \sup_{\mathbf{x} \in \mathbb{R}^n} |\mathbf{g}(\mathbf{x})|,$$

$$\left\|\frac{\partial \mathbf{h}}{\partial \mathbf{u}}(\mathbf{x}, \mathbf{u})\right\| = \sup_{\mathbf{x} \in \mathbb{R}^n, \mathbf{u} \in \mathbb{R}^p} \left|\frac{\partial \mathbf{h}}{\partial \mathbf{u}}(\mathbf{x}, \mathbf{u})\right|,$$

$$\left\|\frac{\partial \mathbf{h}}{\partial \mathbf{x}}(\mathbf{x}, \mathbf{u})\right\| = \sup_{\mathbf{x} \in \mathbb{R}^n, \mathbf{u} \in \mathbb{R}^p} \left|\frac{\partial \mathbf{h}}{\partial \mathbf{x}}(\mathbf{x}, \mathbf{u})\right|.$$

The desired trajectory is $\mathbf{y}_d \in C^1[0, T]$. For any given $\mathbf{y}_d$ to be realizable, it is assumed that there exist $\mathbf{x}_d \in C^1[0, T]$ and $\mathbf{u}_d \in C[0, T]$ such that the following dynamics hold:

$$\dot{\mathbf{x}}_d = \mathbf{f}(\mathbf{x}_d) + \mathbf{g}(\mathbf{x}_d)\mathbf{u}_d, \tag{7.2a}$$

$$\mathbf{y}_d = \mathbf{h}(\mathbf{x}_d, \mathbf{u}_d). \tag{7.2b}$$

Let ω_i denote a variable at the ith iteration, where $\omega \in \{\mathbf{x}, \mathbf{y}, \mathbf{u}\}$. Define the tracking error at the ith iteration as $\mathbf{e}_i = \mathbf{y}_d - \mathbf{y}_i$. The traditional P-type ILC law is constructed as

$$\mathbf{u}_{i+1} = \mathbf{u}_i + \Gamma \mathbf{e}_i, \tag{7.3}$$

where Γ is some suitable learning gain such that

$$\left\|I_m - \Gamma \frac{\partial \mathbf{h}}{\partial \mathbf{u}}\right\| = \sup_{\mathbf{x} \in \mathbb{R}^n, \mathbf{u} \in \mathbb{R}^p} \left|I_m - \Gamma \frac{\partial \mathbf{h}}{\partial \mathbf{u}}\right| \leq \rho_0 < 1, \tag{7.4}$$

I_m is the identity matrix, and the subscript m denotes its dimension.

We say that the P-type learning rule is convergent if $\mathbf{u}_i$ converges to $\mathbf{u}_d$ as the learning iteration tends to infinity. Subsequently, the tracking error $\mathbf{e}_i$ converges to zero as well. It is a well established result that (7.3) is convergent if $\mathbf{f}(\mathbf{x})$ is global Lipschitz continuous in $\mathbf{x}$ (Xu and Tan, 2003). The problem in this chapter is to explore to what extent the learning rule (7.3) can be applied to local Lipschitz systems.

To restrict our discussion, the following assumptions are imposed.

Assumption 7.1   Let D be a compact subset of $\mathbb{R}^n$. For any $\mathbf{z}_1, \mathbf{z}_2 \in D$, there exists a continuous bounding function $\phi(\mathbf{z})$ such that

$$|\mathbf{f}(\mathbf{z}_1) - \mathbf{f}(\mathbf{z}_2)| \leq \phi(\mathbf{z})|\mathbf{z}_1 - \mathbf{z}_2|,$$

where $\mathbf{z} = [\mathbf{z}_1^T, \mathbf{z}_2^T]^T$. Furthermore, if $D \to \mathbb{R}^n$, then $\lim\limits_{|\mathbf{z}| \to \infty} \phi(\mathbf{z}) \to \infty$.

Remark 7.1   If $\phi(\mathbf{z})$ has an upper bound for all $\mathbf{z} \in \mathbb{R}^{2n}$, Assumption 7.1 degenerates to the global Lipschitz condition. Thus, traditional λ-norm analysis can be applied to prove ILC convergence. As $\phi(\mathbf{z})$ is unbounded in $\mathbb{R}^{2n}$, Assumption 7.1 represents a class of local Lipschitz functions, for example high-order polynomials. Under Assumption 7.1, if the system state $\mathbf{x}_i$ is bounded along the iteration axis when controller (7.3) is applied, then the local Lipschitz condition can be treated as a global one in the learning context. This observation motivates the two-step idea to prove ILC convergence. First, show that the state is confined in certain compact set D under learning rule (7.3) by taking advantages of system properties. Next, $\phi(\mathbf{z})$ is bounded on D as it is a continuous function. Therefore, the λ-norm can be utilized to construct a contraction-mapping.

Assumption 7.2   The initial state is reset to the desired initial state at every iteration, that is, $\mathbf{x}_i(0) = \mathbf{x}_d(0)$.

Assumption 7.2 is one of the most commonly used assumptions in ILC literature. For simplicity we assume the perfect resetting condition holds since the initial condition problem is not the main topic in this chapter. As discussed in Chapters 2 and 4, there are many options to relax such a restrictive condition.

7.3   Convergence Properties with Lyapunov Stability Conditions

In this section, we investigate several classes of local Lipschitz systems with the help of the Lyapunov analysis method, and study their P-type ILC convergence.

7.3.1   Preliminary Results

In this subsection, two useful lemmas are introduced. They will be utilized to prove the main results in this chapter.

Lemma 7.1   If $\alpha \in \mathcal{K}_\infty$[1] and convex, then for any $r > 0$ there exist some positive constants k and b such that the following inequality holds:

$$\alpha^{-1}(r) \leq kr + b.$$

---

1 $\mathcal{K}$ is a class of functions $[0, \infty) \to [0, \infty)$ which are zeros at zero, strictly increasing, and continuous. $\mathcal{K}_\infty$ is a subset of $\mathcal{K}$ whose elements are unbounded (Khalil, 2002).

In Lemma 7.1, $\alpha \in \mathcal{K}_\infty$ implies α is a one-to-one mapping from $[0, \infty)$ to $[0, \infty)$. Hence, the inverse function α^{-1} exists. As α is a convex function, α^{-1} has to be concave. Therefore, we can find a linear bounding function for α^{-1}. The constants k and b depend on α. Lemma 7.1 will be used to establish the growth rate of system state as a function of input.

Lemma 7.2   Consider system (7.1) under the P-type learning rule (7.3), and Assumptions 7.1 and 7.2. If the system state satisfies the following inequality:

$$|\mathbf{x}(t)| < l \int_0^t |\mathbf{u}(\tau)|\, d\tau + \eta(t), \tag{7.5}$$

where l is a positive constant and $\eta(t)$ is a bounded continuous function of time, then the P-type learning rule (7.3) is convergent.

Proof: Substituting (7.1b) into the P-type learning rule (7.3) yields

$$\begin{aligned}
\mathbf{u}_{i+1} &= \mathbf{u}_i + \Gamma \mathbf{e}_i \\
&= \mathbf{u}_i + \Gamma(\mathbf{h}(\mathbf{x}_d, \mathbf{u}_d) - \mathbf{h}(\mathbf{x}_i, \mathbf{u}_i)) \\
&= \mathbf{u}_i + \Gamma(\mathbf{h}(\mathbf{x}_d, \mathbf{u}_d) - \mathbf{h}(\mathbf{x}_d, \mathbf{u}_i)) + \Gamma(\mathbf{h}(\mathbf{x}_d, \mathbf{u}_i) - \mathbf{h}(\mathbf{x}_i, \mathbf{u}_i)).
\end{aligned} \tag{7.6}$$

By using the Mean Value Theorem (Khalil, 2002), the last two terms in (7.6) can be respectively written as

$$\mathbf{h}(\mathbf{x}_d, \mathbf{u}_d) - \mathbf{h}(\mathbf{x}_d, \mathbf{u}_i) = \frac{\partial \mathbf{h}}{\partial \mathbf{u}}(\mathbf{x}_d, \hat{\mathbf{u}}_i)(\mathbf{u}_d - \mathbf{u}_i),$$

$$\mathbf{h}(\mathbf{x}_d, \mathbf{u}_i) - \mathbf{h}(\mathbf{x}_i, \mathbf{u}_i) = \frac{\partial \mathbf{h}}{\partial \mathbf{x}}(\hat{\mathbf{x}}_i, \mathbf{u}_i)(\mathbf{x}_d - \mathbf{x}_i),$$

where $\hat{\mathbf{u}}_i = \mathbf{u}_d + \theta_i^\mathbf{u}(\mathbf{u}_i - \mathbf{u}_d), \hat{\mathbf{x}}_i = \mathbf{x}_d + \theta_i^\mathbf{x}(\mathbf{x}_i - \mathbf{x}_d),$ and $\theta_i^\mathbf{u}, \theta_i^\mathbf{x} \in [0, 1]$. Thus, (7.6) can be simplified as

$$\mathbf{u}_{i+1} = \left(I_m - \Gamma\frac{\partial \mathbf{h}}{\partial \mathbf{u}}\right)\mathbf{u}_i - \Gamma\frac{\partial \mathbf{h}}{\partial \mathbf{x}}\mathbf{x}_i + \Gamma\frac{\partial \mathbf{h}}{\partial \mathbf{u}}\mathbf{u}_d + \Gamma\frac{\partial \mathbf{h}}{\partial \mathbf{x}}\mathbf{x}_d. \tag{7.7}$$

Since $\frac{\partial \mathbf{h}}{\partial \mathbf{u}}$ and $\frac{\partial \mathbf{h}}{\partial \mathbf{x}}$ are bounded, denote $d = \|\frac{\partial \mathbf{h}}{\partial \mathbf{u}}\|$ and $c = \|\frac{\partial \mathbf{h}}{\partial \mathbf{x}}\|$. Noticing the convergence condition (7.4) and inequality (7.5), taking the norm on both sides of (7.7) yields

$$\begin{aligned}
|\mathbf{u}_{i+1}| &\leq \left|I_m - \Gamma\frac{\partial \mathbf{h}}{\partial \mathbf{u}}\right||\mathbf{u}_i| + c|\Gamma|\left(l \int_0^t |\mathbf{u}_i(\tau)|\, d\tau + \eta(t)\right) + d|\Gamma \mathbf{u}_d| + c|\Gamma \mathbf{x}_d| \\
&\leq \rho_0|\mathbf{u}_i| + cl|\Gamma| \int_0^t |\mathbf{u}_i(\tau)|\, d\tau + \underbrace{c|\Gamma|\eta(t) + d|\Gamma \mathbf{u}_d| + c|\Gamma \mathbf{x}_d|}_{\text{iteration-invariant}}.
\end{aligned} \tag{7.8}$$

Denote the supremum of the iteration-invariant term in (7.8) as $\Delta = c|\Gamma|\|\eta\| + d|\Gamma|\|\mathbf{u}_d\| + c|\Gamma|\|\mathbf{x}_d\|$, which is a constant. Therefore, taking the λ-norm on both sides of (7.8) yields

$$\|\mathbf{u}_{i+1}\|_\lambda \leq \left(\rho_0 + \frac{cl|\Gamma|}{\lambda}\right)\|\mathbf{u}_i\|_\lambda + \Delta. \tag{7.9}$$

Choose a sufficiently large λ such that $\rho_0 + \frac{cl|\Gamma|}{\lambda} \triangleq \rho_1 < 1$. Therefore, $\|\mathbf{u}_i\|_\lambda$ is bounded for any iteration i, and

$$\|\mathbf{u}_i\|_\lambda \leq \frac{\Delta}{1 - \rho_1}.$$

From the analysis above, $\|\mathbf{u}_i\|$ must be bounded for all iterations as the supremum norm and λ-norm are equivalent. Based on the inequality (7.5), $\|\mathbf{x}_i\|$ must belong to certain compact set D for all iterations. Therefore, $\mathbf{f}(\mathbf{x})$ can be treated as a global Lipschitz function in the learning context. Now utilizing Assumptions 7.1 and 7.2, the convergence of the P-type learning rule (7.3) can be proved by traditional λ-norm analysis (Xu and Tan, 2003). ∎

    The condition (7.5) is more general than the global Lipschitz condition in ILC literature. To demonstrate this point, assume that $\mathbf{f}(\mathbf{x})$ satisfies the global Lipschitz condition, that is for any $\mathbf{z}_1, \mathbf{z}_2 \in \mathbb{R}^n$,

$$|\mathbf{f}(\mathbf{z}_1) - \mathbf{f}(\mathbf{z}_2)| \leq L|\mathbf{z}_1 - \mathbf{z}_2|,$$

where L is the global Lipschitz constant.

    From (7.1a) the solution of $\mathbf{x}(t)$ can be written as

$$\mathbf{x}(t) = \mathbf{x}(0) + \int_0^t (\mathbf{f}(\mathbf{x}(\tau)) + \mathbf{g}(\mathbf{x}(\tau))\mathbf{u}(\tau))\, d\tau.$$

Taking the norm on both sides and applying the global Lipschitz condition on $\mathbf{f}(\mathbf{x})$ yields

$$|\mathbf{x}(t)| \leq |\mathbf{x}(0)| + \int_0^t (|\mathbf{f}(\mathbf{x})| + |\mathbf{g}(\mathbf{x})\mathbf{u}(\tau)|)\, d\tau$$

$$\leq |\mathbf{x}(0)| + \int_0^t |\mathbf{f}(0)|\, d\tau + \int_0^t \|\mathbf{g}(\mathbf{x})\||\mathbf{u}(\tau)|\, d\tau + \int_0^t L|\mathbf{x}(\tau)|\, d\tau.$$

Therefore, applying Gronwall–Bellman's Lemma (Khalil, 2002) yields

$$|\mathbf{x}(t)| \leq e^{Lt}|\mathbf{x}(0)| + \int_0^t e^{L(t-\tau)}\|\mathbf{g}(\mathbf{x})\||\mathbf{u}(\tau)|\, d\tau + \int_0^t e^{L(t-\tau)}|\mathbf{f}(0)|\, d\tau$$

$$\leq l \int_0^t |\mathbf{u}(\tau)|\, d\tau + \eta(t),$$

where $l = \|\mathbf{g}(\mathbf{x})\| \int_0^T e^{L(T-\tau)}\, d\tau$ and $\eta(t) = e^{Lt}|\mathbf{x}(0)| + \int_0^t e^{L(t-\tau)}|\mathbf{f}(0)|\, d\tau$. Therefore, we can conclude that the global Lipschitz condition is a special case of (7.5).

7.3.2 Lyapunov Stable Systems

To explore the applicability of ILC to local Lipschitz systems, the first natural candidate is the Lyapunov stable system. Furthermore, many industry processes are stable due to heat dissipation and frictions. This subsection focuses on Lyapunov stable systems.

Theorem 7.1 Consider system (7.1) under the P-type learning rule (7.3), and Assumptions 7.1 and 7.2. If there exist $\alpha_1, \alpha_2 \in \mathcal{K}_\infty$, a continuously differentiable function $W(\mathbf{x})$, a positive semi-definite function $\alpha_3(\mathbf{x})$, and a positive constant γ, such that

$$\alpha_1(|\mathbf{x}|) \le W(\mathbf{x}) \le \alpha_2(|\mathbf{x}|) \tag{7.10}$$

$$\frac{\partial W}{\partial \mathbf{x}}(\mathbf{f}(\mathbf{x}) + \mathbf{g}(\mathbf{x})\mathbf{u}) \le -\alpha_3(\mathbf{x}) + \gamma|\mathbf{u}|; \tag{7.11}$$

if, in addition, α_1 is convex, then the P-type learning rule (7.3) is convergent.

Proof: From the inequality (7.11), we have

$$W(\mathbf{x}(t)) \le -\int_0^t \alpha_3(\mathbf{x}(\tau))\,d\tau + \gamma \int_0^t |\mathbf{u}(\tau)|\,d\tau + W(\mathbf{x}(0))$$

$$\le \gamma \int_0^t |\mathbf{u}(\tau)|\,d\tau + W(\mathbf{x}(0)). \tag{7.12}$$

By using the comparison function in (7.10), (7.12) can be written as

$$\alpha_1(|\mathbf{x}|) \le \gamma \int_0^t |\mathbf{u}(\tau)|\,d\tau + W(\mathbf{x}(0)). \tag{7.13}$$

Notice that α_1 is convex, from Lemma 7.1 we have the following inequality:

$$|\mathbf{x}| \le \gamma k \int_0^t |\mathbf{u}(\tau)|\,d\tau + kW(\mathbf{x}(0)) + b, \tag{7.14}$$

where k and b are some constants. Hence, (7.14) satisfies the condition in (7.5). Therefore, we can conclude that the P-type learning rule is convergent by using Lemma 7.2. ∎

Take a close look at Equations (7.10) and (7.11): when input $\mathbf{u}$ is set to zero, the unforced system of (7.1a) is Lyapunov stable. Theorem 7.1 provides a sufficient condition for P-type rule convergence. Next we will investigate what kind of systems satisfy (7.10) and (7.11).

Lemma 7.3 If the unforced system of (7.1a) is Lyapunov stable, then there exist a continuously differentiable function $W(\mathbf{x})$, comparison functions $\alpha_1, \alpha_2 \in \mathcal{K}$, and a positive semi-definite function $\alpha_3(\mathbf{x})$, such that

$$\alpha_1(|\mathbf{x}|) \le W(\mathbf{x}) \le \alpha_2(|\mathbf{x}|),$$

$$\frac{\partial W}{\partial \mathbf{x}}(\mathbf{f}(\mathbf{x}) + \mathbf{g}(\mathbf{x})\mathbf{u}) \le -\alpha_3(\mathbf{x}) + \gamma|\mathbf{u}|.$$

Comparing Lemma 7.3 with Theorem 7.1, we can notice that in Lemma 7.3 α_1 and α_2 belong to $\mathcal{K}$ functions instead of $\mathcal{K}_\infty$ functions, and α_1 is not necessarily a convex function.

Proof: As the unforced system of (7.1a) is Lyapunov stable, by the Converse Theorem (Khalil, 2002) there exists a Lyapunov function $V(\mathbf{x})$ such that

$$\overline{\alpha}_1(|\mathbf{x}|) \le V(\mathbf{x}) \le \overline{\alpha}_2(|\mathbf{x}|), \tag{7.15}$$

$$\frac{\partial V}{\partial \mathbf{x}}\mathbf{f}(\mathbf{x}) \le -\overline{\alpha}_3(\mathbf{x}), \tag{7.16}$$

where $\overline{\alpha}_1, \overline{\alpha}_2 \in \mathcal{K}_\infty$ and $\overline{\alpha}_3(\mathbf{x})$ is a positive semi-definite function.

Define a positive and non-decreasing function $\overline{\alpha}_4$ and a class $\mathcal{K}$ function π as follows:

$$\overline{\alpha}_4(r) = \sup_{|\mathbf{x}| \leq r} \left| \frac{\partial V}{\partial \mathbf{x}} \right|,$$

and

$$\pi(r) = \int_0^r \frac{1}{1 + \chi(s)} \, ds,$$

where $\chi(s)$ is a positive and non-decreasing function to be defined later.

Let $W(\mathbf{x}) = \pi(V(\mathbf{x}))$, then

$$\frac{\partial W}{\partial \mathbf{x}} \dot{\mathbf{x}} = \frac{\partial \pi}{\partial V} \frac{\partial V}{\partial \mathbf{x}} \mathbf{f}(\mathbf{x}) + \frac{\partial \pi}{\partial V} \frac{\partial V}{\partial \mathbf{x}} \mathbf{g}(\mathbf{x}) \mathbf{u}$$

$$\leq \frac{-\overline{\alpha}_3(\mathbf{x})}{1 + \chi(V(\mathbf{x}))} + \frac{\overline{\alpha}_4(|\mathbf{x}|)|\mathbf{g}(\mathbf{x})| \, |\mathbf{u}|}{1 + \chi(V(\mathbf{x}))}. \tag{7.17}$$

Define $\chi(V(\mathbf{x})) = \overline{\alpha}_4(\overline{\alpha}_1^{-1}(V(\mathbf{x})))$, which is positive and non-decreasing.

From (7.15), we have $|\mathbf{x}| \leq \overline{\alpha}_1^{-1}(V(\mathbf{x}))$ and $V(\mathbf{x}) \leq \overline{\alpha}_2(|\mathbf{x}|)$. As $\overline{\alpha}_4$ is a positive and non-decreasing function, we have $\overline{\alpha}_4(|\mathbf{x}|) \leq \overline{\alpha}_4(\overline{\alpha}_1^{-1}(V(\mathbf{x})))$. Similarly, $\overline{\alpha}_4(\overline{\alpha}_1^{-1}(V(\mathbf{x}))) \leq \overline{\alpha}_4(\overline{\alpha}_1^{-1}(\overline{\alpha}_2(|\mathbf{x}|)))$. With these inequalities, (7.17) can be written as

$$\frac{\partial W}{\partial \mathbf{x}} \dot{\mathbf{x}} \leq \frac{-\overline{\alpha}_3(\mathbf{x})}{1 + \overline{\alpha}_4(\overline{\alpha}_1^{-1}(V(\mathbf{x})))} + \frac{\overline{\alpha}_4(|\mathbf{x}|)|\mathbf{g}(\mathbf{x})| \, |\mathbf{u}|}{1 + \overline{\alpha}_4(\overline{\alpha}_1^{-1}(V(\mathbf{x})))}$$

$$\leq \frac{-\overline{\alpha}_3(\mathbf{x})}{1 + \overline{\alpha}_4(\overline{\alpha}_1^{-1}(\overline{\alpha}_2(|\mathbf{x}|)))} + \frac{\overline{\alpha}_4(|\mathbf{x}|)|\mathbf{g}(\mathbf{x})| \, |\mathbf{u}|}{1 + \overline{\alpha}_4(|\mathbf{x}|)}$$

$$\leq \frac{-\overline{\alpha}_3(\mathbf{x})}{1 + \overline{\alpha}_4(\overline{\alpha}_1^{-1}(\overline{\alpha}_2(|\mathbf{x}|)))} + |\mathbf{g}(\mathbf{x})| \, |\mathbf{u}|. \tag{7.18}$$

Therefore, set $\alpha_1 = \pi(\overline{\alpha}_1)$, $\alpha_2 = \pi(\overline{\alpha}_2)$, $\alpha_3 = \frac{-\overline{\alpha}_3(\mathbf{x})}{1 + \overline{\alpha}_4(\overline{\alpha}_1^{-1}(\overline{\alpha}_2(|\mathbf{x}|)))}$, and $\gamma = \|\mathbf{g}(\mathbf{x})\|$. This completes the proof. ∎

The constructions of $\pi(r)$ and $W(\mathbf{x})$ are motivated by the techniques in Angeli *et al.* (2000a).

Lemma 7.3 shows that if the unforced system is Lyapunov stable, there exists a continuously differentiable function $W(\mathbf{x})$ satisfying the similar conditions (7.10) and (7.11) in Theorem 7.1. As Theorem 7.1 provides a sufficient condition for P-type learning rule convergence, comparing the results in Lemma 7.3 and the conditions in Theorem 7.1 shows that when α_1 in Lemma 7.3 happens to be a $\mathcal{K}_\infty$ function and convex, all the conditions in Theorem 7.1 are fulfilled. Therefore, we have the following corollary.

Corollary 7.1 If α_1 in Lemma 7.3 is a $\mathcal{K}_\infty$ function and convex, then the P-type learning rule (7.3) is convergent.

The proof of Lemma 7.3 and Corollary 7.1 provide a constructive method to check if the P-type learning rule works for a Lyapunov stable system.

Next, we demonstrate the applications of Corollary 7.1 by two examples, namely, the stable system with quadratic Lyapunov function and the globally exponentially stable system. As the results cannot be obtained by existing analysis method in the literature, they are presented in the form of corollaries.

Corollary 7.2 Consider system (7.1) under the P-type learning rule (7.3) and Assumptions 7.1 and 7.2. If the unforced system of (7.1a) is Lyapunov stable, and admits a quadratic Lyapunov function, that is $V(\mathbf{x}) = \mathbf{x}^T P \mathbf{x}$ where P is a symmetric positive definite matrix, then the P-type learning rule (7.3) is convergent.

Proof: As the unforced system of (7.1a) admits a quadratic Lyapunov function $V(\mathbf{x}) = \mathbf{x}^T P \mathbf{x}$, we can construct two $\mathcal{K}_\infty$ functions $\overline{\alpha}_1$ and $\overline{\alpha}_2$ such that

$$\overline{\alpha}_1(|\mathbf{x}|) \le V(\mathbf{x}) \le \overline{\alpha}_2(|\mathbf{x}|),$$

where $\overline{\alpha}_1(r) = c_1 r^2$, $\overline{\alpha}_2(r) = c_2 r^2$, $c_1 = \lambda_{\min}(P)$, and $c_2 = \lambda_{\max}(P)$. $\overline{\alpha}_4$ can be calculated as follows:

$$\overline{\alpha}_4(r) = \sup_{|x| \le r} \left| \frac{\partial V(\mathbf{x})}{\partial \mathbf{x}} \right| = 2|P|r = c_4 r.$$

Following similar procedures to the proof of Lemma 7.3, $W(\mathbf{x})$ can be constructed by $\pi(V(\mathbf{x}))$. Therefore, $\alpha_1 = \pi(\overline{\alpha}_1)$. Utilizing Lemma 7.3 and Corollary 7.1, it suffices to show that α_1 is a $\mathcal{K}_\infty$ function and convex in order to conclude Corollary 7.2. Calculating α_1, we have

$$\alpha_1(r) = \int_0^{\overline{\alpha}(r)} \frac{1}{1 + \overline{\alpha}_4(\overline{\alpha}_1^{-1}(s))} \, ds$$

$$= \int_0^{c_1 r^2} \frac{1}{1 + \frac{c_4}{\sqrt{c_1}} \sqrt{s}} \, ds. \tag{7.19}$$

From (7.19), we can conclude that $\alpha_1(r)$ is nonnegative, continuous, monotonically increasing, and tends to infinity as $r \to \infty$. Thus, $\alpha_1 \in \mathcal{K}_\infty$.

Differentiating α_1 twice against r yields

$$\frac{d\alpha_1^2}{d^2 r} = \frac{2c_1}{(1 + c_4 r)^2} > 0.$$

Therefore, α_1 is a convex function. This completes the proof. ∎

As a specific example, consider the dynamics below:

$$\dot{x} = -x - x^3 + u, \tag{7.20a}$$

$$y = x + u. \tag{7.20b}$$

The system (7.20) contains a local Lipschitz term $-x^3$, hence all existing ILC theory does not apply to this system. It is straightforward to verify that the unforced system of (7.20a) admits the Lyapunov function $V(x) = x^2$. Thus, by Corollary 7.2, the P-type learning rule is applicable to system (7.20).

Corollary 7.3 Consider system (7.1) under the P-type learning rule (7.3) and Assumptions 7.1 and 7.2. If the unforced system of (7.1a) is globally exponentially stable, then the P-type learning rule (7.3) is convergent.

Proof: As the unforced system of (7.1a) is globally exponentially stable, by the Converse Theorem (Khalil, 2002), there exists a Lyapunov function $V(\mathbf{x})$ such that

$$c_1|\mathbf{x}|^2 \leq V(\mathbf{x}) \leq c_2|\mathbf{x}|^2,$$
$$\frac{\partial V}{\partial \mathbf{x}}\mathbf{f}(\mathbf{x}) \leq -c_3|\mathbf{x}|^2,$$
$$\left|\frac{\partial V}{\partial \mathbf{x}}\right| \leq c_4|\mathbf{x}|,$$

where c_1, c_2, c_3, c_4 are positive constants. Therefore, we can draw the conclusion by using similar arguments to the proof of Corollary 7.2. ∎

It is easy to show that the unforced system of (7.20a) is globally exponentially stable, Corollary 7.3 says that the P-type learning rule is applicable to system (7.20), which is the same as predicted by Corollary 7.2. Corollary 7.3 can be regarded as a special case of Corollary 7.2.

7.3.3 Systems with Stable Local Lipschitz Terms but Unstable Global Lipschitz Factors

All the results presented so far in this section assume Lyapunov stable unforced systems. However, it is widely observed that ILC works for unstable systems as well, especially unstable system that are globally Lipschitz continuous. This motivates us to further explore what kinds of unstable systems ILC is applicable to. The following Lyapunov criterion renders the existing ILC for global Lipschitz systems as a special case.

Theorem 7.2 Consider system (7.1) under the P-type learning rule (7.3) and Assumptions 7.1 and 7.2. If there exist $\alpha_1, \alpha_2 \in \mathcal{K}_\infty$, a continuously differentiable function $W(\mathbf{x})$, and positive constants $\gamma_1, \gamma_2, \gamma_3$, such that

$$\alpha_1(|\mathbf{x}|) \leq W(\mathbf{x}) \leq \alpha_2(|\mathbf{x}|) \tag{7.21}$$
$$\frac{\partial W}{\partial \mathbf{x}}(\mathbf{f}(\mathbf{x}) + \mathbf{g}(\mathbf{x})\mathbf{u}) \leq \gamma_1|\mathbf{x}| + \gamma_2|\mathbf{u}| + \gamma_3; \tag{7.22}$$

if, in addition, α_1 is convex, then the P-type learning rule (7.3) is convergent.

Proof: From (7.22) we have

$$W(\mathbf{x}(t)) \leq \gamma_1 \int_0^t |\mathbf{x}(\tau)|\, d\tau + \gamma_2 \int_0^t |\mathbf{u}(\tau)|\, d\tau + \gamma_3 t + W(\mathbf{x}(0)). \tag{7.23}$$

Noticing Equation (7.21) and Lemma 7.1, we have the following relation:

$$|\mathbf{x}(t)| \leq \gamma_1 k \int_0^t |\mathbf{x}(\tau)|\, d\tau + \gamma_2 k \int_0^t |\mathbf{u}(\tau)|\, d\tau + \gamma_3 kt + b, \tag{7.24}$$

for $t \in [0, T]$, where k depends on α_1, and b is a positive constant that depends on α_1 and $W(\mathbf{x}(0))$.

Taking the λ-norm on both sides of (7.24) yields

$$\|\mathbf{x}\|_\lambda \leq \frac{\gamma_1 k}{\lambda}\|\mathbf{x}\|_\lambda + \frac{\gamma_2 k}{\lambda}\|\mathbf{u}\|_\lambda + \gamma_3 kT + b.$$

Solving $\|\mathbf{x}\|_\lambda$ in the above equation yields

$$\|\mathbf{x}\|_\lambda \leq \left(1 - \frac{\gamma_1 k}{\lambda}\right)^{-1}\left(\frac{\gamma_2 k}{\lambda}\|\mathbf{u}\|_\lambda + \gamma_3 kT + b\right). \tag{7.25}$$

The relation between $\mathbf{u}_{i+1}$ and $\mathbf{u}_i$ has been derived in (7.7). Similarly, taking the λ-norm on both sides of (7.7) yields

$$\|\mathbf{u}_{i+1}\|_\lambda \leq \rho_0\|\mathbf{u}_i\|_\lambda + c|\Gamma|\|\mathbf{x}_i\|_\lambda + d|\Gamma|\|\mathbf{u}_d\| + c|\Gamma|\|\mathbf{x}_d\|, \tag{7.26}$$

where c and d are positive constants defined in the proof of Lemma 7.2.

Substituting (7.25) in (7.26), we can obtain

$$\|\mathbf{u}_{i+1}\|_\lambda \leq \left(\rho_0 + \left(1 - \frac{\gamma_1 k}{\lambda}\right)^{-1}\frac{c\gamma_2 k|\Gamma|}{\lambda}\right)\|\mathbf{u}_i\|_\lambda$$

$$+ \underbrace{c|\Gamma|\left(1 - \frac{\gamma_1 k}{\lambda}\right)^{-1}(\gamma_3 kT + b) + d|\Gamma|\|\mathbf{u}_d\| + c|\Gamma|\|\mathbf{x}_d\|}_{\text{iteration-invariant}}.$$

$$\tag{7.27}$$

Choose a sufficiently large λ such that

$$\rho_0 + \left(1 - \frac{\gamma_1 k}{\lambda}\right)^{-1}\frac{c\gamma_2 k|\Gamma|}{\lambda} = \rho_1 < 1.$$

Hence, $\mathbf{u}_i$ and $\mathbf{x}_i$ are bounded for all iterations. By following a similar line to the proof of Lemma 7.2, we can conclude Theorem 7.2. ∎

From Theorem 7.2 we can conclude that the P-type learning rule is applicable to a system whose local Lipschitz terms are stable and whose global Lipschitz terms can be unstable. Next, we show that Theorem 7.2 includes the global Lipschitz system as a special case. Consider system (7.1), and assume that $\mathbf{f}(\mathbf{x})$ satisfies the global Lipschitz condition with Lipschitz constant L. Let $\pi(r) = \int_0^r \frac{1}{1+\sqrt{s}}\,ds$, $V(\mathbf{x}) = \mathbf{x}^T\mathbf{x}$, and $W(\mathbf{x}) = \pi(V(\mathbf{x}))$. Thus, $W(\mathbf{x})$ can be written as

$$W(\mathbf{x}) = \int_0^{\mathbf{x}^T\mathbf{x}} \frac{1}{1+\sqrt{s}}\,ds.$$

In this case $\bar{\alpha}_1(r) = r^2$. Note that we have already shown that $\pi(\bar{\alpha}_1)$ is a $\mathcal{K}_\infty$ function and convex in the proof to Corollary 7.2.

Differentiating $W(\mathbf{x})$ yields

$$\frac{\partial W}{\partial \mathbf{x}}(\mathbf{f}(\mathbf{x}) + \mathbf{g}(\mathbf{x})\mathbf{u}) = \frac{2\mathbf{x}^T\mathbf{f}(\mathbf{x}) + 2\mathbf{x}^T\mathbf{g}(\mathbf{x})\mathbf{u}}{1+|\mathbf{x}|}$$

$$\leq 2L|\mathbf{x}| + 2\|\mathbf{g}(\mathbf{x})\|\|\mathbf{u}\| + 2|\mathbf{f}(0)|.$$

Set $\gamma_1 = 2L$, $\gamma_2 = 2\|\mathbf{g}(\mathbf{x})\|$, and $\gamma_3 = 2|\mathbf{f}(0)|$. Hence, all global Lipschitz systems satisfy the condition in Theorem 7.2.

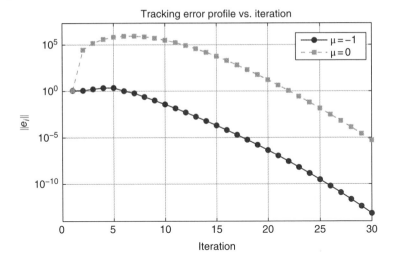

Figure 7.1 Tracking error profiles vs. iteration number for $\mu = -1$ and $\mu = 0$.

Consider the dynamics below:

$$\dot{x} = 2x + \mu x^3 + u, \tag{7.28a}$$

$$y = x + u. \tag{7.28b}$$

The unforced system of (7.28a) has an unstable origin and contains a local Lipschitz term μx^3. It is easy to show that the P-type learning rule is applicable to system (7.28) by Theorem 7.2 when $\mu < 0$. In contrast, all existing ILC theories cannot claim such a conclusion unless $\mu = 0$. To verify the theoretical prediction, the following P-type controller is applied to system (7.28):

$$u_{i+1} = u_i + e_i,$$

where $e_i = y_d - y_i$, and $y_d(t) = 0.1t + \sin(t)$ for $t \in [0, 5]$. The initial state is set to $x_i(0) = 0$ for all iterations, and the control input in the first iteration $u_i(t) = 0$. It can be seen from Figure 7.1 that the P-type learning rule converges for both systems, that is, $\mu = -1$ and $\mu = 0$. A closer examination shows that when $\mu = 0$ the tracking error initially increases up to the magnitude of 10^5, and then exponentially converges to zero. In contrast, the transient performance for the system when $\mu = -1$ is much smoother. This is because the local Lipschitz term $-x^3$ is stable and it prevents the system state from growing to a large magnitude. This observation also suggests that it may help improve the transient performance by stabilizing a system before applying the ILC rule.

7.4 Convergence Properties in the Presence of Bounding Conditions

7.4.1 Systems with Bounded Drift Term

In addition to Assumption 7.1, we further assume that the drift term $\mathbf{f}(\mathbf{x})$ in system (7.1) is bounded for all $\mathbf{x} \in \mathbb{R}^n$. Then, we have the following theorem.

Theorem 7.3 Consider system (7.1) under the P-type learning rule (7.3) and Assumptions 7.1 and 7.2. If $\|\mathbf{f}(\mathbf{x})\| = \sup_{\mathbf{x} \in \mathbb{R}^n} |\mathbf{f}(\mathbf{x})| \leq b_f$ and b_f is a positive constant, then the P-type learning rule (7.3) is convergent.

Proof: From the system dynamics (7.1a), the system state can be written as

$$\mathbf{x}(t) = \mathbf{x}(0) + \int_0^t (\mathbf{f}(\mathbf{x}(\tau)) + \mathbf{g}(\mathbf{x}(\tau))\mathbf{u}(\tau))\, d\tau. \tag{7.29}$$

Taking the norm on both sides of (7.29) and noticing the boundedness assumption on $\mathbf{f}(\mathbf{x})$, a bounding function on $|\mathbf{x}(t)|$ can be obtained:

$$|\mathbf{x}(t)| \leq \|\mathbf{g}(\mathbf{x})\| \int_0^t |\mathbf{u}(\tau)|\, d\tau + b_f t + |\mathbf{x}(0)|. \tag{7.30}$$

Note that the bounding function on $|\mathbf{x}(t)|$ in (7.30) has the same form as in (7.5). Thus, we can conclude that the P-type learning rule (7.3) is convergent based on Lemma 7.2. ∎

To verify the result in Theorem 7.3, consider the following system:

$$\dot{x} = \sin(x^2) + u,$$

$$y = x + u.$$

Note that the local Lipschitz term $\sin(x^2)$ is bounded for all $x \in \mathbb{R}$. Let the desired control input be $u_d(t) = 0.1t + 1.5 \sin(2t)$, the initial state $x_i(0) = 0$, and $u_1(t) = 0$ for $t \in [0, 5]$. Applying the P-type learning rule

$$u_{i+1} = u_i + e_i,$$

it can be seen from Figure 7.2 that the tracking error increases slightly and then exponentially converges to zero.

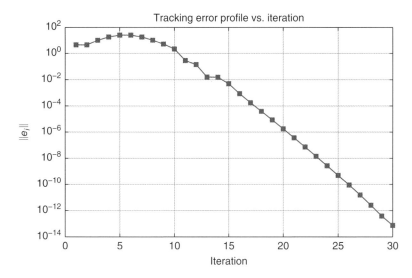

Figure 7.2 Tracking error profiles vs. iteration number for system with bounded local Lipschitz term.

7.4.2 Systems with Bounded Control Input

In almost all practical systems, the inputs always have finite power due to physical limitations. In this subsection, we apply control saturation to a class of uniformly bounded energy bounded state (UBEBS) systems (Angeli *et al.*, 2000b), and explore the applicability of the P-type ILC rule to UBEBS systems.

Definition 7.1 The system (7.1) is UBEBS if there exist class $\mathcal{K}_\infty$ functions α_1, α_2, and some positive constant a such that

$$|\mathbf{x}(t)| \leq \int_0^t \alpha_1(|\mathbf{u}(\tau)|)\, d\tau + \alpha_2(|\mathbf{x}(0)|) + a,$$

for any $t \in [0, T]$.

The definition of UBEBS shares the similar but more general definitions than ISS (Input to State Stability) (Jiang and Wang, 2001; Khalil, 2002) and iISS (integral Input to State Stability) (Angeli *et al.*, 2000a; Sontag, 2006). The system (7.1) is said to be ISS if the state satisfies

$$|\mathbf{x}(t)| \leq \beta(|\mathbf{x}(0)|, t) + \gamma \|\mathbf{u}\|,$$

where $\beta(|\mathbf{x}(0)|, t)$ is a class $\mathcal{KL}$ function[2], and γ is a positive constant.
The system (7.1) is said to be iISS if the state satisfies

$$\alpha_1(|\mathbf{x}(t)|) \leq \beta(|\mathbf{x}(0)|, t) + \int_0^t \alpha_2(|\mathbf{u}(\tau)|)\, d\tau,$$

where α_1 and α_2 are class $\mathcal{K}$ functions.
Based on the definitions of UBEBS, ISS and iISS, we can conclude that UBEBS systems include ISS and iISS systems as special cases. Both ISS and iISS require that the system has a stable origin. However, a UBEBS system does not require a stable origin. Therefore, the results derived in this subsection can be applied to a large class of systems when control saturation is imposed.
The P-type learning rule with control saturation is proposed as follows:

$$\mathbf{v}_{i+1} = \mathbf{u}_i + \Gamma \mathbf{e}_i, \tag{7.31a}$$

$$\mathbf{u}_{i+1} = \mathrm{sat}(\mathbf{v}_{i+1}, \mathbf{u}^*), \tag{7.31b}$$

where Γ is chosen such that (7.4) is satisfied, the saturation function $\mathrm{sat}(\cdot, \mathbf{u}^*)$ is defined component-wise, the saturation bound

$$\mathbf{u}^* = \begin{bmatrix} \underline{u}_1, \overline{u}_1 \\ \underline{u}_2, \overline{u}_2 \\ \vdots \quad \vdots \\ \underline{u}_p, \overline{u}_p \end{bmatrix},$$

and $\underline{u}_j, \overline{u}_j$ denote the lower and upper limits of the jth input channel and $1 \leq j \leq p$.

---

2 $\beta(s, t)$ is said to belong to class $\mathcal{KL}$ if for each fixed t, the mapping $\beta(s, t)$ belongs to class $\mathcal{K}$ with respect to s, and for each fixed s, the mapping $\beta(s, t)$ is decreasing with respect to t and $\beta(s, t) \to 0$ as $t \to \infty$.

As the desired trajectory $\mathbf{y}_d$ in (7.2) should be realizable, $\mathbf{u}_d$ must not exceed the saturation bound, that is $\mathbf{u}_d = \text{sat}(\mathbf{u}_d, \mathbf{u}^*)$.

Applying the control saturation (7.31b), the input never exceeds the saturation bound. Therefore, the upper bound on state $|\mathbf{x}(t)|$ can be obtained from Definition 7.1. Denote the upper bound of $|\mathbf{x}(t)|$ by $b_x = b_x(\mathbf{x}(0), \mathbf{u}^*, T)$. As such, $\mathbf{f}(\mathbf{x})$ can be regarded as a global Lipschitz function in $\mathbf{x}$ with Lipschitz constant L, and

$$L = \sup_{\|\mathbf{z}_1\| < b_x, \|\mathbf{z}_2\| < b_x} \phi(\mathbf{z}).$$

Theorem 7.4 Consider system (7.1) under the P-type learning rule (7.31) and Assumptions 7.1 and 7.2. If the system (7.1) is UBEBS, then the P-type learning rule (7.31) with control saturation is convergent.

Proof: As $\mathbf{g}(\mathbf{x})$ is a continuously differentiable function and $\|\mathbf{x}\| = b_x$, the following inequality holds:

$$|\mathbf{g}(\mathbf{x}_d) - \mathbf{g}(\mathbf{x}_i)| \le L_g |\mathbf{x}_d - \mathbf{x}_i|, \tag{7.32}$$

where L_g is a constant.

Let $\delta \mathbf{x}_i = \mathbf{x}_d - \mathbf{x}_i$ and $\delta \mathbf{u}_i = \mathbf{u}_d - \mathbf{u}_i$. Note that $\delta \mathbf{x}_i(0) = 0$ by Assumption 7.2, thus

$$\delta \mathbf{x}_i(t) = \int_0^t \left(\mathbf{f}(\mathbf{x}_d) - \mathbf{f}(\mathbf{x}_i) + \mathbf{g}(\mathbf{x}_d)\mathbf{u}_d - \mathbf{g}(\mathbf{x}_i)\mathbf{u}_i \right) d\tau$$

$$= \int_0^t \left(\mathbf{f}(\mathbf{x}_d) - \mathbf{f}(\mathbf{x}_i) + \mathbf{g}(\mathbf{x}_d)\mathbf{u}_d - \mathbf{g}(\mathbf{x}_i)\mathbf{u}_d + \mathbf{g}(\mathbf{x}_i)\mathbf{u}_d - \mathbf{g}(\mathbf{x}_i)\mathbf{u}_i \right) d\tau. \tag{7.33}$$

Taking the norm on both sides of (7.33) yields

$$|\delta \mathbf{x}_i(t)| \le \int_0^t \left((L + L_g \|\mathbf{u}_d\|) |\delta \mathbf{x}_i| + \|\mathbf{g}(\mathbf{x}_i)\| |\delta \mathbf{u}_i| \right) d\tau. \tag{7.34}$$

Applying Gronwall–Bellman's Lemma to (7.34) yields

$$|\delta \mathbf{x}_i(t)| \le \int_0^t e^{(L + L_g \|\mathbf{u}_d\|)(t-\tau)} \|\mathbf{g}(\mathbf{x}_i)\| |\delta \mathbf{u}_i(\tau)| d\tau. \tag{7.35}$$

Taking the λ-norm of (7.35) we obtain

$$\|\delta \mathbf{x}_i\|_\lambda \le \frac{b_g}{\lambda - l} \|\delta \mathbf{u}_i\|_\lambda, \tag{7.36}$$

where $l = L + L_g \|\mathbf{u}_d\|$ and $b_g = \|\mathbf{g}(\mathbf{x})\|$.

Next, calculate the relation between $\delta \mathbf{u}_{i+1}$ and $\delta \mathbf{u}_i$:

$$|\delta \mathbf{u}_{i+1}| = |\mathbf{u}_d - \text{sat}(\mathbf{v}_{i+1}, \mathbf{u}^*)|$$

$$\le |\mathbf{u}_d - \mathbf{v}_{i+1}|$$

$$= |\mathbf{u}_d - \mathbf{u}_i - \Gamma \mathbf{e}_i|$$

$$= \left| \left(I_m - \Gamma \frac{\partial \mathbf{h}}{\partial \mathbf{u}} \right) \delta \mathbf{u}_i + \frac{\partial \mathbf{h}}{\partial \mathbf{x}} \delta \mathbf{x}_i \right|$$

$$\le \rho_0 |\delta \mathbf{u}_i| + c |\delta \mathbf{x}_i|, \tag{7.37}$$

where c is defined in the proof to Lemma 7.2.

Note that (7.7) is utilized in deriving (7.37). Taking the λ-norm on (7.37) yields

$$\|\delta \mathbf{u}_{i+1}\|_\lambda \le \rho_0 \|\delta \mathbf{u}_i\|_\lambda + c\|\delta \mathbf{x}_i\|_\lambda. \tag{7.38}$$

Substituting (7.36) in (7.38), we obtain

$$\|\delta \mathbf{u}_{i+1}\|_\lambda \le \left(\rho_0 + \frac{cb_g}{\lambda - l}\right)\|\delta \mathbf{u}_i\|_\lambda. \tag{7.39}$$

Choose λ sufficiently large such that (7.39) is a contraction-mapping. Therefore, $\|\delta \mathbf{u}_i\|$ converges to zero in the iteration domain. Consequently, $\|\delta \mathbf{x}_i\|$ converges to zero as well. This completes the proof. ∎

To demonstrate the application of Theorem 7.4, we consider the switched reluctance motor (SRM) control problem, which is one of the benchmark problems in ILC literature (Sahoo *et al.*, 2004). In an SRM, the rotor is made of steel laminations without conductors or permanent magnets, and only the stator presents windings. Due to the simple mechanical structure, the SRM is cheaper to manufacture compared with other ac and dc motors. However, the presence of nonlinearity makes SRM a challenging control problem.

According to Spong *et al.* (1987), an m-phase SRM can be modeled by the following dynamics:

$$\dot{\theta}_e = \omega_e,$$

$$\dot{\omega}_e = J^{-1}\left(\sum_{j=1}^m T_j(\theta_e, \eta_j) - f\omega_e\right),$$

$$T_j = \frac{\psi_s}{h_j^2(\theta_e)}\frac{d\,h_j(\theta_e)}{d\,\theta_e}\left(1 - [1 + \eta_j h_j(\theta_e)]e^{-\eta_j h_j(\theta_e)}\right),$$

$$\dot{\eta}_j = \left(\frac{\partial \psi_j}{\partial \eta_j}\right)^{-1}\left(-R\eta_j + u_j - \frac{\partial \psi_j}{\partial \theta_e}\omega_e\right),$$

$$\psi_j = \psi_s\left(1 - e^{\eta_j h_j(\theta_e)}\right),$$

$$h_j(\theta_e) \approx h_0 + b_1 \sin(N_r\theta_e - 2\pi(j-1)/m),$$

where $j = 1, \ldots, m$ denotes the phase, $m = 3$, θ_e is the electrical angular position of rotor, ω_e is the angular speed, $N_r = 4$ is the number of rotor poles, η_j is the stator current of phase j, u_j is the voltage applied on phase j, ψ_j is the flux-linkage of phase j, $\psi_s = 1.2\mathrm{Wb}$ is the flux constant, $J = 0.0016\mathrm{kgm}^2$ is the inertia of the rotor, $R = 4\Omega$ is resistance of phase j, $f = 0.2$ is the friction coefficient, $h_0 = 0.0545$ and $b_1 = 0.0454$ are two constants.

To illustrate the P-type learning rule, consider only the current to torque loop. Let the desired torque profile be $T_d(t) \in C[0, 0.04]$ as follows:

$$T_d(t) = \begin{cases} 0 & 0 \le t < 0.05 \\ \frac{3}{\tau^2}(t - 0.005)^2 - \frac{2}{\tau^3}(t - 0.005)^3 & 0.005 \le t < 0.015 \\ 1 & 0.015 \le t < 0.025 \\ 1 - \frac{3}{\tau^2}(t - 0.025)^2 + \frac{2}{\tau^3}(t - 0.025)^3 & 0.025 \le t < 0.035 \\ 0 & 0.035 \le t \le 0.04 \end{cases}$$

where $\tau = 0.01$. Figure 7.3 shows the desired torque profile.

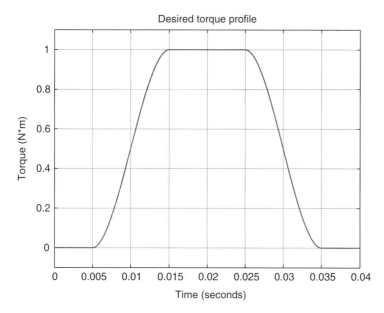

Figure 7.3 Desired torque profile.

Due to the symmetric structure, apply the learning rule to the first phase only, and the controller is chosen as

$$\eta_{1,i+1} = \text{sat}(\eta_{1,i} + \gamma e_i, \eta^*),$$

where i denotes the iteration number, learning gain $\gamma = 2$, tracking error $e_i = T_d - T_{1,i}$, and saturation bound $\eta^* = [-20, 20]$. The application of control saturation is reasonable as in any electrical drives a current limiter is always incorporated to prevent the circuit from current overflow or overheating.

In the simulation study, the initial conditions are chosen as $\theta_e = 0, \omega_e = 0$, and $\eta_{1,0} = 0$. Figure 7.4 shows the tracking error profiles in the iteration domain. After 37 iterations of learning, the tracking error is reduced to 10^{-3}.

7.5 Application of P-type Rule in MAS with Local Lipschitz Uncertainties

As mentioned in the Introduction, it is possible to apply P-type rules in the multi-agent problem setup. In this chapter, we have developed several criteria for P-type rules to work for local Lipschitz systems. In this section, we demonstrate that the P-type ILC law is applicable to MAS under certain local Lipschitz nonlinearities.

Consider a network consisting of four heterogeneous agents for the consensus tracking problem. The agents' dynamics are governed by

$$\begin{cases} \dot{x}_{i,1} = -x_{i,1}^3 + u_{i,1}, \\ y_{i,1} = 2x_{i,1} + u_{i,1}, \end{cases}$$
$$\begin{cases} \dot{x}_{i,2} = \sin(x_{i,2}^2) + u_{i,2}, \\ y_{i,2} = x_{i,2} + u_{i,2}, \end{cases}$$

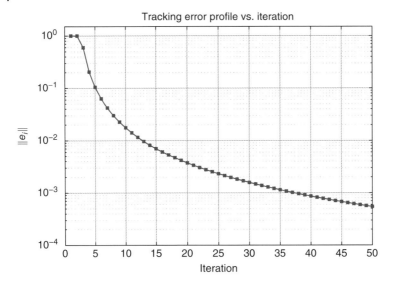

Figure 7.4 Tracking error profiles vs. iteration number under control saturation.

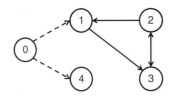

Figure 7.5 Communication topology among agents in the network.

$$
\begin{cases}
\dot{x}_{i,3} = -x_{i,3}^2 \mathrm{sign}(x_{i,3}) + u_{i,3}, \\
y_{i,3} = 3x_{i,3} + u_{i,3},
\end{cases}
$$
$$
\begin{cases}
\dot{x}_{i,4} = \arctan(x_{i,4}^2) + u_{i,4}, \\
y_{i,4} = 2x_{i,4} + u_{i,4}.
\end{cases}
$$

Note that four different local Lipschitz terms are respectively involved in the agents' models, which belong to the categories discussed in previous sections, that is, the drift terms for agents 1 and 3 are stable, meanwhile, the drift terms for agents 2 and 4 are bounded.

Let the desired reference trajectory be $y_d(t) = t + 2\sin(t)$, $t \in [0, 5]$, and the initial state for each agent $x_{i,j}(0) = 0, j = 1, 2, 3, 4$.

The communication graph among followers and the leader is depicted in Figure 7.5. The virtual leader is labeled by vertex 0 in the communication graph, and it has edges (dashed arrows) to agents 1 and 4. The communication among followers is depicted by solid arrows. Notice that the communication graph among followers is not connected. However, the communication graph including the leader contains a spanning tree with the leader being the root. Based on the results from previous chapters, the

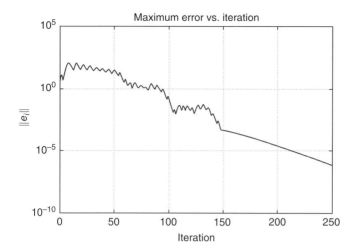

Figure 7.6 Maximal tracking error profiles vs. iteration number for MAS with local Lipschitz uncertainties.

communication is sufficient for the network to achieve consensus tracking. The Laplacian for the follower agents is

$$
L = \begin{bmatrix} 1 & -1 & 0 & 0 \\ 0 & 1 & -1 & 0 \\ -1 & -1 & 2 & 0 \\ 0 & 0 & 0 & 0 \end{bmatrix},
$$

and $D = \mathrm{diag}(1, 0, 0, 1)$. Applying the following P-type learning rule:

$$
u_{i+1,j}(t) = u_{i,j}(t) + \gamma \xi_{i,j}(t), \ u_{0,j}(t) \equiv 0, \tag{7.40}
$$

where $\gamma = 0.5$ is the learning gain and $\xi_{i,j}$ is the distributed error for agent j as defined in Chapter 2. The convergence of the maximal tracking error $\|\mathbf{e}_i\|$ is shown in Figure 7.6, where $\|\mathbf{e}_i\| \triangleq \max_{j=1,2,3,4}\{\sup_{t\in[0,5]} |e_{i,j}(t)|\}$.

7.6 Conclusion

This chapter explores the applicability of the P-type ILC rule to local Lipschitz systems. In the current ILC literature, contraction-mapping based ILC is only applicable to global Lipschitz systems. In contrast, a number of sufficient conditions for local Lipschitz systems in the form of Lyapunov criteria are developed in this chapter. By using a similar proof approach, it is further shown that the P-type ILC can be applied to systems whose local Lipschitz terms are bounded, and UBEBS systems under control saturation. Several examples are provided to verify the new results. In particular, P-type rules are successfully applied to the MAS consensus tracking problem in the presence of some local Lipschitz nonlinearities.

8

Synchronization for Nonlinear Multi-agent Systems by Adaptive Iterative Learning Control

8.1 Introduction

As we discussed in earlier chapters, there are two main frameworks for iterative learning control, namely contraction-mapping (CM) and composite energy function (CEF)-based approaches. CM-based ILC is usually preferred due to its simplicity in implementation, little system information requirement, and applicability to non-affine-in-input systems. However, in the current literature, CM-based ILC is only applicable to global Lipschitz systems. In Chapter 7, CM-based ILC is extended to several classes of local Lipschitz systems by combining Lyapunov and CM analysis methods. Consequently, the algorithms developed in Chapters 2–5 can be applied to a large range of systems. In contrast, the CEF-based method uses full state information to design estimation rules and ILC controllers. This method can be applied if the system dynamics is linear in parameters and full state information is available for feedback or nonlinear compensation. As the current state information is utilized in the controller, the transient performance is usually better than for the CM-based method.

Yang and Xu (2012) demonstrates promising results of the CEF method in the consensus tracking problem for first-order parametric agents. Li and Li (2013) apply a similar approach to second-order agents. The focus of this chapter is to design ILC schemes for agents that have more general nonlinear dynamics. The CEF framework is used to analyze the performance of synchronization algorithms. Whereas in Li and Li (2013), the nonparametric uncertainties are assumed to be bounded and suppressed by sliding mode control, in this chapter it is assumed that the nonparametric uncertainties satisfy the local Lipschitz like condition and could be unbounded. They are carefully handled by some robust terms. In Yang and Xu (2012) and Li and Li (2013), the problem formulations assume constant and known input gains, whereas in this chapter, time-varying input gains with uncertainties are considered. Moreover, the developed methods are not restricted to the first-order or second-order systems, but can also be applied to any high-order systems.

In a traditional ILC setting (tracking problem), the desired trajectory is used to design updating laws. However, in the multi-agent coordination problem, it is very common that the desired trajectory is only available to a subset of followers. Hence, only local measurement (the extended error) can be utilized in the controller design. Although there is a clear relationship between the extended error and the consensus tracking error, it is not straightforward to apply the CEF-based design framework. In this chapter, the desired trajectory is treated as an unknown time-varying (iteration-invariant)

Iterative Learning Control for Multi-agent Systems Coordination, First Edition.
Shiping Yang, Jian-Xin Xu, Xuefang Li, and Dong Shen.
© 2017 John Wiley & Sons Singapore Pte. Ltd. Published 2017 by John Wiley & Sons Singapore Pte. Ltd.

parametric uncertainty so that each follower tries to learn. This desired trajectory is combined with other parametric uncertainties that ILC can handle easily. On the other hand, nonlinear uncertainties that are not global Lipschitz continuous will be handled by robust control. The proposed algorithm enables all the followers to learn the parametric uncertainties and deal with lumped uncertainties based on local information from their neighborhoods. Convergence analysis using an appropriate CEF shows that the proposed ILC algorithms can achieve consensus tracking in the presence of parametric and lumped uncertainties under some appropriate convergence conditions.

The rest of this chapter is organized as follows. Section 8.2 provides one preliminary result in graph theory and the problem formulation. Next, ILC control laws are designed for the first-order MAS in Section 8.3. Section 8.4 extends the obtained results to the more general settings: high-order systems. Illustrative examples are presented in Section 8.5 to demonstrate efficacy of the proposed algorithms, followed by conclusions in Section 8.6.

8.2 Preliminaries and Problem Description

8.2.1 Preliminaries

Graph theory is an instrumental tool for characterizing the communication topology in the multi-agent setting. Unlike the previous chapters, an undirected graph $\mathcal{G}(\mathcal{V}, \mathcal{E})$ is adopted to model the communication among followers. That is, the information flow among followers is symmetric. The following lemma is a well known result in multi-agent coordination, and it is an important result needed in the construction of an appropriate CEF.

Lemma 8.1 (Hong *et al.*, 2006)   If the undirected graph $\mathcal{G}$ is connected, and D is any nonnegative diagonal matrix with at least one of the diagonal entries being positive, then $H = L + D$ is symmetric positive definite, where L is the Laplacian matrix of $\mathcal{G}$.

8.2.2 Problem Description for First-Order Systems

The problem formulation starts with simple first-order systems. This simplifies the design of the ILC updating law. Section 8.4 shows how to generalize the obtained results to high-order systems.

Consider a group of N heterogeneous agents. At the ith iteration, the dynamics of the jth agent, $(j = 1, 2, \ldots, N)$, take the following form:

$$\dot{x}_{i,j}(t) = \theta_j^0(t)\xi_j^0(t, x_{i,j}) + \eta_j(t, x_{i,j}) + b_j(t)u_{i,j}(t), \tag{8.1}$$

where $x_{i,j} \in \mathbb{R}^1$ is the system state, $u_{i,j} \in \mathbb{R}^1$ is the input, $\theta_j^0 : [0, T] \to \mathbb{R}^{1 \times n_j}$ is an unknown time-varying function, $\xi_j^0 : [0, T] \times \mathbb{R}^1 \to \mathbb{R}^{n_j}$ is a state-dependent known function, $\eta_j : [0, T] \times \mathbb{R}^1 \to \mathbb{R}^1$ represents the unknown lumped system uncertainties or disturbances. The parameter $b_j \in \mathbb{R}^1$ is the time-varying input gain, and it is either positive or negative. Without loss of generality, it is assumed that $0 < \underline{b_j} \le b_j \le \overline{b_j}$. In addition, $\theta_j^0(\cdot)$, $\xi_j^0(\cdot, \cdot)$, $\eta_j(\cdot, \cdot)$, and $b_j(\cdot)$ are continuous functions with respect to their arguments. For simplicity of the presentation, for a time-varying function $f(t)$, the argument t is dropped when no confusion arises.

The desired trajectory (virtual leader) satisfies $x_d \in C^1[0, T]$, hence, $\dot{x}_d$ is continuous. Meanwhile, the virtual leader's trajectory $x_d(t)$, for $t \in (0, T]$ is only accessible to a small portion of the followers. Furthermore, we assume that the upper bound of the desired trajectory x_d and its initial condition x_0 are known to each follower. The supremum norm of x_d is denoted as b_d.

Remark 8.1 Under certain scenarios, the initial states of the agents are easily obtainable compared with the states during the task execution. For example, a group of unmanned aerial vehicles (UAVs) take off simultaneously from an air base and perform a formation task for surveillance. Before the UAVs take off, the relative geometric positions among them can be prescribed by human operators. Whereas, during the task execution, the leader's position and velocity are not available to some of the followers due to communication or sensor limitations. Therefore, it is emphasized here that $x_d(t)$, for $t \in (0, T]$ is only accessible to a small portion of the followers.

Remark 8.2 For many practical systems, there are limitations due to physical constraints. For instance, the range space of an industry manipulator, the thrust on a fixed wing aircraft, or the speed of a vehicle, are all limited. As the controller requires the upper bound of x_d, which is global information, we shall minimize the usage of global information. Depending on the specific control task, the system's operation range may be used to estimate b_d.

Denote the tracking error for the jth agent at the ith iteration as

$$e_{i,j} = x_d - x_{i,j}. \tag{8.2}$$

A follower can only measure or observe the state information within its neighborhood (local information). This local information is called the extended tracking error and it is defined as follows:

$$\epsilon_{i,j} \triangleq \sum_{k \in \mathcal{N}_j} a_{j,k}(x_{i,k} - x_{i,j}) + d_j(x_d - x_{i,j}), \tag{8.3}$$

where $d_j = 1$ if the jth agent knows the desired trajectory, and $d_j = 0$ otherwise.

The control objective is to design a set of distributed controllers such that all the followers can perfectly track the desired trajectory in the presence of parametric and lumped uncertainties, that is, $\lim_{i \to \infty} e_{i,j} = 0$, for $j = 1, 2, \ldots, N$.

Noting the fact that $L\mathbf{1} = 0$ and using (8.2), the extended tracking error (8.3) can be written in the stack vector form,

$$\epsilon_i = -L\mathbf{x}_i + D\mathbf{e}_i = L(x_d\mathbf{1} - \mathbf{x}_i) + D\mathbf{e}_i$$
$$= (L + D)\mathbf{e}_i = H\mathbf{e}_i, \tag{8.4}$$

where $D = \mathrm{diag}(d_1, \ldots, d_N)$, $\mathbf{x}_i$, $\mathbf{e}_i$, and ϵ_i are the column stack vectors of $x_{i,j}$, $e_{i,j}$, and $\epsilon_{i,j}$. If Assumption 8.2 (introduced later) holds, Lemma 8.1 shows that H is symmetric positive definite. This important feature will be used to construct an appropriate CEF.

Remark 8.3   The relationship (8.4) plays a crucial role to ensure consensus tracking with local information. If ϵ_i converges, the relationship (8.4) shows that the tracking error $\mathbf{e}_i$ will also converge as H is positive definite. This also supports the key idea of MAS: by sharing information among the communication networks, each agent does not have to equip a lot of sensors to measure needed signals.

To simplify the discussion, the following assumptions are imposed.

Assumption 8.1   For any $j = 1, 2, \ldots, N$, the nonlinear function $\eta_j(t, x)$ satisfies the following condition:

$$|\eta_j(t, z_1) - \eta_j(t, z_2)| \leq \phi_j(\mathbf{z})|z_1 - z_2|, t \in [0, T], \tag{8.5}$$

where $\mathbf{z} = [z_1, z_2]^T$, and $\phi_j : \mathbb{R}^2 \to \mathbb{R}_{\geq 0}$ is a known continuous function and it is radially unbounded.[1]

Remark 8.4   Assumption 8.1 is much weaker than the global Lipschitz condition as $\phi_j(\cdot)$ might go to infinity when the state of the jth agent is unbounded. Thus the CM-based ILC design framework cannot be applied in general (see Chapter 7). Although this condition is more generic and many nonlinear functions satisfy this condition, it is more strict than the local Lipschitz condition as the inequality (8.5) holds for all $\mathbf{z} \in \mathbb{R}^2$.

Assumption 8.2   The communication graph $\mathcal{G}$ is undirected and connected, and at least one of the followers can access the leader's trajectory.

Remark 8.5   Assumption 8.2 assumes that the leader is globally reachable from all followers. This assumption is a necessary requirement for a leader-follower consensus tracking problem. If there is an isolated agent, it is not possible that the agent can track the leader's trajectory as there is no information to correct its control action.

Assumption 8.3   The initial state of all followers are reset to be at x_0 at every iteration, that is, $x_{i,j}(0) = x_d(0)$.

Remark 8.6   Assumption 8.3 is the well known identical initialization condition (*iic*), which is one of the fundamental problems in the ILC literature. It has been used in many multi-agent coordination problems, for example the formation problems considered in Ahn and Chen (2009) and Liu and Jia (2012). Without *iic*, no matter how ILC repeats, perfect tracking can never be achieved. Many modifications have been dedicated to the relaxation of *iic*, but they require either extra system information, or additional control mechanisms. For example the initial state learning in Yang *et al.* (2012). This fundamental issue has been explored over the past 30 years without much progress. In fact, the *iic* problem is essentially the initial condition problem in ordinary differential equations (ODE). The solution trajectory of an ODE is determined by its initial condition. Therefore, different initial conditions will yield distinct solution trajectories. Hence, without

---

1 A function $V : \mathbb{R}^n \to \mathbb{R}_{\geq 0}^1$ is radially unbounded if $|\mathbf{x}| \to \infty \Rightarrow V(\mathbf{x}) \to \infty$ (Khalil, 2002, Chapter 4).

iic, it is impossible for the control system to generate a solution trajectory that is identical to the reference $x_d(t)$. That is the main reason why *iic* is indispensable for perfect tracking. For more discussion, please see Remark 2.4 in Chapter 2.

From the tracking error definition in (8.2), we can find the error dynamics as follows:

$$
\begin{aligned}
\dot{e}_{i,j} &= \dot{x}_d - \dot{x}_{i,j}, \\
&= \dot{x}_d - \theta_j^0 \xi_j^0(t, x_{i,j}) - \eta_j(t, x_{i,j}) - b_j u_{i,j} \\
&= \dot{x}_d - \eta_j(t, x_d) - \theta_j^0 \xi_j^0(t, x_{i,j}) - b_j u_{i,j} + \left(\eta_j(t, x_d) - \eta_j(t, x_{i,j}) \right) \\
&= b_j \theta_j \xi_j(t, x_{i,j}) + \left(\eta_j(t, x_d) - \eta_j(t, x_{i,j}) \right) - b_j u_{i,j},
\end{aligned}
\tag{8.6}
$$

where

$$
\theta_j \triangleq [-b_j^{-1} \theta_j^0, \; b_j^{-1}(\dot{x}_d - \eta_j(t, x_d))] \in \mathbb{R}^{1 \times (n_j+1)},
\tag{8.7}
$$

$$
\xi_j(t, x_{i,j}) \triangleq [(\xi_j^0(t, x_{i,j}))^T, \; 1]^T \in \mathbb{R}^{(n_j+1) \times 1}.
\tag{8.8}
$$

Note that since $b_j \neq 0$, (8.7) and (8.8) are well defined.

The error dynamics (8.6) contain parametric uncertainty $\theta_j^0 \xi_j^0(t, x_{i,j})$, the lumped uncertainty $\eta_j(t, x_{i,j})$, and the derivative of unknown desired trajectory $\dot{x}_d$. Equation (8.6) shows that by lumping $\dot{x}_d - \eta_j(t, x_d)$ into parametric uncertainties, it is possible to incorporate adaptive ILC with robust ILC to deal with parametric uncertainties $\theta_j \xi_j(t, x_{i,j})$ and $\eta_j(t, x_d) - \eta_j(t, x_{i,j})$ with the help of Assumption 8.1, provided that $e_{i,j}$ is available for each follower. However, $e_{i,j}$ cannot be used to design ILC schemes because only a small portion of followers can access the leader's trajectory. Remark 8.3 discussed the possibility that by using local information, consensus tracking might be achieved when an ILC law is designed appropriately.

8.3 Controller Design for First-Order Multi-agent Systems

This section contains two parts. The first part shows how to design an ILC law to ensure robust consensus tracking for the MAS discussed in Section 8.2.2. The second part discusses how to design an ILC updating law for system (8.1) if Assumption 8.3 is relaxed.

8.3.1 Main Results

In the following convergence analysis, let Δ represent the difference operator over two consecutive iterations. Specifically, $\Delta \omega_i = \omega_i - \omega_{i-1}$, where $\omega_i \in \{E_i, V_i\}$, and E_i, V_i are defined later.

Obviously, the proposed ILC updating law for the jth agent needs to ensure consensus tracking, reject lumped uncertainties, and learn parametric uncertainties. Therefore, it has three components:

$$
u_{i,j} = \underbrace{b_j^{-1} \gamma \epsilon_{i,j}}_{\text{consensus tracking}} + \underbrace{b_j^{-1} \left(\overline{\phi}_j(x_{i,j}) \right)^2 \epsilon_{i,j}}_{\text{robust control}} + \underbrace{\hat{\theta}_{i,j} \xi_j(t, x_{i,j})}_{\text{parameter learning}},
\tag{8.9}
$$

where

$$\overline{\phi}_j(x_{i,j}) = \sup_{|z| \le b_d} \phi_j(z, x_{i,j}), \tag{8.10}$$

γ is a positive constant to be designed, and $\hat{\theta}_{i,j}$ is an estimation of θ_j at the ith iteration. It is updated as follows

$$\hat{\theta}_{i,j}(t) = \hat{\theta}_{i-1,j}(t) + \kappa \epsilon_{i,j} \left(\xi_j(t, x_{i,j}) \right)^T,$$
$$\hat{\theta}_{0,j}(t) = 0, \quad \forall t \in [0, T], \tag{8.11}$$

where $\kappa > 0$ is the parameter learning gain.

The updating laws (8.9) can be rewritten in a stack vector form,

$$\mathbf{u}_i = \underline{B}^{-1} \left(\gamma I + \left(\overline{\Phi}(\mathbf{x}_i) \right)^2 \right) \epsilon_i + \hat{\Theta}_i \xi(t, \mathbf{x}_i), \tag{8.12}$$

where $\underline{B} \triangleq \mathrm{diag}(\underline{b}_1, \dots, \underline{b}_N)$; ϵ_i, $\mathbf{u}_i$ and $\xi(t, \mathbf{x}_i)$ are the column stack vectors of $\epsilon_{i,j}$, $u_{i,j}$ and $\xi_j(t, x_{i,j})$ respectively, and

$$\overline{\Phi}(\mathbf{x}_i) \triangleq \mathrm{diag}(\overline{\phi}_1(x_{i,1}), \overline{\phi}_2(x_{i,2}) \dots, \overline{\phi}_N(x_{i,N})),$$
$$\hat{\Theta}_i \triangleq \mathrm{diag}(\hat{\theta}_{i,1}, \hat{\theta}_{i,2}, \dots, \hat{\theta}_{i,N}).$$

For convenience, we introduce the following notations:

$$\eta_d \triangleq \left[\eta_1(t, x_d), \ \eta_2(t, x_d), \ \cdots, \ \eta_N(t, x_d) \right]^T,$$
$$\eta(t, \mathbf{x}_i) \triangleq \left[\eta_1(t, x_{i,1}), \ \eta_2(t, x_{i,2}), \ \cdots, \ \eta_N(t, x_{i,N}) \right]^T,$$
$$B \triangleq \mathrm{diag}(b_1, b_2 \dots, b_N),$$
$$\Theta \triangleq \mathrm{diag}(\theta_1, \theta_2, \dots, \theta_N),$$
$$\tilde{\Theta}_i \triangleq \Theta - \hat{\Theta}_i.$$

This leads to the following closed loop error dynamics:

$$\dot{\mathbf{e}}_i = \left(\eta_d - \eta(t, \mathbf{x}_i) \right) + B\underline{B}^{-1} \left(\gamma I + \left(\overline{\Phi}(\mathbf{x}_i) \right)^2 \right) \epsilon_i + \tilde{\Theta}_i \xi(t, \mathbf{x}_i). \tag{8.13}$$

The following CEF is introduced:

$$E_i(t) = \underbrace{V_i(\mathbf{e}_i)}_{\text{consensus tracking}} + \underbrace{\frac{1}{2\kappa} \int_0^t \mathrm{Trace} \left(\left(\tilde{\Theta}_i(\tau) \right)^T B \tilde{\Theta}_i(\tau) \right) d\tau}_{\text{parameter learning}}, \tag{8.14}$$

where $V_i(\mathbf{e}_i) \triangleq \frac{1}{2} \mathbf{e}_i^T H \mathbf{e}_i$. The CEF contains the information related to consensus tracking and parameter learning. It is defined for each time instant over $[0, T]$.

The first result is stated in Theorem 8.1.

Theorem 8.1 Assume that Assumptions 8.1–8.3 hold for the multi-agent system (8.1). The closed loop system consisting of (8.1) and the updating laws (8.9)–(8.11), can ensure that the tracking error $e_{i,j}(t)$ converges to zero point-wisely ($j = 1, 2, \ldots, N$) for any $t \in [0, T]$ along the iteration axis, if

$$\gamma \geq \frac{1}{4\underline{\sigma}(H^2)} + \alpha, \tag{8.15}$$

for some positive constant α. Moreover, $u_{i,j} \in \mathcal{L}^2[0, T]$ for any $j = 1, 2, \ldots, N, i \in \mathbb{N}$.

Proof: see Appendix B.4 ∎

Remark 8.7 In the proof of Theorem 8.1, it shows that $E_{i+1}(t) \leq E_i(t) - V_i(e_i(t))$ for each time instant over $[0, T]$. After showing the uniform boundedness of $E_i(t)$ over time, the point-wise convergence of the tracking error can be ensured. The uniform boundedness of $E_i(t)$ for any $i \in \mathbb{N}, t \in [0, T]$ shows that the control input is $\mathcal{L}^2$ bounded uniformly for all $i \in \mathbb{N}$.

Remark 8.8 As pointed out by Xu and Tan (2002a), the parameter updating law (8.11) cannot ensure the uniform boundedness of the $\hat{\Theta}_i, i \in \mathbb{N}$, only the point-wise convergence of the tracking error can be ensured and the input signal is $\mathcal{L}^2$ bounded. When an appropriate projection method is used for the updating law (8.11), it is possible to get stronger results (uniform convergence of the tracking error and uniform boundedness of the input signal). However, the projection method always requires extra information: the upper and lower bounds of unknown time-varying parameters.

Remark 8.9 In the parameter updating rule (8.11), the derivative of the desired trajectory $\dot{x}_d$ is treated as a completely unknown variable, though some of the followers are able to know $\dot{x}_d$. This available information can potentially be used to improve the performance (for example, transient response along iteration domain). Let Ω_0 be the set of agents who can obtain the leader's information, that is, $\Omega_0 \triangleq \{j \in \mathcal{V} | d_j = 1\}$, which leads to another parameterization:

$$\theta_j^a \triangleq \begin{cases} [-b_j^{-1}\theta_j^0, -b_j^{-1}\eta_j(t, x_d), b_j^{-1}], & \text{if } j \in \Omega_0, \\ [-b_j^{-1}\theta_j^0, b_j^{-1}(\dot{x}_d - \eta_j(t, x_d))], & \text{otherwise,} \end{cases}$$

and

$$\xi_j^a(t, x_{i,j}) \triangleq \begin{cases} \left[(\xi_j^0(t, x_{i,j}))^T, 1, \dot{x}_d\right]^T, & \text{if } j \in \Omega_0, \\ [(\xi_{i,j}^0(t, x_{i,j}))^T, 1]^T, & \text{otherwise.} \end{cases}$$

Under this parameterization, the parameter updating law (8.11) is applicable if θ_j and ξ_j are replaced by θ_j^a and ξ_j^a respectively. It is interesting to observe that when more information is available, the number of unknown parameters increases, leading to a more complicated controller.

8.3.2 Extension to Alignment Condition

Although Assumption 8.3 is a necessary condition to ensure perfect tracking performance, it is not easy to achieve *iic* in general. It is widely observed in the industry that many motion systems start from the position in which they stopped in the previous iteration (Xu and Qu, 1998; Xu and Xu, 2004). For example, consider an industry manipulator performing the pick and place task repetitively. The starting position is always the final position in the previous task execution. This motivates the widely used alignment condition in the area of ILC.

Assumption 8.4 For any $i \in \mathbb{N}$ and $j = 1, 2, \ldots, N$, the system (8.1) satisfies $x_{i+1,j}(0) = x_{i,j}(T)$. Moreover, the desired trajectory also satisfies $x_d(0) = x_d(T)$.

With this relaxed assumption, we can obtain the following weaker result.

Corollary 8.1 Assume that Assumptions 8.1, 8.2 and 8.4 hold for the multi-agent system (8.1). The closed loop system consisting of (8.1) and the updating laws (8.9)–(8.11) satisfying (8.15), can ensure that the tracking error $e_{i,j}(t)$ converges in the sense of $\mathcal{L}^2[0, T]$ norm, that is,

$$\lim_{i \to \infty} \int_0^T \left(e_{i,j}(\tau) \right)^2 d\tau = 0. \tag{8.16}$$

Moreover, $u_{i,j} \in \mathcal{L}^2[0, T]$ for any $j = 1, 2, \ldots, N, i \in \mathbb{N}$.

Proof: see Appendix B.5. ∎

Remark 8.10 As Assumption 8.3 is relaxed by Assumption 8.4, Corollary 8.1 obtains a weaker convergence result compared with that in Theorem 8.1, namely, from point-wise to $\mathcal{L}^2$ convergence.

8.4 Extension to High-Order Systems

With some nontrivial modifications of the proposed methods in Section 8.3, the learning controllers can be applied to high-order systems. The results are first derived under *iic*, then extended to the imperfect initial conditions with the inital rectifying action.

Consider the jth agent modeled by the following dynamics:

$$\dot{x}_{i,j_k} = x_{i,j_{k+1}}, \quad k = 1, 2, \ldots, n - 1,$$
$$\dot{x}_{i,j_n} = \theta_j^0 \xi_j^0(t, \mathbf{x}_{i,j}) + \eta_j(t, \mathbf{x}_{i,j}) + b_j(t) u_{i,j}, \tag{8.17}$$

where $\mathbf{x}_{i,j}$ is the column stack vector of $x_{i,j_k}, k = 1, \ldots, n$.

The desired trajectory $\mathbf{x}_d = [x_{d_1}, x_{d_2}, \ldots, x_{d_n}]^T$ is generated by

$$\dot{x}_{d_k} = x_{d_{k+1}}, \quad k = 1, 2, \ldots, n - 1,$$
$$\dot{x}_{d_n} = \dot{x}_d.$$

Define the following two auxiliary variables:

$$s_{i,j} \triangleq \sum_{k=1}^n c_k x_{i,j_k}, \text{ and } s_d \triangleq \sum_{k=1}^n c_k x_{d_k},$$

where $c_n = 1$, and

$$\lambda^{n-1} + c_{n-1} \lambda^{n-2} + \cdots + c_1 = 0 \tag{8.18}$$

is a stable polynomial. For instance, c_i can be chosen the same coefficient as the polynomial $(\lambda + 1)^{n-1}$.

Let the tracking error for the jth agent at the ith iteration be $e_{i,j} = s_d - s_{i,j}$. Since the the polynomial (8.18) is stable, and $\mathbf{x}_{i,j}(0) = \mathbf{x}_d(0)$, hence, if $e_{i,j} = 0$, perfect tracking is achieved, that is, $\mathbf{x}_{i,j}(t) = \mathbf{x}_d(t)$, $t \in [0, T]$. The corresponding extended tracking error is defined as

$$\epsilon_{i,j} \triangleq \sum_{k \in \mathcal{N}_j} a_{j,k}(s_{i,k} - s_{i,j}) + d_j(s_d - s_{i,j}). \tag{8.19}$$

Similarly, the extended tracking error $\epsilon_{i,j}$ in (8.19) can be written in the compact matrix form,

$$\epsilon_i = -L\mathbf{s}_i + D\mathbf{e}_i = H\mathbf{e}_i, \tag{8.20}$$

where ϵ_i, $\mathbf{e}_i$, and $\mathbf{s}_i$ are the column stack vectors of $\epsilon_{i,j}$, $e_{i,j}$, and $s_{i,j}$. Notice that the definitions of $e_{i,j}$ and $s_{i,j}$ are rather different from the ones in Section 8.3.

The error dynamics now become

$$
\begin{aligned}
\dot{e}_{i,j} &= \sum_{k=1}^{n} c_k \dot{x}_{d_k} - \sum_{k=1}^{n} c_k \dot{x}_{i,j_k} \\
&= \sum_{k=1}^{n-1} c_k(x_{d_{k+1}} - x_{i,j_{k+1}}) + \dot{x}_d - \theta_j^0 \xi_j^0(t, \mathbf{x}_{i,j}) - \eta_j(t, \mathbf{x}_{i,j}) - b_j u_{i,j} \\
&= b_j \theta_j \xi_j(t, \mathbf{x}_{i,j}) + \eta_j(t, \mathbf{x}_d) - \eta_j(t, \mathbf{x}_{i,j}) - b_j u_{i,j},
\end{aligned} \tag{8.21}
$$

where

$$\theta_j \triangleq \left[-b_j^{-1}\theta_j^0, \ b_j^{-1}\left(\sum_{k=1}^{n-1} c_k x_{d_{k+1}} + \dot{x}_d - \eta_j(t, \mathbf{x}_d) \right), \ -b_j^{-1} \right],$$

and

$$\xi_j(t, \mathbf{x}_{i,j}) \triangleq \left[\left(\xi_j^0(t, \mathbf{x}_{i,j}) \right)^T, \ 1, \ \sum_{k=1}^{n-1} c_k x_{i,j_{k+1}} \right]^T.$$

The proposed ILC controller and updating rules are

$$u_{i,j} = \underline{b}_j^{-1} \left(\gamma + \left(\overline{\phi}_j(\mathbf{x}_{i,j}) \right)^2 \right) \epsilon_{i,j} + \hat{\theta}_{i,j} \xi_j(t, \mathbf{x}_{i,j}), \tag{8.22}$$

$$\hat{\theta}_{i,j} = \hat{\theta}_{i-1,j} + \kappa \epsilon_{i,j} \left(\xi_j(t, \mathbf{x}_{i,j}) \right)^T, \ \hat{\theta}_{0,j} = 0, \tag{8.23}$$

where $\overline{\phi}_j(\mathbf{x}_{i,j}) = \sup_{|\mathbf{z}| \le b_d} \phi_j(\mathbf{z}, \mathbf{x}_{i,j})$, and $\hat{\theta}_{i,j}$ is the estimate of θ_j at the ith iteration.

To analyze the convergence properties of the proposed algorithm. The following result is required. Denote $\delta \mathbf{x}_{i,j} = \mathbf{x}_d - \mathbf{x}_{i,j}$, and $\mathbf{c}^T = [c_1, \dots, c_n]$. Then we have $e_{i,j} = \mathbf{c}^T \delta \mathbf{x}_{i,j}$.

Lemma 8.2   Under Assumption 8.3, $\mathbf{x}_{i,j}(0) = \mathbf{x}_d(0)$, we have

$$|\delta \mathbf{x}_{i,j}| \le r|e_{i,j}|,$$

where $r = \|e^{Ct}\mathbf{g}\|$,

$$C = \begin{bmatrix} 0 & 1 & 0 & \cdots & \cdots \\ 0 & 0 & 1 & 0 & \cdots \\ \vdots & \vdots & \vdots & \ddots & \vdots \\ 0 & -c_1 & -c_2 & \cdots & -c_{n-1} \end{bmatrix},$$

and

$$\mathbf{g} = \begin{bmatrix} 0 \\ 0 \\ \vdots \\ 1 \end{bmatrix}.$$

Proof: Taking the derivative of $e_{i,j}$ yields $\dot{e}_{i,j} = \mathbf{c}^T \delta\dot{\mathbf{x}}_{i,j}$. Rewrite $\delta\dot{\mathbf{x}}_{i,j}$ in the linear matrix form, one has

$$\delta\dot{\mathbf{x}}_{i,j} = C\delta\mathbf{x}_{i,j} + \mathbf{g}\dot{e}_{i,j}.$$

Noticing (Assumption 8.3) that $\delta\mathbf{x}_{i,j}(0) = 0$, the above differential equation renders the following solution:

$$\delta\mathbf{x}_{i,j} = e^{Ct}\delta\mathbf{x}_{i,j}(0) + \int_0^t e^{C(t-\tau)}\mathbf{g}\dot{e}_{i,j}(\tau)\,d\tau$$

$$= \int_0^t e^{C(t-\tau)}\mathbf{g}\dot{e}_{i,j}(\tau)\,d\tau.$$

Taking the norm operation of the both sides yields,

$$|\delta\mathbf{x}_{i,j}| \leq \left(\max_{t\in[0,T]} |e^{Ct}\mathbf{g}|\right)\left|\int_0^t \dot{e}_{i,j}(\tau)\,d\tau\right| = r|e_{i,j}|.$$

∎

Remark 8.11 The characteristic polynomial of C is $\lambda(\lambda^{n-1} + c_{n-1}\lambda^{n-2} + \ldots + c_1) = 0$. Notice that the polynomial in (8.18) is stable, therefore, zero is one of the eigenvalues of C, and all the other eigenvalues are stable. Hence, $\|e^{Ct}\|$ is finite for all $t \in [0, \infty)$.

Define

$$\eta(t, \mathbf{x}_d) \triangleq \left[\eta_1(t, \mathbf{x}_d), \eta_2(t, \mathbf{x}_d), \ldots, \eta_N(t, \mathbf{x}_d)\right]^T,$$

$$\eta(t, \mathbf{x}_i) \triangleq \left[\eta_1(t, \mathbf{x}_{i,1}), \eta_2(t, \mathbf{x}_{i,2}), \ldots, \eta_N(t, \mathbf{x}_{i,N})\right]^T.$$

With the help of Lemma 8.2, the Equation (B.10) in Appendix B.4 for the first-order case now becomes

$$|\epsilon_i^T(\eta(t, \mathbf{x}_d) - \eta(t, \mathbf{x}_i))| \leq \epsilon_i^T\left(\overline{\Phi}(\mathbf{x}_i)\right)^2 \epsilon_i + \frac{1}{4\underline{\sigma}(H)^2}r^2\epsilon_i^T\epsilon_i. \tag{8.24}$$

Theorem 8.2 Assume that Assumptions 8.1–8.3 hold for the high-order multi-agent system (8.17). The closed loop system consisting of (8.17) and the updating rules

(8.22)–(8.23), can ensure that the tracking error $e_{i,j}(t)$ approaches to zero point-wisely ($j = 1, 2, \ldots, N$) for any $t \in [0, T]$ along iteration axis, if

$$\gamma \geq \frac{r^2}{4\underline{\sigma}(H^2)} + \alpha,$$

for some positive constant α. Moreover, $u_{i,j} \in \mathcal{L}^2[0, T]$ for any $j = 1, 2, \ldots, N, i \in \mathbb{N}$.

Sketch of proof: The error dynamics (8.21) and controllers (8.22), (8.23) are very similar to the ones in the first-order case, except that the parameterizations become more complex and the error term definitions are different. Consider the CEF below:

$$E_i(t) = V_i(\mathbf{e}_i) + \frac{1}{2\kappa} \int_0^t \text{Trace}\left(\tilde{\Theta}_i^T B \tilde{\Theta}_i\right) \, d\tau,$$

together with (8.24), and follow the similar proof procedures in the previous section, we conclude Theorem 8.2. ∎

Theorem 8.2 depends on the *iic*, which makes the results sensitive to the initial conditions. Assume that the initial state of each agent cannot be manipulated and is reset to a fixed state at every iteration, but not the same as the desired state $\mathbf{x}_d(0)$. By extending the concept of the initial rectifying action (Sun and Wang, 2002), it is possible to make the results more robust to the CEF-based ILC.

Assumption 8.5 The initial state of the jth follower is reset to a fixed initial state at every iteration, that is, $\mathbf{x}_{i,j}(0) = \mathbf{x}_j(0) \neq \mathbf{x}_d(0)$ for all $i \in \mathbb{N}$.

Define matrices $A, \overline{A} \in \mathbb{R}^{n \times n}$ below:

$$A = \begin{bmatrix} 0 & 1 & 0 & \cdots & 0 \\ 0 & 0 & 1 & \cdots & 0 \\ \vdots & \vdots & \vdots & \ddots & \vdots \\ 0 & \cdots & \cdots & \cdots & 0 \end{bmatrix},$$

and $\overline{A} = A - \mathbf{g}K$ is a stable matrix, where $K \in \mathbb{R}^{1 \times n}$ and $\mathbf{g}$ is defined in Lemma 8.2. K can be interpreted as the state feedback gain. The structure of A is decided by the structure of system model (8.17). Let the extended tracking error be

$$\hat{e}_{i,j} \triangleq \sum_{k \in \mathcal{N}_j} a_{j,k}(s_{i,k} - s_{i,j}) + d_j(s_d - s_{i,j}) - \sum_{k \in \mathcal{N}_j} a_{j,k} \mathbf{c}^T e^{\overline{A}t}(\mathbf{x}_k(0) - \mathbf{x}_j(0))$$

$$- d_j \mathbf{c}^T e^{\overline{A}t}(\mathbf{x}_d(0) - \mathbf{x}_j(0)). \tag{8.25}$$

It is worth noting that the extended tracking error in (8.25) is a distributed measurement. Hence, it can be used in the distributed controller design.

To design and analyze the distributed version of initial rectifying action in the CEF framework, define the following auxiliary variables:

$$\hat{\mathbf{x}}_{d,j} \triangleq \mathbf{x}_d - e^{\overline{A}t}(\mathbf{x}_d(0) - \mathbf{x}_j(0)),$$

$$\hat{s}_{d,j} \triangleq \mathbf{c}^T \hat{\mathbf{x}}_{d,j} = s_d - \mathbf{c}^T e^{\overline{A}t}(\mathbf{x}_d(0) - \mathbf{x}_j(0)),$$

$$\hat{e}_{i,j} \triangleq \hat{s}_{d,j} - s_{i,j}.$$

The dynamics of $\hat{e}_{i,j}$ can be obtained as

$$
\begin{aligned}
\dot{\hat{e}}_{i,j} &= \dot{\hat{s}}_{d,j} - \dot{s}_{i,j} \\
&= \dot{\hat{s}}_{d,j} - \sum_{k=1}^{n-1} c_k x_{i,j_{k+1}} - \theta_j^0 \xi_j^0(t, \mathbf{x}_{i,j}) - \eta_j(t, \mathbf{x}_{i,j}) - b_j u_{i,j} \\
&= b_j \theta_j \xi_j(t, \mathbf{x}_{i,j}) + \eta_j(t, \hat{\mathbf{x}}_{d,j}) - \eta_j(t, \mathbf{x}_{i,j}) - b_j u_{i,j},
\end{aligned}
\tag{8.26}
$$

where the new parameterizations are

$$
\theta_j \triangleq \left[-b_j^{-1} \theta_j^0, \; b_j^{-1} \left(\dot{\hat{s}}_{d,j} - \eta_j(t, \hat{\mathbf{x}}_{d,j}) \right), \; -b_j^{-1} \right],
$$

and

$$
\xi_j(t, \mathbf{x}_{i,j}) \triangleq \left[\left(\xi_j^0(t, \mathbf{x}_{i,j}) \right)^T, \; 1, \; \sum_{k=1}^{n-1} c_k x_{i,j_{k+1}} \right]^T.
$$

The dynamics in (8.26) are identical to (8.21) except that the error definition and parameterizations are different. Analogous to the controller design in the *iic* case, the proposed ILC controller and updating rules are

$$
u_{i,j} = \underline{b}_j^{-1} \left(\gamma + \left(\overline{\phi}_j(\mathbf{x}_{i,j}) \right)^2 \right) \hat{e}_{i,j} + \hat{\theta}_{i,j} \xi_j(t, \mathbf{x}_{i,j}),
\tag{8.27}
$$

$$
\hat{\theta}_{i,j} = \hat{\theta}_{i-1,j} + \kappa \hat{e}_{i,j} \left(\xi_j(t, \mathbf{x}_{i,j}) \right)^T, \; \hat{\theta}_{0,j} = 0,
\tag{8.28}
$$

where $\overline{\phi}_j(\mathbf{x}_{i,j}) = \sup_{|\mathbf{z}| \leq \hat{b}_d} \phi_j(\mathbf{z}, \mathbf{x}_{i,j})$, $\hat{b}_d = b_d + |\mathbf{x}_d(0) - \mathbf{x}_j(0)|$, and $\hat{\theta}_{i,j}$ is the estimate of θ_j at the ith iteration.

To analyze the convergence properties, $\hat{e}_{i,j}$ should be expressed in terms of $\hat{e}_{i,j}$. With the help of auxiliary variables, we have

$$
\begin{aligned}
\hat{e}_{i,j} &= \sum_{k \in \mathcal{N}_j} a_{j,k} (\hat{s}_{d,j} - s_{i,j} - (\hat{s}_{d,k} - s_{i,k}) - \hat{s}_{d,j} + \hat{s}_{d,k}) - \sum_{k \in \mathcal{N}_j} a_{j,k} \mathbf{c}^T e^{\overline{A}t} (\mathbf{x}_k(0) - \mathbf{x}_j(0)) \\
&\quad + d_j(s_d - s_{i,j}) - d_j \mathbf{c}^T e^{\overline{A}t} (\mathbf{x}_d(0) - \mathbf{x}_j(0)) \\
&= \sum_{k \in \mathcal{N}_j} a_{j,k} (\hat{e}_{i,j} - \hat{e}_{i,k}) + d_j \hat{e}_{i,j}.
\end{aligned}
$$

Rewriting the above equation in compact matrix form yields

$$
\hat{e}_i = H \hat{e}_i,
\tag{8.29}
$$

where $\hat{e}_i$ and $\hat{e}_i$ are the column stack vectors of $\hat{e}_{i,j}$ and $\hat{e}_{i,j}$, respectively.

Based on the definition of $\hat{\mathbf{x}}_{d,j}$, it is guaranteed that $\hat{\mathbf{x}}_{d,j}(0) = \mathbf{x}_j(0)$. Define $\delta \hat{\mathbf{x}}_{i,j} = \hat{\mathbf{x}}_{d,j} - \mathbf{x}_{i,j}$, following the idea in Lemma 8.2, we have $|\delta \hat{\mathbf{x}}_{i,j}| \leq r |\hat{e}_{i,j}|$.

With the above developments and analogy to Theorem 8.2, we have the following results.

Theorem 8.3   Assume that Assumptions 8.1, 8.2, and 8.5 hold for the high-order multi-agent system (8.17). The closed loop system consisting of (8.17) and the updating rules (8.27)–(8.28), can ensure that the tracking error $\hat{e}_{i,j}(t)$ approaches to zero point-wisely ($j = 1, 2, \ldots, N$) for any $t \in [0, T]$ along iteration axis, if

$$\gamma \geq \frac{r^2}{4\underline{\sigma}(H^2)} + \alpha,$$

for some positive constant α. Therefore,

$$\lim_{i \to \infty} \mathbf{x}_{i,j}(t) = \mathbf{x}_d(t) - e^{\overline{A}t}(\mathbf{x}_d(0) - \mathbf{x}_j(0)).$$

Moreover, $u_{i,j} \in \mathcal{L}^2[0, T]$ for any $j = 1, 2, \ldots, N, i \in \mathbb{N}$.

The convergence results can be proved analogously to Theorems 8.1 and 8.2 by studying the following CEF:

$$E_i(t) = V_i(\hat{\mathbf{e}}_i) + \frac{1}{2\kappa} \int_0^t \text{Trace}\left(\tilde{\Theta}_i^T B \tilde{\Theta}_i\right) d\tau.$$

As discussed in Remark 8.6, perfect tracking can never be achieved when *iic* is violated. From Theorem 8.3, it can be seen that the final trajectory of each individual agent can be adjusted by the matrix K in $\overline{A}$.

8.5 Illustrative Example

To illustrate the applications of the developed algorithms, consider a group of four agents. The communication graph among followers and leader is depicted in Figure 8.1. Vertex 0 represents the virtual leader, and the dashed lines stand for the communication links between leader and followers, that is, only agents 1 and 4 can access the state information of the leader. The solid lines denote the communication links between followers, and the communication graph among followers is connected. Then the Laplacian is

$$L = \begin{bmatrix} 2 & -1 & 0 & -1 \\ -1 & 3 & -1 & -1 \\ 0 & -1 & 1 & 0 \\ -1 & -1 & 0 & 2 \end{bmatrix},$$

and $D = \text{diag}(1, 0, 0, 1)$. Hence, the smallest singular value of H is $\underline{\sigma}(H) = 0.3249$.

In the simulation study, we first demonstrate the results for first-order systems under *iic* and alignment conditions. Next, we illustrate the results for high-order systems.

Figure 8.1 Communication among agents in the network.

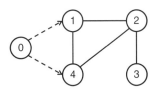

8.5.1 First-Order Agents

Let the four followers be modeled by the following dynamics:

$$\dot{x}_{i,1} = \sin(t)x_{i,1}^2 + (1 + 0.1\sin^2(t))u_{i,1} + \eta_1(t, x_{i,1}),$$
$$\dot{x}_{i,2} = \cos(t)x_{i,2}^3 + (1 + 0.2\sin^2(t))u_{i,2} + \eta_2(t, x_{i,2}),$$
$$\dot{x}_{i,3} = e^{-t}x_{i,3}^2 + (1 + 0.3\sin^2(t))u_{i,3} + \eta_3(t, x_{i,3}),$$
$$\dot{x}_{i,4} = -tx_{i,4}^3 + (1 + 0.4\sin^2(t))u_{i,4} + \eta_4(t, x_{i,4}),$$

where the unknown disturbance $\eta_j(t, x_{i,j}) = x_{i,j}^2 \sin(j \cdot t)$ for $j = 1, 2, 3, 4$. Notice that the disturbance $\eta_j(t, x_{i,j})$ satisfies the Assumption 8.1, that is,

$$|\eta_j(t, x_d) - \eta_j(t, x_{i,j})| \le \phi_j(x_d, x_{i,j})|x_d - x_{i,j}|,$$

where $\phi_j(x_d, x_{i,j}) = |x_d| + |x_{i,j}|$.

The agent dynamics have the same form as in (8.1). So the learning rules (8.9) and (8.11) can be applied directly.

Identical Initialization Condition

Choose the desired trajectory $x_d = \sin^3(t)$, $t \in [0, 5]$, and let all the agents satisfy the *iic*, that is, $x_{i,j}(0) = x_d(0)$. Set $\gamma = 7.4$, $\kappa = 5$, and $\bar{\phi}_j(x_{i,j}) = 1 + |x_{i,j}|$.

Figure 8.2 shows the trajectory profiles at the 1st and 50th iterations. At the 1st iteration, the followers' trajectories have some deviations from the leader's. Due to the current error feedback in the controller, the transient response is much better than the

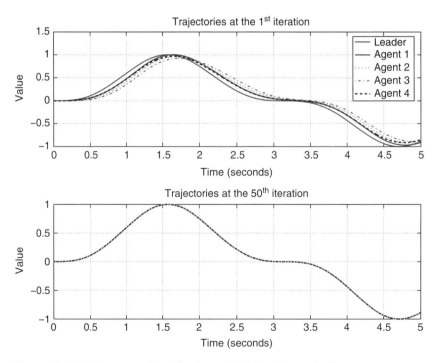

Figure 8.2 The trajectory profiles at the 1st and 50th iterations under *iic*.

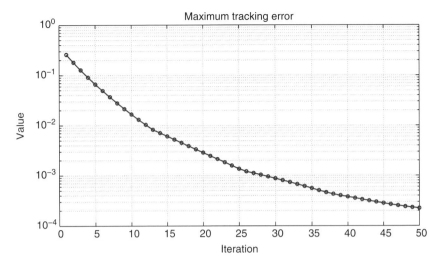

Figure 8.3 Maximum tracking error vs. iteration number under *iic*.

typical contraction-mapping-based controllers. The followers' trajectories almost overlap with the leader's at the 50th iteration. Figure 8.3 shows the maximum tracking error versus iteration number. The maximum tracking error has been reduced to 0.0086% of the one at the 1st iteration, where the maximum tracking error at the *i*th iteration is defined as $\max\limits_{j=1,2,3,4} \max\limits_{t\in[0,T]} |x_d - x_{i,j}|$.

Alignment Condition
Choose the desired trajectory $x_d = \sin^3(t)$, $t \in [0, \pi]$. x_d is a closed orbit since $x_d(0) = x_d(\pi)$. The initial condition for the follower agents are $x_{1,1}(0) = 0.2$, $x_{1,2}(0) = 0.4$, $x_{1,3}(0) = 0.6$, $x_{1,4}(0) = -0.4$. Set $\gamma = 7.4$, $\kappa = 5$, and $\overline{\phi}_j(x_{i,j}) = 1 + |x_{i,j}|$.
Figure 8.4 shows the trajectory profiles at the 1st and 50th iterations. At the 1st iteration, the followers' trajectories have large deviations from the leader's, especially at the starting time. Due to the alignment condition, the controllers (8.9) and (8.11) can still work. The trajectories almost overlap with the leader's at the 50th iteration. Figure 8.5 shows the maximum tracking error versus iteration number. The maximum tracking error has been reduced to 1.52% of the one at the 1st iteration.

8.5.2 High-Order Agents

Consider the agent dynamics as follows:

$$\begin{bmatrix} \dot{x}_{i,j_1} \\ \dot{x}_{i,j_2} \end{bmatrix} = \begin{bmatrix} 0 & 1 \\ 0 & 0 \end{bmatrix} \begin{bmatrix} x_{i,j_1} \\ x_{i,j_2} \end{bmatrix} + \begin{bmatrix} 0 \\ \frac{1+0.1*j\sin^2 t}{m_j l_j^2 + I_j} \end{bmatrix} \begin{bmatrix} 0 \\ u_{i,j} - gl_j\cos(x_{i,j_1}) \end{bmatrix}$$
$$+ \begin{bmatrix} 0 \\ \eta_j(t, \mathbf{x}_{i,j}) \end{bmatrix}, \tag{8.30}$$

where x_{i,j_1} is the position, x_{i,j_2} is the velocity, m_j is the mass, l_j is the length of the rigid-body, I_j is the moment of inertia, $u_{i,j}$ is the control input, and $\eta_j(t, \mathbf{x}_{i,j}) = x_{i,j_1}^2 \sin(j \cdot t)$ is the disturbance. The plant parameters are assumed to be unknown. In the simulation example, they are specified in Table 8.1.

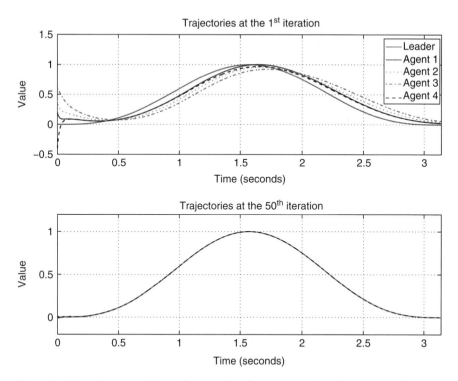

Figure 8.4 The trajectory profiles at the 1st and 50th iterations under alignment condition.

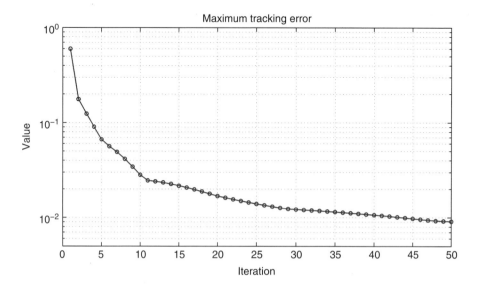

Figure 8.5 Maximum tracking error vs. iteration number under alignment condition.

Table 8.1 Agent Parameters.

Agent	m (kg)	l (m)	I (kg $\cdot$ m^2)	b
1	1.5	0.8	0.48	$0.60(1 + .1\sin^2 t)$
2	2	0.9	0.81	$0.38(1 + .2\sin^2 t)$
3	1.8	1	0.9	$0.37(1 + .3\sin^2 t)$
4	1.7	0.7	0.42	$0.62(1 + .4\sin^2 t)$

The agent dynamics are slightly different from the system in (8.17); with some mod-ifications of the parameterization, controllers (8.22) and (8.23) are still applicable. In this part of simulation study, *iic* is assumed. The leader's trajectory is chosen as $x_{d_1}(t) = \sin^3 t$ for $t \in [0, 5]$, and the controller parameters are chosen as $c = 0.5$, $\gamma = 10$, $\kappa = 15$, and $\overline{\phi}_j(\mathbf{x}_{i,j}) = 1 + |x_{i,j_1}|$.

Let $\mathbf{c} = [c, 1]^T$, and $c = 0.5$. The auxiliary variable $s_{i,j} = cx_{i,j_1} + x_{i,j_2}$, and $s_d = cx_{d_1} + x_{d_2}$. Then, the error dynamics are

$$\dot{e}_{i,j} = cx_{d_2} - cx_{i,j_1} + \dot{x}_d - \eta_j(t, \mathbf{x}_{i,j}) - b_j(u_{i,j} - gl_j\cos(x_{i,j_1}))$$
$$= b_j\theta_j\xi_{i,j} + \eta_j(t, \mathbf{x}_d) - \eta_j(t, \mathbf{x}_{i,j}) - b_j u_{i,j},$$

where

$$\theta_j \triangleq [gl_j, \ b_j^{-1}(cx_{d_2} + \dot{x}_d - \eta_j(t, \mathbf{x}_d)), \ -b_j^{-1}],$$

and

$$\xi_{i,j} \triangleq [\cos(x_{i,j_1}), \ 1, \ cx_{i,j_2}]^T.$$

From Lemma 8.2, it can be shown that $r = \max\limits_{t \in [0,5]} |e^{Ct}\mathbf{g}| < 2$, where

$$C = \begin{bmatrix} 0 & 1 \\ 0 & -.5 \end{bmatrix} \text{ and } \mathbf{g} = \begin{bmatrix} 0 \\ 1 \end{bmatrix}.$$

In the simulation study, the following control laws are applied:

$$u_{i,j} = \underline{b}_j^{-1}\left(\gamma + \left(\overline{\phi}_j(\mathbf{x}_{i,j})\right)^2\right)\epsilon_{i,j} + \hat{\theta}_{i,j}\xi_j(t, \mathbf{x}_{i,j}),$$
$$\hat{\theta}_{i,j} = \hat{\theta}_{i-1,j} + \kappa\epsilon_{i,j}\left(\xi_j(t, \mathbf{x}_{i,j})\right)^T, \hat{\theta}_{0,j}(t) = 0,$$

where $\gamma = 10$, $\kappa = 15$, and $\overline{\phi}_j(\mathbf{x}_{i,j}) = 1 + |x_{i,j_1}|$. $5 \leq \hat{\theta}_{i,j_1} \leq 10$, $-20 \leq \hat{\theta}_{i,j_1} \leq 20$, and $-4 \leq \hat{\theta}_{i,j_3} \leq -1$.

Under *iic*, Figures 8.6 and 8.7 describe the position and velocity trajectories of all agents at the 1st and 50th iterations. At the 1st iteration, neither positions nor velocities of the followers match those of the leader. At the 50th iteration the followers' trajectories overlap with the leader's.

Define the maximum position error at the *i*th iteration as $\max\limits_{j=1,2,\dots,4} \|x_{d_1} - x_{i,j_1}\|$, and the maximum velocity error is defined analogously with the position error. The max-imum position and velocity tracking error profiles are shown in Figure 8.8. Based on the simulation results, the maximum position tracking error at the 50th iteration has

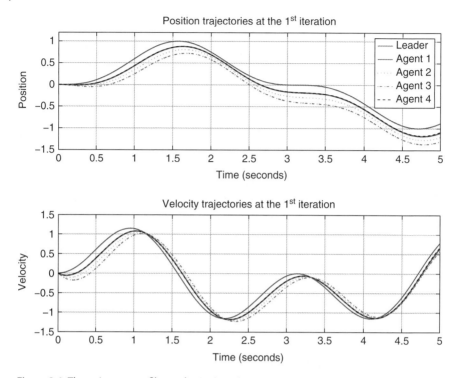

Figure 8.6 The trajectory profiles at the 1st iteration.

been reduced to 0.05% of the one at the 1st iteration, meanwhile, the corresponding maximum velocity tracking error has been reduce to 1.93%.

To demonstrate the effectiveness of the initial rectifying action, let Assumption 8.5 hold, and the initial states of agents are chosen as $x_1(0) = [0.4, -0.5]^T$, $x_2(0) = [0.5, -0.8]^T$, $x_3(0) = [0.8, 0.5]^T$, and $x_4(0) = [0.6, -0.4]^T$. Obviously the initial states are not at the desired states. Matrix K is chosen such that

$$\overline{A} = \begin{bmatrix} 0 & 1 \\ -25 & -10 \end{bmatrix},$$

which has two eigenvalues located at -5 in the complex plane. By using the controllers (8.27) and (8.28), Figures 8.9 and 8.10 depict the trajectory profiles at the 1st and 20th iterations respectively. It can be seen from Figure 8.10 that the learning controllers with initial rectifying action demonstrate satisfactory performance even under imperfect initial conditions.

8.6 Conclusion

Adaptive ILC algorithms are developed for a synchronization problem, which is formulated for a group of heterogeneous agents. Their dynamics are of a general nonlinear form that does not satisfy the global Lipschitz condition. Although this group of agents

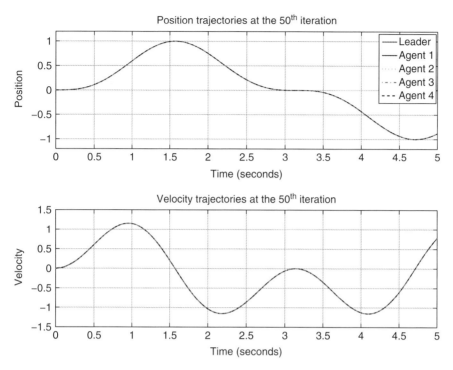

Figure 8.7 The trajectory profiles at the 50th iteration.

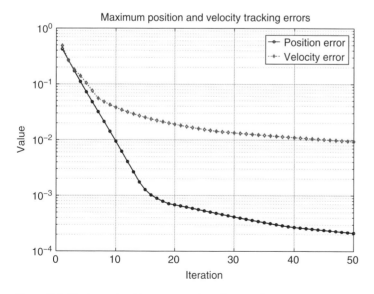

Figure 8.8 Maximum tracking errors vs. iteration number.

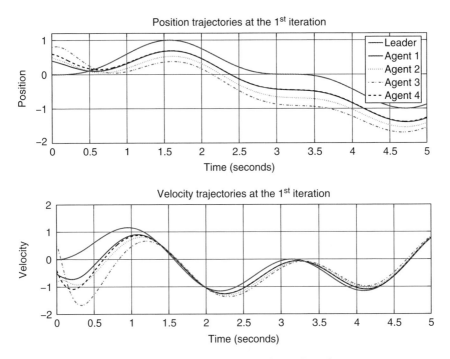

Figure 8.9 The trajectory profiles at the 1st iteration with initial rectifying action.

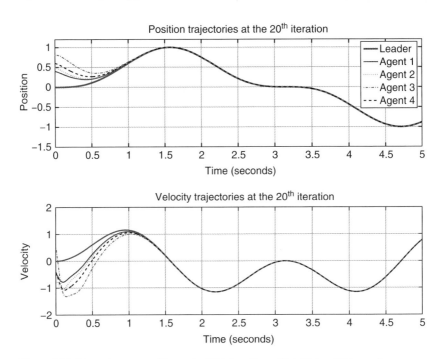

Figure 8.10 The trajectory profiles at the 20th iteration with initial rectifying action.

are connected through communications, the desired trajectory is only known to a few of agents in the system. By incorporating the desired trajectories as a part of the parametric uncertainty, the proposed ILC algorithm combines the parameter learning with robust control to handle both parametric uncertainties and lumped uncertainties. Meanwhile, the composite energy function plays an important role to ensure that the proposed ILC algorithm can synchronize all agents' trajectories to the desired one. Extensive synchronization examples verify the correctness of the developed methods.

9

Distributed Adaptive Iterative Learning Control for Nonlinear Multi-agent Systems with State Constraints

9.1 Introduction

As discussed in the previous chapters, there are many works investigating consensus tracking problems of MAS under the framework of ILC (Yang *et al.*, 2014, 2015; Yang and Xu, 2016; Meng *et al.*, 2012, 2013a,b, 2015a; Meng and Moore, 2016). However, in many practical problems, consensus tracking with constraints needs to be considered. Convex input constraints were discussed in Qiu *et al.* (2016) and Johansson *et al.* (2008) due to practical limitations on driving forces for agents, where the input constraints were handled by using optimization techniques. In contrast to the above two works, this chapter will investigate the consensus problem of a class of nonlinear MAS with state or output constraints under the framework of adaptive ILC.

When considering MAS in the real world, we find that almost all real systems are subject to constraints in one way or another. Such constraints may arise due to various kinds of practical limitations or due to requirements of safe operation. It is well known that the actuators of real systems are usually saturated because there is a limit to the driving force, like voltages or currents, which cannot be arbitrarily large. In many system operations, the states or outputs are also bounded within certain ranges. For example, the motorcade, consisting of several vehicles, is a typical MAS. Clearly, no matter how a vehicle moves, it should stay on the road for safety. In other words, the position of the vehicle is limited to be within the range of the road. Meanwhile, the speed of any vehicle is also limited due to safe driving requirements, although the speed limit may vary in different cases. For such a case, the state or the output is bounded within certain ranges. Similar examples include formation control performance of robots in a given site, a group dance performance on stage, coordination search by UAVs in a designated area, and so on. Another example comes from communication limitations due to physical transmission bandwidth. In this case, in order to guarantee good communication results, requirements are commonly imposed on the data packets, which can be regarded as a kind of constraint on the output of each agent. Moreover, there might also exist measurement limitations due to the usage of simple and cheap devices. When the output exceeds the measurement range, one would not get the real output. This requires that the systems should satisfy certain constraints in turn to get good observations.

Based on the above observations, it is of great value to consider the learning control of MAS with state constraints. To be specific, how to design a coordinative learning scheme to ensure state boundedness while achieving the desired consensus or formation

Iterative Learning Control for Multi-agent Systems Coordination, First Edition.
Shiping Yang, Jian-Xin Xu, Xuefang Li, and Dong Shen.
© 2017 John Wiley & Sons Singapore Pte. Ltd. Published 2017 by John Wiley & Sons Singapore Pte. Ltd.

objectives, is of interest. In other control topics, such as adaptive control, the barrier Lyapunov function (BLF) has been proposed to address such problems (Liu and Tong, 2016; Ren *et al.*, 2010; Li *et al.*, 2014). In the field of ILC, Xu *et al* also presented some pioneering results on systems with state or output constraints (Xu and Jin, 2013; Jin and Xu, 2013). However, when a MAS is considered, where only local information can be obtained, the approach to dealing with the state or output constraints in the system is still an open topic.

This chapter considers distributed adaptive ILC design and analysis for nonlinear MAS with state constraints, where the states are required to be bounded in a predefined zone. For the first-order model of an agent, five control schemes are proposed in turn to address the consensus problem comprehensively from both theoretical and practical viewpoints. The first scheme is the original adaptive algorithm with conventional parameter learning processes, where a sign function is introduced to compensate for the unknown uncertainties. To ensure the prior boundedness of learning parameters, the second control scheme replaces the conventional learning processes with projection-based ones. On the other hand, the sign function may make the control signal discontinuous as it switches between positive and negative. Thus, in the third control scheme, the sign function is replaced by a hyperbolic tangent function in order to generate a smooth control signal. To ensure zero-error consensus, a fast decreasing parameter sequence is also introduced into this scheme. However, from the practical application viewpoint, this is not quite realistic, thus in the fourth control scheme, this decreasing parameter is replaced by a fixed one, which can only ensure the boundedness consensus. To guarantee the boundedness of learning parameters, we also adopt the forgetting factor mechanism in the fourth scheme, which may result in additional consensus error. Therefore, a simple dead-zone-like scheme (the fifth scheme) is also given, meaning that the parameter learning process would stop so long as the consensus error converges into a prior given bound. The convergence of each control scheme is analyzed by introducing a general γ-type barrier Lyapunov function.

The rest of this chapter is arranged as follows. Section 9.2 formulates the problem and introduces a γ-type barrier Lyapunov function. Controller design and convergence analyses are presented in Section 9.3. Section 9.4 provides the illustrative simulations for all the proposed controllers. Lastly, Section 9.5 draws conclusions.

9.2 Problem Formulation

Consider a multi-agent system formulated by N ($N > 2$) agents, one of which is described by the following first-order SISO (single input, single output) nonlinear system:

$$\dot{x}_{i,j} = \theta(t)^T \xi_j(x_{i,j}, t) + b_j(t) u_{i,j}, \tag{9.1}$$

where i is the iteration number and j is the agent number. $\theta^T \xi_j(x_{i,j}, t)$ is parametric uncertainty, where $\theta(t)$ is unknown time-varying parameter while $\xi_j(x_{i,j}, t)$ is a known time-varying function. $b_j(t) \triangleq b_j(x_{i,j}, t)$ is unknown time-varying control gain. In the following, denote $\xi_{i,j} \triangleq \xi(x_{i,j}, t)$ and $b_j \triangleq b_j(x_{i,j}, t)$ where no confusion arises.

Let the desired trajectory (virtual leader) be x_r, which satisfies

$$\dot{x}_r = \theta^T \xi(x_r, t) + b_r u_r \tag{9.2}$$

and similarly, $\xi_r \triangleq \xi(x_r, t)$, $b_r \triangleq b_r(x_r, t)$.

The following assumptions are required for (9.1) and (9.2).

Assumption 9.1 Assume that the control gain b_j does not change its sign. Meanwhile, it has lower and upper bound. Without loss of any generality, we assume that $0 < b_{\min} \leq b_j \leq b_{\max}$, where $b_{\min}$ is assumed to be known.

Assumption 9.2 Each agent satisfies the alignment condition, that is, $x_{i,j}(0) = x_{i-1,j}(T)$. In addition, the desired trajectory is spatially closed, that is, $x_r(0) = x_r(T)$.

Denote the tracking error of the jth agent to the desired trajectory as $e_{i,j} \triangleq x_{i,j} - x_r$. Note that not all agents can get information of the desired trajectory. Thus, the tracking error $e_{i,j}$ is only available to a subset of agents within the neighborhood of the virtual leader. Meanwhile, all agents can access the information of their neighbors within a specified distance. Therefore, for the jth agent, denote its extended observation error as follows:

$$z_{i,j} = \varepsilon_j(x_{i,j} - x_r) + \sum_{l \in \mathcal{N}_j}(x_{i,j} - x_{i,l}), \tag{9.3}$$

where ε_j denotes the access of the jth agent to the desired trajectory, that is, $\varepsilon_j = 1$ if agent j has direct access to the full information of the desired trajectory, otherwise $\varepsilon_j = 0$. Note that the above definition could also be formulated as

$$z_{i,j} = \varepsilon_j(x_{i,j} - x_r) + \sum_{l=1}^{N} a_{jl}(x_{i,j} - x_{i,l}). \tag{9.4}$$

The control objective is to design a set of distributed controllers such that all the agents can perfectly track the desired trajectory in the presence of parametric uncertainties, that is,

$$\lim_{i \to \infty} e_{i,j} = 0, \quad \forall j = 1, \cdots, N, \tag{9.5}$$

and ensure the prior given boundedness of the state of all agents for all iterations.

To obtain a compact form of the MAS, denote $\bar{e}_i, \bar{x}_i$ and $\bar{z}_i$ as the stack of tracking error and extended observation error for all agents, that is,

$$\bar{e}_i = [e_{i,1}, \cdots, e_{i,N}]^T,$$
$$\bar{x}_i = [x_{i,1}, \cdots, x_{i,N}]^T,$$
$$\bar{z}_i = [z_{i,1}, \cdots, z_{i,N}]^T.$$

Noting the fact that $L\mathbf{1} = 0$ and using the definition of $e_{i,j}$, the relationship between the extended observation error and the tracking error is given as follows:

$$\begin{aligned}
\bar{z}_i &= L\bar{x}_i + D\bar{e}_i \\
&= L(\bar{x}_i - \mathbf{1}x_r) + D\bar{e}_i \\
&= (L + D)\bar{e}_i, \tag{9.6}
\end{aligned}$$

where $D = \text{diag}\{\varepsilon_1, \cdots, \varepsilon_N\}$ and $\mathbf{1} = [1, \cdots, 1]^T \in \mathbb{R}^N$. Let $H = L + D$.

The following assumption is made on the connection of agents.

Assumption 9.3 The undirected graph $\mathcal{G}$ is connected.

Remark 9.1 Assumption 9.3 assumes that the virtual leader actually is reachable for each agent no matter whether the virtual leader lies in its neighborhood or not. Here by reachable we mean there is a path from the virtual leader to the agent possibly passing through several other agents. This assumption is a necessary requirement for the leader-follower consensus tracking problem. If there is an isolated agent, it is not possible to ensure the agent tracks the leader's trajectory since no information of the virtual leader could be obtained by the agent.

Then one has the positive property of H from the following lemma.

Lemma 9.1 (Hong *et al.*, 2006) If the undirected graph $\mathcal{G}$ is connected, and M is any nonnegative diagonal matrix with at least one of the diagonal entries being positive, then $L + M$ is symmetric positive definite, where L is the Laplacian matrix of $\mathcal{G}$.

Under 9.3, it is seen that $L + D$ is positive from Lemma 9.1. Let us denote its minimum and maximum singular values as $\sigma_{\min}(L + D)$ and $\sigma_{\max}(L + D)$.

To ensure output boundedness, we need a general class of barrier Lyapunov function (BLF) satisfying the following definition.

Definition 9.1 We call a BLF $V(t) = V(\gamma^2(t), k_b)$ "γ-type BLF" if all the following conditions hold

- $V \to \infty$ if and only if $\gamma^2 \to k_b^2$, where k_b is a certain fixed parameter in V, provided $\gamma^2(0) < k_b^2$.
- $V \to \infty$ if and only if $\frac{\partial V}{\partial \gamma^2} \to \infty$.
- If $\gamma^2 < k_b^2$, then $\frac{\partial V}{\partial \gamma^2} \geq C$, where $C > 0$ is a constant.
- $\lim_{k_b \to \infty} V(\gamma^2(t), k_b) = \frac{1}{2}\gamma^2(t)$.

Remark 9.2 The first term of the definition is to ensure the boundedness of γ^2 as long as the BLF is finite, so it is fundamental. The second term is given to show the boundedness of the BLF by making use of $\frac{\partial V}{\partial \gamma^2}$ in the controller design. The third term provides flexibility of the BLF as can be seen in the latter proof of our main results. From the last term, the defined γ-type BLF can be regarded as a special case of the conventional quadratic Lyapunov function, in the sense that they are mathematically equivalent when $k_b \to \infty$.

Remark 9.3 Two typical examples of the so-called γ-type BLF are given as follows. The first one is of log-type,

$$V(t) = \frac{k_b^2}{2} \log\left(\frac{k_b^2}{k_b^2 - \gamma^2(t)}\right), \tag{9.7}$$

and the other one is of tan-type,

$$V(t) = \frac{k_b^2}{\pi} \tan\left(\frac{\pi \gamma^2(t)}{2k_b^2}\right).$$ (9.8)

By direct calculation, one can find that all the terms of a γ-type BLF are satisfied.

In the following, to simplify the notation, the time and state-dependence of the system will be omitted whenever no confusion would arise.

9.3 Main Results

9.3.1 Original Algorithms

Before we propose the distributed learning algorithms, we first give some auxiliary functions to deal with the state constraints. Let

$$\gamma_{i,j} = z_{i,j} = \varepsilon_j(x_{i,j} - x_r) + \sum_{l=1}^{N} a_{jl}(x_{i,j} - x_{i,l})$$ (9.9)

with its stabilizing function

$$\sigma_{i,j} = \left(\varepsilon_j \dot{x}_r + \sum_{l=1}^{N} a_{jl} \dot{x}_{i,l}\right) - \lambda_{i,j}^{-1} \mu_j \gamma_{i,j},$$ (9.10)

where

$$\lambda_{i,j} \triangleq \lambda_{i,j}(t) = \frac{1}{\gamma_{i,j}} \frac{\partial V_{i,j}}{\partial \gamma_{i,j}}, \quad V_{i,j} = V(\gamma_{i,j}^2, k_{b_j}),$$ (9.11)

and V here is the γ-type BLF. $k_{b_j} > 0$ is the constraint for $\gamma_{i,j}$, $\forall i$. μ_j is a positive constant to be designed later. Under the definition of γ-type BLF, since $\lim_{k_b \to \infty} V(\gamma^2, k_b) = \frac{1}{2}\gamma^2$, it is evident that $\lim_{k_{b_j} \to \infty} \lambda_{i,j} = 1$.

The distributed control law is designed as follows:

$$(C_1): \quad u_{i,j} = \hat{u}_{i,j} - \frac{1}{b_{\min}} \hat{\theta}_{i,j}^T \xi_{i,j} \text{sign}\left(\lambda_{i,j} \gamma_{i,j} \hat{\theta}_{i,j}^T \xi_{i,j}\right)$$

$$- \frac{1}{b_{\min}(\varepsilon_j + d_j^{in})} \sigma_{i,j} \text{sign}\left(\lambda_{i,j} \gamma_{i,j} \sigma_{i,j}\right)$$ (9.12)

with iterative updating laws

$$\hat{u}_{i,j} = \hat{u}_{i-1,j} - q_j \lambda_{i,j} \gamma_{i,j},$$ (9.13)

$$\hat{\theta}_{i,j} = \hat{\theta}_{i-1,j} + p_j \lambda_{i,j} \gamma_{i,j} \xi_{i,j},$$ (9.14)

where $q_j > 0$ and $p_j > 0$ are design parameters, $\forall j = 1, \cdots, N$; and $\text{sign}(\cdot)$ is a sign function, defined as follows

$$\text{sign}(\chi) = \begin{cases} +1, & \text{if } \chi > 0, \\ 0, & \text{if } \chi = 0, \\ -1, & \text{if } \chi < 0. \end{cases}$$ (9.15)

The initial values of the iterative update laws are set to be zero, that is, $\hat{u}_{0,j} = 0$, $\hat{\theta}_{0,j} = 0$, $\forall j = 1, \cdots, N$.

Remark 9.4 Consider the log-type BLF define in (9.7), and notice that

$$\lambda_{i,j} = \frac{k_{b_j}^2}{k_{b_j}^2 - \gamma_{i,j}^2}.$$

Thus the controller C_1 and its iterative updating laws will become

$$(C_1') : \; u_{i,j} = \hat{u}_{i,j} - \text{sign}\left(k_{b_j}^2 - \gamma_{i,j}^2\right)\left[\frac{1}{b_{\min}}\hat{\theta}_{i,j}^T\xi_{i,j}\text{sign}(\gamma_{i,j}\hat{\theta}_{i,j}^T\xi_{i,j})\right.$$

$$\left. + \frac{1}{b_{\min}(\varepsilon_j + d_j^{in})}\sigma_{i,j}\text{sign}\left(\gamma_{i,j}\sigma_{i,j}\right)\right], \tag{9.16}$$

$$\hat{u}_{i,j} = \hat{u}_{i-1,j} - \frac{k_{b_j}^2 q_j \gamma_{i,j}}{k_{b_j}^2 - \gamma_{i,j}^2}, \tag{9.17}$$

$$\hat{\theta}_{i,j} = \hat{\theta}_{i-1,j} + \frac{k_{b_j}^2 p_j \gamma_{i,j}\xi_{i,j}}{k_{b_j}^2 - \gamma_{i,j}^2}. \tag{9.18}$$

The term $\text{sign}\left(k_{b_j}^2 - \gamma_{i,j}^2\right)$ can be further removed from the controller C_1' as it is always equal to 1; this can be seen from the following technical analysis that the boundedness of the BLF is guaranteed and so is the inequality $\gamma_{i,j}^2 < k_{b_j}^2$. On the other hand, if the tan-type BLF (9.8) is selected, it is found that

$$\lambda_{i,j} = \frac{1}{\cos^2\left(\frac{\pi\gamma_{i,j}^2}{2k_{b_j}^2}\right)}.$$

Then the controller C_1 and its updating laws will become

$$(C_1'') : \; u_{i,j} = \hat{u}_{i,j} - \frac{1}{b_{\min}}\hat{\theta}_{i,j}^T\xi_{i,j}\text{sign}(\gamma_{i,j}\hat{\theta}_{i,j}^T\xi_{i,j}) - \frac{1}{b_{\min}(\varepsilon_j + d_j^{in})}\sigma_{i,j}\text{sign}\left(\gamma_{i,j}\sigma_{i,j}\right), \tag{9.19}$$

$$\hat{u}_{i,j} = \hat{u}_{i-1,j} - \frac{q_j\gamma_{i,j}}{\cos^2\left(\frac{\pi\gamma_{i,j}^2}{2k_{b_j}^2}\right)}, \tag{9.20}$$

$$\hat{\theta}_{i,j} = \hat{\theta}_{i-1,j} + \frac{p_j\gamma_{i,j}\xi_{i,j}}{\cos^2\left(\frac{\pi\gamma_{i,j}^2}{2k_{b_j}^2}\right)}, \tag{9.21}$$

since $\text{sign}(\lambda_{i,j}) = 1$ is always true. It is interesting to see that the differences between (C_1') and (C_1'') only lie in the iterative updating laws.

Now we have the following theorem on the consensus results of (9.1) under state constraints.

Theorem 9.1 Assume that Assumptions 9.1–9.3 hold for the multi-agent system (9.1). The closed loop system consisting of (9.1) and the control update algorithms

(9.12)–(9.14), can ensure that: (1) the tracking error $e_{i,j}(t)$ converges to zero uniformly as the iteration number i tends to infinity, $\forall j = 1, \cdots, N$; (2) the system state $x_{i,j}$ is bounded by a predefined constraint, that is, $|x_{i,j}| < k_s$, will always be guaranteed for all iterations i and all agents j provided that $|\gamma_{1,j}| < k_{b_j}$ over $[0, T]$ for $j = 1, \cdots, N$.

Remark 9.5   Let us take the agent $e_{i,j}$ for example and let there be a predefined bound such that $|x_{i,j}| < k_s$. Since $|x_r| < k_s$, we can select an upper bound $k_{b_j} < k_s - |x_r|$ for $e_{i,j}$ to guarantee the boundedness requirement. That is, if we could ensure that $|e_{i,j}| < k_{b_j} < k_s - |x_r|$ for all iterations, then it is evident that $|x_{i,j}| < k_{b_j} + |x_r| < k_s - |x_r| + |x_r| = k_s$ for all iterations. Further, from the relationship between the extended observation error and tracking error (9.6), the upper bound of $|e_{i,j}|$ can be derived by imposing the boundedness condition on $|z_{i,j}|$, as can be seen from the following analysis.

Remark 9.6   As can be seen from the algorithms (9.12)–(9.14), if the state approximates the given constraint bound, that is, $x_{i,j} \to k_s$, then $\lambda_{i,j} \to \infty$ from the definition of the BLF. In other words, the controller (9.12) will be dominated by the pure ILC part $\hat{u}_{i,j}$ as iteration number increases. This shows the inherent effect of the learning process in the proposed algorithms.

Proof of Theorem 9.1: The proof consists of five parts. First, we investigate the decreasing property of the given barrier composite energy function (BCEF) in the iteration domain at $t = T$. Then finiteness of the given BCEF is shown by verification on its derivative, and the boundedness of related parameters is also given here. Next, we give the proof of the convergence of extended observation errors. In the fourth part, the verification of the constraints of state $x_{i,j}$ for all iterations is demonstrated. Last, uniform consensus tracking is shown based on the previous analysis.

Define the following BCEF:

$$E_i(t) = \sum_{j=1}^{N} E_{i,j}(t) = \sum_{j=1}^{N} (V_{i,j}^1(t) + V_{i,j}^2(t) + V_{i,j}^3(t)), \tag{9.22}$$

$$V_{i,j}^1(t) = V(\gamma_{i,j}^2(t), t), \tag{9.23}$$

$$V_{i,j}^2(t) = \frac{(\varepsilon_j + d_j^{in})}{2p_j} \int_0^t (\hat{\theta}_{i,j} - \theta)^T (\hat{\theta}_{i,j} - \theta) d\tau, \tag{9.24}$$

$$V_{i,j}^3(t) = \frac{(\varepsilon_j + d_j^{in})}{2q_j} \int_0^t b_j \hat{u}_{i,j}^2 d\tau. \tag{9.25}$$

Part I. Difference of $E_i(t)$

Denote $\Delta E_i(t) \triangleq E_i(t) - E_{i-1}(t)$ as the difference of the BCEF along the iteration axis. Take time $t = T$, and let us examine the difference of $E_i(T)$, that is,

$$\Delta E_i(T) = \sum_{j=1}^{N} \Delta E_{i,j}(T)$$

$$= \sum_{j=1}^{N} (\Delta V_{i,j}^1(T) + \Delta V_{i,j}^2(T) + \Delta V_{i,j}^3(T)).$$

Here $\Delta \eta_i \triangleq \eta_i - \eta_{i-1}$, where η_i denotes arbitrary variable of i.

We first examine the first term $\Delta V_{i,j}^1(T)$. Noticing (9.23), we have

$$\Delta V_{i,j}^1(T) = V_{i,j}^1(T) - V_{i-1,j}^1(T)$$

$$= V_{i,j}^1(0) - V_{i-1,j}^1(T) + \int_0^T \frac{1}{\gamma_{i,j}} \frac{\partial V_{i,j}^1}{\partial \gamma_{i,j}} \gamma_{i,j} \dot{\gamma}_{i,j} d\tau$$

$$= V(\gamma_{i,j}^2(0), 0) - V(\gamma_{i-1,j}^2(T), T) + \int_0^T \frac{1}{\gamma_{i,j}} \frac{\partial V(\gamma_{i,j}^2)}{\partial \gamma_{i,j}} \gamma_{i,j} \dot{\gamma}_{i,j} d\tau. \tag{9.26}$$

Note that

$$\gamma_{i,j}(0) = z_{i,j}(0)$$

$$= \varepsilon_j(x_{i,j}(0) - x_r(0)) + \sum_{l=1}^N a_{jl}(x_{i,j}(0) - x_{i,l}(0))$$

$$= \varepsilon_j(x_{i-1,j}(T) - x_r(T)) + \sum_{l=1}^N a_{jl}(x_{i-1,j}(T) - x_{i-1,l}(T))$$

$$= \gamma_{i-1,j}(T),$$

thus $V(\gamma_{i,j}^2(0), 0) = V(\gamma_{i-1,j}^2(T), T)$. This further yields

$$\Delta V_{i,j}^1(T) = \int_0^T \lambda_{i,j} \gamma_{i,j} \dot{\gamma}_{i,j} d\tau$$

$$= \int_0^T \lambda_{i,j} \gamma_{i,j} \dot{z}_{i,j} d\tau$$

$$= \int_0^T \lambda_{i,j} \gamma_{i,j} \left[(\varepsilon_j + d_j^{in}) \dot{x}_{i,j} - \left(\varepsilon_j \dot{x}_r + \sum_{l=1}^N a_{jl} \dot{x}_{i,l} \right) \right] d\tau$$

$$= \int_0^T \lambda_{i,j} \gamma_{i,j} \left[(\varepsilon_j + d_j^{in}) \theta^T \xi_{i,j} + (\varepsilon_j + d_j^{in}) b_j u_{i,j} - \sigma_{i,j} - \lambda_{i,j}^{-1} \mu_j \gamma_{i,j} \right] d\tau$$

$$= \int_0^T \lambda_{i,j} \gamma_{i,j} \left(-(\varepsilon_j + d_j^{in}) \tilde{\theta}_{i,j}^T \xi_{i,j} + (\varepsilon_j + d_j^{in}) \hat{\theta}_{i,j}^T \xi_{i,j} + (\varepsilon_j + d_j^{in}) b_j u_{i,j} - \sigma_{i,j} \right) d\tau$$

$$- \int_0^T \mu_j \gamma_{i,j}^2 d\tau, \tag{9.27}$$

where $\tilde{\theta}_{i,j} = \hat{\theta}_{i,j} - \theta$ and (9.10) is used.

From the controller (9.12), one has

$$(\varepsilon_j + d_j^{in}) b_j u_{i,j} = (\varepsilon_j + d_j^{in}) b_j \left[\hat{u}_{i,j} - \frac{1}{b_{min}} \hat{\theta}_{i,j}^T \xi_{i,j} \mathrm{sign} \left(\lambda_{i,j} \gamma_{i,j} \hat{\theta}_{i,j}^T \xi_{i,j} \right) \right.$$

$$\left. - \frac{1}{b_{min}(\varepsilon_j + d_j^{in})} \sigma_{i,j} \mathrm{sign} \left(\lambda_{i,j} \gamma_{i,j} \sigma_{i,j} \right) \right]$$

$$= (\varepsilon_j + d_j^{in}) b_j \hat{u}_{i,j} - (\varepsilon_j + d_j^{in}) \frac{b_j}{b_{min}} \hat{\theta}_{i,j}^T \xi_{i,j} \mathrm{sign} \left(\lambda_{i,j} \gamma_{i,j} \hat{\theta}_{i,j}^T \xi_{i,j} \right)$$

$$- \frac{b_j}{b_{min}} \sigma_{i,j} \mathrm{sign} \left(\lambda_{i,j} \gamma_{i,j} \sigma_{i,j} \right). \tag{9.28}$$

Notice that the following inequalities are true:

$$(\varepsilon_j + d_j^{in})\lambda_{i,j}\gamma_{i,j}\hat{\theta}_{i,j}^T\xi_{i,j} - (\varepsilon_j + d_j^{in})\frac{b_j}{b_{\min}}\lambda_{i,j}\gamma_{i,j}\hat{\theta}_{i,j}^T\xi_{i,j}\text{sign}\left(\lambda_{i,j}\gamma_{i,j}\hat{\theta}_{i,j}^T\xi_{i,j}\right)$$

$$\leq (\varepsilon_j + d_j^{in})\lambda_{i,j}\gamma_{i,j}\hat{\theta}_{i,j}^T\xi_{i,j} - (\varepsilon_j + d_j^{in})|\lambda_{i,j}\gamma_{i,j}\hat{\theta}_{i,j}^T\xi_{i,j}| \leq 0 \tag{9.29}$$

and

$$-\lambda_{i,j}\gamma_{i,j}\sigma_{i,j} - \frac{b_j}{b_{\min}}\lambda_{i,j}\gamma_{i,j}\sigma_{i,j}\text{sign}\left(\lambda_{i,j}\gamma_{i,j}\sigma_{i,j}\right)$$

$$\leq \lambda_{i,j}\gamma_{i,j}\sigma_{i,j} - |\lambda_{i,j}\gamma_{i,j}\sigma_{i,j}| \leq 0 \tag{9.30}$$

where 9.1 is used.

Therefore, the difference of $\Delta V_{i,j}^1(T)$ now becomes

$$\Delta V_{i,j}^1(T) \leq \int_0^T \left[-\lambda_{i,j}\gamma_{i,j}(\varepsilon_j + d_j^{in})\tilde{\theta}_{i,j}^T\xi_{i,j} + \lambda_{i,j}\gamma_{i,j}(\varepsilon_j + d_j^{in})b_j\hat{u}_{i,j} - \mu_j\gamma_{i,j}^2\right]d\tau. \tag{9.31}$$

Next, we move on to the second term $V_{i,j}^2(T)$.

$$\Delta V_{i,j}^2(T) = \frac{\varepsilon_j + d_j^{in}}{2p_j}\int_0^T (\hat{\theta}_{i,j} - \theta)^T(\hat{\theta}_{i,j} - \theta)d\tau - \frac{\varepsilon_j + d_j^{in}}{2p_j}\int_0^T (\hat{\theta}_{i-1,j} - \theta)^T(\hat{\theta}_{i-1,j} - \theta)d\tau$$

$$= \frac{\varepsilon_j + d_j^{in}}{2p_j}\int_0^T (\hat{\theta}_{i,j} - \hat{\theta}_{i-1,j})^T(\hat{\theta}_{i,j} + \hat{\theta}_{i-1,j} - 2\theta)d\tau$$

$$= -\frac{\varepsilon_j + d_j^{in}}{2p_j}\int_0^T (\hat{\theta}_{i,j} - \hat{\theta}_{i-1,j})^T(\hat{\theta}_{i,j} - \hat{\theta}_{i-1,j})d\tau$$

$$+ \frac{\varepsilon_j + d_j^{in}}{p_j}\int_0^T (\hat{\theta}_{i,j} - \hat{\theta}_{i-1,j})^T(\hat{\theta}_{i,j} - \theta)d\tau$$

$$\leq \frac{\varepsilon_j + d_j^{in}}{p_j}\int_0^T (\hat{\theta}_{i,j} - \theta)^T(\hat{\theta}_{i,j} - \hat{\theta}_{i-1,j})d\tau$$

$$= (\varepsilon_j + d_j^{in})\int_0^T \lambda_{i,j}\gamma_{i,j}\tilde{\theta}_{i,j}^T\xi_{i,j}d\tau, \tag{9.32}$$

where (9.14) is used for the last equality.

Then, for the last term $\Delta V_{i,j}^3(T)$, we have

$$\Delta V_{i,j}^3(T) = \frac{\varepsilon_j + d_j^{in}}{2q_j}b_j\left[\int_0^T \hat{u}_{i,j}^2 d\tau - \int_0^T \hat{u}_{i-1,j}^2 d\tau\right]$$

$$= \frac{\varepsilon_j + d_j^{in}}{2q_j}b_j\int_0^T \left(\hat{u}_{i,j} + \hat{u}_{i-1,j}\right)\left(\hat{u}_{i,j} - \hat{u}_{i-1,j}\right)d\tau$$

$$= -\frac{\varepsilon_j + d_j^{in}}{2q_j}b_j\int_0^T \left(\hat{u}_{i,j} - \hat{u}_{i-1,j}\right)^2 d\tau + \frac{\varepsilon_j + d_j^{in}}{q_j}b_j\int_0^T \hat{u}_{i,j}\left(\hat{u}_{i,j} - \hat{u}_{i-1,j}\right)d\tau$$

$$\leq \frac{\varepsilon_j + d_j^{in}}{q_j}b_j\int_0^T \hat{u}_{i,j}\left(\hat{u}_{i,j} - \hat{u}_{i-1,j}\right)d\tau$$

$$= -(\varepsilon_j + d_j^{in})b_j\int_0^T \lambda_{i,j}\gamma_{i,j}\hat{u}_{i,j}d\tau, \tag{9.33}$$

where (9.13) is used in the last equality.

Consequently, combing (9.31)–(9.33) results in

$$\Delta E_{i,j}(T) \le - \int_0^T \mu_j \gamma_{i,j}^2 d\tau \tag{9.34}$$

which further yields

$$\Delta E_i(T) = \sum_{j=1}^N \Delta E_{i,j}(T)$$

$$\le - \sum_{j=1}^N \int_0^T \mu_j \gamma_{i,j}^2 d\tau. \tag{9.35}$$

Thus the decreasing property of the BCEF in the iteration domain at $t = T$ is obtained.

Part II. Finiteness of $E_i(t)$ and involved quantities

The finiteness of $E_i(t)$ will be proved for the first iteration and then generalized to the following. For this point, we first give the expressions of $\dot{E}_i(t)$, and then show the finiteness of $E_1(t)$.

For any iteration index i, we have

$$\dot{E}_i(t) = \sum_{j=1}^N \dot{E}_{i,j}(t)$$

$$= \sum_{j=1}^N (\dot{V}_{i,j}^1(t) + \dot{V}_{i,j}^2(t) + \dot{V}_{i,j}^3(t)). \tag{9.36}$$

Similar to the derivations in Part I, for $\dot{V}_{i,j}^1(t)$, we have

$$\dot{V}_{i,j}^1(t) \le -\lambda_{i,j}\gamma_{i,j}(\varepsilon_j + d_j^{in})\tilde{\theta}_{i,j}^T\xi_{i,j} + \lambda_{i,j}\gamma_{i,j}(\varepsilon_j + d_j^{in})b_j\hat{u}_{i,j} - \mu_j\gamma_{i,j}^2. \tag{9.37}$$

For $\dot{V}_{i,j}^2(t)$, one has

$$\frac{2p_j}{\varepsilon_j + d_j^{in}} \dot{V}_{i,j}^2(t) = (\hat{\theta}_{i,j} - \theta)^T(\hat{\theta}_{i,j} - \theta)$$

$$= \theta^T\theta - 2\theta^T\hat{\theta}_{i,j} + \hat{\theta}_{i,j}^T\hat{\theta}_{i,j}$$

$$= \theta^T\theta - 2\theta^T\hat{\theta}_{i-1,j} - 2p_j\lambda_{i,j}\gamma_{i,j}\theta^T\xi_{i,j}$$

$$+ \hat{\theta}_{i-1,j}^T\hat{\theta}_{i-1,j} + p_j^2\lambda_{i,j}^2\gamma_{i,j}^2\xi_{i,j}^T\xi_{i,j} + 2p_j\hat{\theta}_{i-1,j}^T\lambda_{i,j}\gamma_{i,j}\xi_{i,j}$$

$$\le \theta^T\theta - 2\theta^T\hat{\theta}_{i-1,j} + \hat{\theta}_{i-1,j}^T\hat{\theta}_{i-1,j}$$

$$- 2p_j\lambda_{i,j}\gamma_{i,j}\theta^T\xi_{i,j} + 2p_j\hat{\theta}_{i,j}^T\lambda_{i,j}\gamma_{i,j}\xi_{i,j}$$

$$= \theta^T\theta - 2\theta^T\hat{\theta}_{i-1,j} + \hat{\theta}_{i-1,j}^T\hat{\theta}_{i-1,j} + 2p_j\tilde{\theta}_{i,j}^T\lambda_{i,j}\gamma_{i,j}\xi_{i,j}.$$

Further, for $\dot{V}_{i,j}^3(t)$, we have

$$\frac{2q_j}{(\varepsilon_j + d_j^{in})b_j}\dot{V}_{i,j}^3(t) = \hat{u}_{i,j}^2$$

$$= (\hat{u}_{i-1,j} - q_j\lambda_{i,j}\gamma_{i,j})^2$$

$$= \hat{u}_{i-1,j}^2 - 2q_j\hat{u}_{i-1,j}\lambda_{i,j}\gamma_{i,j} + q_j^2\lambda_{i,j}^2\gamma_{i,j}^2$$

$$\leq \hat{u}_{i-1,j}^2 - 2q_j\hat{u}_{i,j}\lambda_{i,j}\gamma_{i,j}.$$

Combining the above three inequalities of $\dot{V}_{i,j}^1$, $\dot{V}_{i,j}^2$, and $\dot{V}_{i,j}^3$ together leads to

$$\dot{E}_{i,j} \leq -\mu_j\gamma_{i,j}^2 + \frac{\varepsilon_j + d_j^{in}}{2p_j}(\theta^T\theta - 2\theta^T\hat{\theta}_{i-1,j} + \hat{\theta}_{i-1,j}^T\hat{\theta}_{i-1,j}) + \frac{\varepsilon_j + d_j^{in}}{2q_j}b_j\hat{u}_{i-1,j}^2$$

$$= -\mu_j\gamma_{i,j}^2 + \frac{\varepsilon_j + d_j^{in}}{2p_j}(\theta - \hat{\theta}_{i-1,j})^T(\theta - \hat{\theta}_{i-1,j}) + \frac{\varepsilon_j + d_j^{in}}{2q_j}b_j\hat{u}_{i-1,j}^2. \qquad (9.38)$$

Now, we move on to verify the finiteness of $E_1(t)$. Meanwhile, it can be derived from (9.35) that the finiteness or boundedness of $E_i(T)$ is ensured for each iteration provided that $E_1(T)$ is finite.

Because the initial values of updating laws (9.13) and (9.14) are set to be zero, that is, $\hat{u}_{0,j} = 0$ and $\hat{\theta}_{0,j} = 0$, it is found that from (9.38) that

$$\dot{E}_{1,j} \leq -\mu_j\gamma_{1,j}^2 + \frac{\varepsilon_j + d_j^{in}}{2p_j}\theta^T\theta. \qquad (9.39)$$

It is evident $\dot{E}_{1,j}(t)$ is bounded over $[0, T]$, $\forall j$. Hence, the boundedness of $E_{1,j}(t)$ over $[0, T]$ is also obtained, $\forall j$. In particular, when $t = T, E_{1,j}(T)$ is bounded. Noticing $E_1(T) = \sum_{j=1}^N E_{1,j}(T)$, one concludes that $E_1(T)$ is bounded.

Now we are in the position of checking the boundedness property of $E_i(t)$ for $i \geq 2$, the parameter estimation $\hat{\theta}_{i,j}$ and the control signal $\hat{u}_{i,j}$.

According to the definition of $E_i(t)$ and the boundedness of $E_i(T)$, the boundedness of $V_{i,j}^2(T)$ and $V_{i,j}^3(T)$ are guaranteed for all iterations. That is, for any $i \in \mathbb{Z}^+$, there are finite constants $M_1 > 0$ and $M_2 > 0$ such that

$$\int_0^t (\hat{\theta}_{i,j} - \theta)^T(\hat{\theta}_{i,j} - \theta)d\tau \leq \int_0^T (\hat{\theta}_{i,j} - \theta)^T(\hat{\theta}_{i,j} - \theta)d\tau \leq M_1 < \infty \qquad (9.40)$$

and

$$\int_0^t b_j\hat{u}_{i,j}^2 d\tau \leq \int_0^T b_j\hat{u}_{i,j}^2 d\tau \leq M_2 < \infty. \qquad (9.41)$$

Hence the boundedness of $\hat{\theta}_{i,j}$ and $\hat{u}_{i,j}$ is guaranteed directly.

Recalling the differential of $E_{i,j}$ in (9.38), we have

$$
E_{i,j}(t) = E_{i,j}(0) + \int_0^t \left(-\mu_j \gamma_{i,j}^2 + \frac{\varepsilon_j + d_j^{in}}{2p_j} (\theta - \hat{\theta}_{i-1,j})^T (\theta - \hat{\theta}_{i-1,j}) + \frac{\varepsilon_j + d_j^{in}}{2q_j} b_j \hat{u}_{i-1,j}^2 \right) d\tau
$$

$$
\leq E_{i,j}(0) + \frac{\varepsilon_j + d_j^{in}}{2p_j} \int_0^t (\theta - \hat{\theta}_{i-1,j})^T (\theta - \hat{\theta}_{i-1,j}) d\tau + \frac{\varepsilon_j + d_j^{in}}{2q_j} \int_0^t b_j \hat{u}_{i-1,j}^2 d\tau
$$

$$
\leq E_{i,j}(0) + \frac{\varepsilon_j + d_j^{in}}{2p_j} M_1 + \frac{\varepsilon_j + d_j^{in}}{2q_j} M_2. \tag{9.42}
$$

Meanwhile, from the alignment condition, $E_{i,j}(0) = E_{i-1,j}(T)$ is also bounded. Therefore, it is evident $E_{i,j}(t)$ is bounded over $[0, T]$. And so is the amount $E_i(t)$.

Part III. Convergence of extended observation errors

We recall that

$$
\Delta E_i(T) \leq -\sum_{j=1}^N \int_0^T \mu_j \gamma_{i,j}^2 d\tau. \tag{9.43}
$$

Hence, for $t = T$,

$$
E_i(T) = E_1(T) + \sum_{k=2}^i \Delta E_k(t)
$$

$$
\leq E_1(T) - \sum_{k=2}^i \sum_{j=1}^N \int_0^T \mu_j \gamma_{k,j}^2 d\tau
$$

$$
\leq E_1(T) - \sum_{k=2}^i \sum_{j=1}^N \int_0^T \mu_j \gamma_{k,j}^2 d\tau. \tag{9.44}
$$

Since $E_i(T)$ is positive and $E_1(T)$ is bounded, we conclude that $\sum_{k=2}^i \sum_{j=1}^N \int_0^T \mu_j \gamma_{k,j}^2 d\tau$ is finite. Hence, we conclude that $\gamma_{i,j}$ converges to zero asymptotically in the sense of L_T^2-norm, as $i \to \infty$, namely,

$$
\lim_{i \to \infty} \int_0^T \mu_j \gamma_{i,j}^2 d\tau = 0, \quad j = 1, \cdots, N. \tag{9.45}
$$

Moreover, as $\gamma_{i,j} = z_{i,j}$ from (9.9), we have actually obtained the convergence of extended observation errors $\bar{z}_i$ in the sense of L_T^2-norm, namely, $\lim_{i \to \infty} \int_0^T \|\bar{z}_i\|_2^2 d\tau = 0$.

Part IV. Constraints verification on states

In part III, we showed that $E_i(t)$ is bounded over $[0, T]$ for all iterations. So it is guaranteed that $V_{i,j}^1(t)$, that is, $V(\gamma_{i,j}^2(t), t)$, is bounded over $[0, T]$ for all iterations and all agents. According to the definition of the so-called γ-type BLF, we can conclude that $|\gamma_{i,j}| < k_{b_j}$ holds over $[0, T]$, $\forall j = 1, \cdots, N$, $\forall i \in \mathbb{Z}^+$. Noticing that $\gamma_{i,j} = z_{i,j}$, we have $|z_{i,j}| < k_{b_j}$, $\forall j = 1, \cdots, N$. Denote $k_m \triangleq \max_j k_{b_j}$. Then it is evident that $\|\bar{z}_i\| \leq N k_m$, $\forall i \in \mathbb{Z}^+$.

On the other hand, the relationship between $\bar{z}_i$ and $\bar{e}_i$ in (9.6) leads to $\bar{e}_i = H^{-1}\bar{z}_i$. This further yields

$$
\|\bar{e}_i\| \leq \sigma_{\max}(H^{-\infty})\|\bar{z}_i\| \leq \frac{1}{\sigma_{\min}(H)} N k_m \tag{9.46}
$$

for all iteration numbers $i \in \mathbb{Z}^+$.

Then for the constraints imposed on states, $|x_{i,j}| < k_s$, we can set $k_m = (k_s - |x_r|)\sigma_{\min}(H)/N$. In this case, the tracking error will be bounded as follows:

$$|e_{i,j}| \leq \|\bar{e}_i\|$$

$$\leq \frac{1}{\sigma_{\min}(H)}Nk_m$$

$$\leq \frac{1}{\sigma_{\min}(H)}N(k_s - |x_r|)\frac{\sigma_{\min}(H)}{N}$$

$$= k_s - |x_r|.$$

This further yields that $|x_{i,j} - x_r| = |e_{i,j}| \leq k_s - |x_r|$, and therefore $|x_{i,j}| \leq k_s - |x_r| + |x_r| = k_s$. That is, the state constraint of each agent is satisfied.

In addition, the unknown function $\xi_{i,j}$ is bounded as its argument $x_{i,j}$ has been shown to be bounded. Incorporating with the result that $\gamma_{i,j}$ and $\lambda_{i,j}$ are bounded, and noting the control law (9.12), we can conclude that the input profile $u_{i,j}$ is also bounded.

Part V. Uniform consensus tracking

In part IV, it was shown that $|\gamma_{i,j}|$ is bounded by k_{b_j} for all iterations. Recall that $\gamma_{i,j}$ also converges to zero in the sense of L^2_T-norm, which was also shown in Part IV. Then we can conclude that $\gamma_{i,j} \to 0$ uniformly as $i \to \infty$, $\forall j = 1, \cdots, N$. In other words, $z_{i,j} \to 0$ uniformly as $i \to \infty$, $\forall j = 1, \cdots, N$. Then $\bar{z}_i \to 0$. Meanwhile, $\bar{z}_i = H\bar{e}_i$ and H is a positive matrix. Hence, $\bar{e}_i \to 0$ uniformly as $i \to \infty$. In conclusion, uniform consensus tracking is proved. The proof is completed. ∎

Remark 9.7 In the dynamics of each agent (9.1), no lumped uncertainty is taken into account to make our idea clear (Xu, 2011). As a matter of fact, when there exists a lumped uncertainty, we could append an additional robust term to the controller so that the lumped uncertainties could be compensated. To be specific, if a lumped uncertainty exists, (9.1) is reformulated as $\dot{x}_{i,j} = \theta^T\xi_j(x_{i,j}, t) + b_j(t)u_{i,j} + \alpha_j(x_{i,j}, t)$. We assume the lumped uncertainty is bounded by a known function $\alpha'(x_{i,j}, t)$: $|\alpha_j(x_{i,j}, t)| \leq \alpha'(x_{i,j}, t)$. Then the latter term could be added to the controller (9.12) similar to $\sigma_{i,j}$. The following steps are still true. Moreover, if the lumped uncertainty is bounded with unknown coefficient ω: $|\alpha_j(x_{i,j}, t)| \leq \omega\alpha'(x_{i,j}, t)$, then an additional parameter estimation process could be established and the robust compensation term is again appended to the controller with the help of the estimated parameter. The details are omitted in the interests of saving space.

9.3.2 Projection Based Algorithms

In the last subsection, we propose the original control algorithms where no projection is given to the updating laws. Consequently, we have to show the boundedness of the updating laws (9.13) and (9.14). However, it is only guaranteed that $\hat{u}_{i,j}$ and $\hat{\theta}_{i,j}$ are bounded, but the possible maxima are unknown. In practical applications, this may exceed the limitation of physical devices. In this subsection, we give projection-based algorithms as an alternative choice for engineering requirements.

The control law is designed as follows:

$$(C_2): \quad u_{i,j} = \hat{u}_{i,j} - \frac{1}{b_{\min}}\hat{\theta}^T_{i,j}\xi_{i,j}\text{sign}\left(\lambda_{i,j}\gamma_{i,j}\hat{\theta}^T_{i,j}\xi_{i,j}\right) - \frac{1}{b_{\min}(\varepsilon_j + d^{in}_j)}\sigma_{i,j}\text{sign}\left(\lambda_{i,j}\gamma_{i,j}\sigma_{i,j}\right), \quad (9.47)$$

with projection-based iterative updating laws

$$\hat{u}_{i,j} = \mathcal{P}_u(\hat{u}_{i-1,j}) - q_j \lambda_{i,j} \gamma_{i,j}, \tag{9.48}$$

$$\hat{\theta}_{i,j} = \mathcal{P}_\theta(\hat{\theta}_{i-1,j}) + p_j \lambda_{i,j} \gamma_{i,j} \xi_{i,j}, \tag{9.49}$$

where $\mathcal{P}_u$ and $\mathcal{P}_\theta$ indicate projections which are defined as follows:

$$\mathcal{P}_u[\hat{u}] = \begin{cases} \hat{u}, & |\hat{u}| \le \overline{u}, \\ \text{sign}(\hat{u})\overline{u}, & |\hat{u}| > \overline{u} \end{cases} \tag{9.50}$$

and

$$\mathcal{P}_\theta[\theta] = [\mathcal{P}_\theta[\hat{\theta}_1], \cdots, \mathcal{P}_\theta[\hat{\theta}_\ell]]^T, \tag{9.51}$$

$$\mathcal{P}_\theta[\hat{\theta}_m] = \begin{cases} \hat{\theta}_m, & |\hat{\theta}_m| \le \overline{\theta}_m, \\ \text{sign}(\hat{\theta}_m)\overline{\theta}_m, & |\hat{\theta}_m| > \overline{\theta}_m, m = 1, \cdots, \ell \end{cases} \tag{9.52}$$

with $|u_r|_{\sup} \le \overline{u}$ and $|\theta_m|_{\sup} \le \overline{\theta}_m, m = 1, \cdots, \ell$. The initial values of the iterative update laws are set to be zero, that is, $\hat{u}_{0,j} = 0, \hat{\theta}_{0,j} = 0, \forall j = 1, \cdots, N. p_j > 0$ and $q_j > 0$ are design parameters.

Remark 9.8 The projection is added to the iterative learning laws (9.48) and (9.49) to guarantee the boundedness where $\overline{u}$ and $\overline{\theta}_m$ are predefined upper bounds. In practical applications, these upper bounds could be obtained or estimated from specific environments. An intuitive case is that the control signals of the real control system are bounded by certain hardware limits that can be used as the upper bounds. When such information cannot be directly obtained, we can conduct preliminary tests to estimated them. With the help of such projection, the convergence analysis becomes be much easier, as can be seen in the following.

Now we can formulate the following theorem for control scheme (C_2).

Theorem 9.2 Assume that Assumptions 9.1–9.3 hold for the multi-agent system (9.1). The closed loop system consisting of (9.1) and the control update algorithms (9.47)–(9.49), can ensure that: (1) the tracking error $e_{i,j}(t)$ converges to zero uniformly as the iteration number i tends to infinity, $\forall j = 1, \cdots, N$; (2) the system state $x_{i,j}$ is bounded by a predefined constraint, that is, $|x_{i,j}| < k_s$, will always be guaranteed for all iterations i and all agents j provided that $|\gamma_{1,j}| < k_{b_j}$ over $[0, T]$ for $j = 1, \cdots, N$.

Proof: The main proof of this theorem is similar to that of Theorem 9.1 with minor modifications. Use the same BCEF given by (9.22)–(9.25).

The derivations from (9.26) to (9.31) are kept. Modifications are made to (9.32) and (9.33) as follows:

$$\Delta V_{i,j}^2(T) = \frac{\varepsilon_j + d_j^{in}}{2p_j} \int_0^T (\hat{\theta}_{i,j} - \theta)^T(\hat{\theta}_{i,j} - \theta)d\tau - \frac{\varepsilon_j + d_j^{in}}{2p_j} \int_0^T (\hat{\theta}_{i-1,j} - \theta)^T(\hat{\theta}_{i-1,j} - \theta)d\tau$$

$$\leq \frac{\varepsilon_j + d_j^{in}}{2p_j} \int_0^T (\hat{\theta}_{i,j} - \theta)^T (\hat{\theta}_{i,j} - \theta) d\tau$$

$$- \frac{\varepsilon_j + d_j^{in}}{2p_j} \int_0^T (\mathcal{P}_\theta(\hat{\theta}_{i-1,j}) - \theta)^T (\mathcal{P}_\theta(\hat{\theta}_{i-1,j}) - \theta) d\tau$$

$$= \frac{\varepsilon_j + d_j^{in}}{2p_j} \int_0^T (\hat{\theta}_{i,j} - \mathcal{P}_\theta(\hat{\theta}_{i-1,j}))^T (\hat{\theta}_{i,j} + \mathcal{P}_\theta(\hat{\theta}_{i-1,j}) - 2\theta) d\tau$$

$$\leq \frac{\varepsilon_j + d_j^{in}}{p_j} \int_0^T (\hat{\theta}_{i,j} - \theta)^T (\hat{\theta}_{i,j} - \mathcal{P}_\theta(\hat{\theta}_{i-1,j})) d\tau$$

$$= (\varepsilon_j + d_j^{in}) \int_0^T \lambda_{i,j} \gamma_{i,j} \tilde{\theta}_{i,j}^T \xi_{i,j} d\tau, \tag{9.53}$$

and

$$\Delta V_{i,j}^3(T) = \frac{\varepsilon_j + d_j^{in}}{2q_j} b_j \left[\int_0^T \hat{u}_{i,j}^2 d\tau - \int_0^T \hat{u}_{i-1,j}^2 d\tau \right]$$

$$\leq \frac{\varepsilon_j + d_j^{in}}{2q_j} b_j \left[\int_0^T \hat{u}_{i,j}^2 d\tau - \int_0^T \mathcal{P}_u(\hat{u}_{i-1,j})^2 d\tau \right]$$

$$= \frac{\varepsilon_j + d_j^{in}}{2q_j} b_j \int_0^T \left(\hat{u}_{i,j} + \mathcal{P}_u(\hat{u}_{i-1,j}) \right) \left(\hat{u}_{i,j} - \mathcal{P}_u(\hat{u}_{i-1,j}) \right) d\tau$$

$$\leq \frac{\varepsilon_j + d_j^{in}}{q_j} b_j \int_0^T \hat{u}_{i,j} \left(\hat{u}_{i,j} - \mathcal{P}_u(\hat{u}_{i-1,j}) \right) d\tau$$

$$= -(\varepsilon_j + d_j^{in}) b_j \int_0^T \lambda_{i,j} \gamma_{i,j} \hat{u}_{i,j} d\tau. \tag{9.54}$$

As a consequence, the decreasing property of (9.34) and (9.35) is still true.

The boundedness proof in Part II would be much simpler. This is because the introduced projection could guarantee a prior boundedness of the differential of $\dot{E}_{i,j}$. To be specific, the estimation of $\dot{V}_{i,j}^1$ does not change, but the estimations of $\dot{V}_{i,j}^2$ and $\dot{V}_{i,j}^3$ now become

$$\frac{2p_j}{\varepsilon_j + d_j^{in}} \dot{V}_{i,j}^2(t) = (\hat{\theta}_{i,j} - \theta)^T (\hat{\theta}_{i,j} - \theta)$$

$$= \theta^T \theta - 2\theta^T \hat{\theta}_{i,j} + \hat{\theta}_{i,j}^T \hat{\theta}_{i,j}$$

$$= \theta^T \theta - 2\theta^T \mathcal{P}_\theta(\hat{\theta}_{i-1,j}) - 2p_j \lambda_{i,j} \gamma_{i,j} \theta^T \xi_{i,j}$$

$$+ \mathcal{P}_\theta(\hat{\theta}_{i-1,j})^T \mathcal{P}_\theta(\hat{\theta}_{i-1,j}) + p_j^2 \lambda_{i,j}^2 \gamma_{i,j}^2 \xi_{i,j}^T \xi_{i,j}$$

$$+ 2p_j \mathcal{P}_\theta(\hat{\theta}_{i-1,j})^T \lambda_{i,j} \gamma_{i,j} \xi_{i,j}$$

$$\leq Q_1 - 2p_j \lambda_{i,j} \gamma_{i,j} \theta^T \xi_{i,j} + 2p_j \hat{\theta}_{i,j}^T \lambda_{i,j} \gamma_{i,j} \xi_{i,j}$$

$$= Q_1 + 2p_j \tilde{\theta}_{i,j}^T \lambda_{i,j} \gamma_{i,j} \xi_{i,j},$$

where $Q_1 \triangleq \theta^T \theta - 2\theta^T P_\theta(\hat{\theta}_{i-1,j}) P_\theta(\hat{\theta}_{i-1,j})^T P_\theta(\hat{\theta}_{i-1,j})$ is bounded for all iterations and all agents, and

$$\frac{2q_j}{(\varepsilon_j + d_j^{in})b_j} \dot{V}_{i,j}^3(t) = \hat{u}_{i,j}^2$$

$$= (P_u(\hat{u}_{i-1,j}) - q_j \lambda_{i,j} \gamma_{i,j})^2$$
$$= P_u(\hat{u}_{i-1,j})^2 - 2q_j P_u(\hat{u}_{i-1,j}) \lambda_{i,j} \gamma_{i,j} + q_j^2 \lambda_{i,j}^2 \gamma_{i,j}^2$$
$$\leq Q_2 - 2q_j \hat{u}_{i,j} \lambda_{i,j} \gamma_{i,j},$$

where $Q_2 = P_u(\hat{u}_{i-1,j})^2$ is also bounded.

Thus, we have

$$\dot{E}_{i,j} \leq -\mu_j \gamma_{i,j}^2 + \frac{\varepsilon_j + d_j^{in}}{2p_j} Q_1 + \frac{\varepsilon_j + d_j^{in}}{2q_j} b_j Q_2. \tag{9.55}$$

It is seen that $\dot{E}_{i,j}$ is bounded. The subsequent proof steps are similar to those of Theorem 9.1 and are thus omitted. This completes the proof. ∎

9.3.3 Smooth Function Based Algorithms

Generally speaking, the sign function used in control schemes (C_1) and (C_2) make them discontinuous control schemes, which further yields the problem of existence and uniqueness of solutions. Further, it may also cause chattering that might excite high-frequency unmodeled dynamics. This motivates us to seek an appropriate smooth approximation function of the sign function used in (9.12) and (9.47).

We take the hyperbolic tangent function in this paper to approximate the sign function. The following lemma demonstrates the property of the hyperbolic tangent function.

Lemma 9.2 (Polycarpous and Ioannouq, 1996) The following inequality holds for any $\epsilon > 0$ and for any $\rho \in \mathbb{R}$:

$$0 \leq |\rho| - \rho \tanh\left(\frac{\rho}{\epsilon}\right) \leq \delta\epsilon \tag{9.56}$$

where δ is a constant that satisfies $\delta = e^{-(\delta+1)}$, that is, $\delta = 0.2785$.

Now we can propose the third control scheme as follows:

$$(C_3) : \quad u_{i,j} = \hat{u}_{i,j} - \frac{1}{b_{\min}} \hat{\theta}_{i,j}^T \xi_{i,j} \tanh\left(\frac{\lambda_{i,j} \gamma_{i,j} \hat{\theta}_{i,j}^T \xi_{i,j}}{\eta_i}\right)$$

$$- \frac{1}{b_{\min}(\varepsilon_j + d_j^{in})} \sigma_{i,j} \tanh\left(\frac{\lambda_{i,j} \gamma_{i,j} \sigma_{i,j}}{\eta_i}\right) \tag{9.57}$$

with iterative updating laws

$$\hat{u}_{i,j} = \hat{u}_{i-1,j} - q_j \lambda_{i,j} \gamma_{i,j}, \tag{9.58}$$
$$\hat{\theta}_{i,j} = \hat{\theta}_{i-1,j} + p_j \lambda_{i,j} \gamma_{i,j} \xi_{i,j}, \tag{9.59}$$

where $q_j > 0$ and $p_j > 0$ are design parameters, $\forall j = 1, \cdots, N$. In addition, the parameter used in (9.57) is given as $\eta_i = \frac{1}{i^\nu}$ with $\nu \geq 2$. Then it is obvious $\sum_{i=1}^{\infty} \eta_i \leq 2$. The initial values of the iterative update laws are set to be zero, that is, $\hat{u}_{0,j} = 0$, $\hat{\theta}_{0,j} = 0$, $\forall j = 1, \cdots, N$.

Remark 9.9 The parameter η_i used in the hyperbolic tangent function can be predefined before the algorithms are applied. This parameter decides the compensation error of the last terms on the right hand side (RHS) of (9.57). Here a fast convergent sequence of $\{\eta_i\}$ is selected to achieve zero-error consensus tracking. However, as η_i converges to zero, the plot of the hyperbolic tangent function will approximate the sign function more and more accurately. In other words, after enough iterations, the hyperbolic tangent function almost coincides with the sign function. We will relax this condition in the next subsection.

The convergence theorem for smooth function based algorithms is given as follows.

Theorem 9.3 Assume that Assumptions 9.1–9.3 hold for the multi-agent system (9.1). The closed loop system consisting of (9.1) and the control update algorithms (9.57)–(9.59), can ensure that: (1) the tracking error $e_{i,j}(t)$ converges to zero uniformly as the iteration number i tends to infinity, $\forall j = 1, \cdots, N$; (2) the system state $x_{i,j}$ is bounded by a predefined constraint, that is, $|x_{i,j}| < k_s$, will always be guaranteed for all iterations i and all agents j provided that $|\gamma_{1,j}| < k_{b_j}$ over $[0, T]$ for $j = 1, \cdots, N$.

Proof: We give the proof by making modifications to that of Theorem 9.1. To be specific, several steps in Part I, Part II, and Part III need modifications while Parts IV and V are identical to the ones in the proof of Theorem 9.1. The BCEF (9.22)–(9.25) is also adopted in this case. In the following, we give a brief proof of Parts I – III.

Part I. Difference of $E_i(t)$

In a similar way to the first part in the proof of Theorem 9.1, the difference of $V_{i,j}^1$ can be obtained:

$$\Delta V_{i,j}^1(T) = \int_0^T \lambda_{i,j} \gamma_{i,j}(-(\varepsilon_j + d_j^{in})\tilde{\theta}_{i,j}^T \xi_{i,j} + (\varepsilon_j + d_j^{in})\hat{\theta}_{i,j}^T \xi_{i,j}$$
$$+ (\varepsilon_j + d_j^{in})b_j u_{i,j} - \sigma_{i,j})d\tau - \int_0^T \mu_j \gamma_{i,j}^2 d\tau. \tag{9.60}$$

Substituting the controller (9.57) we have

$$(\varepsilon_j + d_j^{in})b_j u_{i,j} = (\varepsilon_j + d_j^{in})b_j \hat{u}_{i,j} - (\varepsilon_j + d_j^{in})\frac{b_j}{b_{min}}\hat{\theta}_{i,j}^T \xi_{i,j} \tanh\left(\frac{\lambda_{i,j}\gamma_{i,j}\hat{\theta}_{i,j}^T \xi_{i,j}}{\eta_i}\right)$$
$$- \frac{b_j}{b_{min}}\sigma_{i,j} \tanh\left(\frac{\lambda_{i,j}\gamma_{i,j}\sigma_{i,j}}{\eta_i}\right). \tag{9.61}$$

Using lemma 9.2, we can conclude that

$$(\varepsilon_j + d_j^{in})\lambda_{i,j}\gamma_{i,j}\hat{\theta}_{i,j}^T\xi_{i,j} - (\varepsilon_j + d_j^{in})\frac{b_j}{b_{min}}\lambda_{i,j}\gamma_{i,j}\hat{\theta}_{i,j}^T\xi_{i,j} \tanh\left(\frac{\lambda_{i,j}\gamma_{i,j}\hat{\theta}_{i,j}^T\xi_{i,j}}{\eta_i}\right)$$

$$\leq \frac{b_j}{b_{min}}(\varepsilon_j + d_j^{in})\left[|\lambda_{i,j}\gamma_{i,j}\hat{\theta}_{i,j}^T\xi_{i,j}| - \lambda_{i,j}\gamma_{i,j}\hat{\theta}_{i,j}^T\xi_{i,j}\tanh\left(\frac{\lambda_{i,j}\gamma_{i,j}\hat{\theta}_{i,j}^T\xi_{i,j}}{\eta_i}\right)\right]$$

$$\leq \frac{b_j}{b_{min}}(\varepsilon_j + d_j^{in})\delta\eta_i \leq \bar{\delta}\eta_i \tag{9.62}$$

and

$$- \lambda_{i,j}\gamma_{i,j}\sigma_{i,j} - \frac{b_j}{b_{min}}\lambda_{i,j}\gamma_{i,j}\sigma_{i,j}\tanh\left(\frac{\lambda_{i,j}\gamma_{i,j}\sigma_{i,j}}{\eta_i}\right)$$

$$\leq \frac{b_j}{b_{min}}\left[|\lambda_{i,j}\gamma_{i,j}\sigma_{i,j}| - \lambda_{i,j}\gamma_{i,j}\sigma_{i,j}\tanh\left(\frac{\lambda_{i,j}\gamma_{i,j}\sigma_{i,j}}{\eta_i}\right)\right]$$

$$\leq \frac{b_j}{b_{min}}\delta\eta_i \leq \bar{\delta}\eta_i \tag{9.63}$$

where $\bar{\delta}$ is a constant satisfying $\bar{\delta} > \frac{b_j}{b_{min}}(\varepsilon_j + d_j^{in})\delta$ in consideration of Assumption 9.1. Therefore, the difference of $V_{i,j}^1$ now becomes

$$\Delta V_{i,j}^1(T) \leq \int_0^T \left[-\lambda_{i,j}\gamma_{i,j}(\varepsilon_j + d_j^{in})\tilde{\theta}_{i,j}^T\xi_{i,j} + \lambda_{i,j}\gamma_{i,j}(\varepsilon_j + d_j^{in})b_j\hat{u}_{i,j} - \mu_j\gamma_{i,j}^2 + 2\bar{\delta}\eta_i\right]d\tau. \tag{9.64}$$

By the same derivations of $\Delta V_{i,j}^2(T)$ and $\Delta V_{i,j}^3(T)$ in (9.32) and (9.33), the difference of $E_{i,j}(T)$ and $E_i(T)$ are formed as

$$\Delta E_{i,j} \leq -\int_0^T \mu_j\gamma_{i,j}^2 d\tau + 2\bar{\delta}\eta_i T \tag{9.65}$$

and

$$\Delta E_i \leq -\sum_{j=1}^N \int_0^T \mu_j\gamma_{i,j}^2 d\tau + 2N\bar{\delta}\eta_i T. \tag{9.66}$$

Part II. Finiteness of $E_i(t)$ and involved quantities

Similar to the derivation of the difference $\Delta V_{i,j}^1$ in (9.64), we have the differential of $V_{i,j}^1$,

$$\dot{V}_{i,j}^1(T) \leq -\lambda_{i,j}\gamma_{i,j}(\varepsilon_j + d_j^{in})\tilde{\theta}_{i,j}^T\xi_{i,j} + \lambda_{i,j}\gamma_{i,j}(\varepsilon_j + d_j^{in})b_j\hat{u}_{i,j} - \mu_j\gamma_{i,j}^2 + 2\bar{\delta}\eta_i. \tag{9.67}$$

Since the iterative updating laws (9.58)–(9.59) are identical to (9.13)–(9.14), the differentials of $V_{i,j}^2$ and $V_{i,j}^3$ keep the same form as that given in the proof of Theorem 9.1. As a consequence, there is an additional term appended to (9.38), which becomes

$$\dot{E}_{i,j} = -\mu_j\gamma_{i,j}^2 + \frac{\varepsilon_j + d_j^{in}}{2p_j}(\theta - \hat{\theta}_{i-1,j})^T(\theta - \hat{\theta}_{i-1,j}) + \frac{\varepsilon_j + d_j^{in}}{2q_j}b_j\hat{u}_{i-1,j}^2 + 2\bar{\delta}\eta_i. \tag{9.68}$$

Note that $2\bar{\delta}\eta_i$ is bounded, so this term would not affect the essential boundedness of $\dot{E}_{i,j}$. Thus following similar steps, it is easy to show that $E_{1,j}(t)$ and $E_1(t)$ are bounded over $[0, T]$. In particular, $E_{1,j}(T)$ is bounded.

Now we are in a position to check the boundedness of $E_i(T)$. To show this, recall (9.65), and we find that, $\forall j = 1, \cdots, N$,

$$E_{i,j}(T) = E_1(T) + \sum_{k=2}^{i} \Delta E_{k,j}(T)$$

$$\leq E_1(T) + 2T\bar{\delta} \sum_{k=2}^{i} \eta_k - \sum_{k=2}^{i} \int_0^T \mu_j \gamma_{k,j}^2 d\tau$$

$$\leq E_1(T) + 4\bar{\delta}T - \sum_{k=2}^{i} \int_0^T \mu_j \gamma_{k,j}^2 d\tau \tag{9.69}$$

where $\sum_{k=1}^{\infty} \eta_k \leq 2$ is used. Thus it is evident that $E_{i,j}(T)$ is bounded for all iterations and all agents, and so is $E_i(T)$. From here, the subsequent steps of Part II in the proof of Theorem 9.1 can be imitated to prove the boundedness of $E_i(t)$, $t \in [0, T]$.

Part III. Convergence of extended observation errors

The proof is an immediate conclusion of the following derivations:

$$E_i(T) = E_1(T) + \sum_{k=2}^{i} \Delta E_k(t)$$

$$\leq E_1(T) - \sum_{k=2}^{i} \sum_{j=1}^{N} \int_0^T \mu_j \gamma_{k,j}^2 d\tau + 2N\bar{\delta}T \sum_{k=2}^{i} \eta_k$$

$$\leq E_1(T) + 4N\bar{\delta}T - \sum_{k=2}^{i} \sum_{j=1}^{N} \int_0^T \mu_j \gamma_{k,j}^2 d\tau. \tag{9.70}$$

Since $E_1(T) + 4N\bar{\delta}T$ is finite and $E_i(T)$ is positive, we have $\lim_{i \to \infty} \int_0^T \gamma_{i,j}^2 d\tau = 0$, $\forall j = 1, \cdots, N$. ∎

9.3.4 Alternative Smooth Function Based Algorithms

As explained in Remark 9.9, the fast convergent η_i make the smooth hyperbolic tangent function close to the sign function, which is discontinuous. Thus, it may be not quite favorable to engineers. We are interested in whether a fixed parameter η, possibly small, is able to ensure convergence. And if so, how does it perform? The answer to this question is given in this subsection.

The controller (9.57) is modified by replacing the iteration-varying parameter η_i with a fixed one η. However, in this case, there would be a constant compensation error in the difference and differential expression of $V_{i,j}^1(t)$. Hence, it is hard to derive that the difference of $E_i(T)$ will be negative even after sufficient iterations. As a result, only a bounded convergence rather than zero-error convergence can be obtained in this case. The details are discussed in the following.

Specifically, the algorithms are formulated as follows:

$$(C_4) : \quad u_{i,j} = \hat{u}_{i,j} - \frac{1}{b_{\min}} \hat{\theta}_{i,j}^T \xi_{i,j} \tanh\left(\frac{\lambda_{i,j}\gamma_{i,j}\hat{\theta}_{i,j}^T\xi_{i,j}}{\eta}\right)$$

$$- \frac{1}{b_{\min}(\varepsilon_j + d_j^{in})} \sigma_{i,j} \tanh\left(\frac{\lambda_{i,j}\gamma_{i,j}\sigma_{i,j}}{\eta}\right) \tag{9.71}$$

with iterative updating laws

$$\hat{u}_{i,j} = \hat{u}_{i-1,j} - q_j\lambda_{i,j}\gamma_{i,j} - \phi\hat{u}_{i,j}, \tag{9.72}$$

$$\hat{\theta}_{i,j} = \hat{\theta}_{i-1,j} + p_j\lambda_{i,j}\gamma_{i,j}\xi_{i,j} - \varphi\hat{\theta}_{i,j}, \tag{9.73}$$

where $q_j > 0$, $p_j > 0$, $\forall j = 1, \cdots, N$, $\phi > 0$ and $\varphi > 0$ are design parameters. The initial values of the iterative update laws are set to be zero, that is, $\hat{u}_{0,j} = 0$, $\hat{\theta}_{0,j} = 0$, $\forall j = 1, \cdots, N$.

Remark 9.10　Compared with the iterative updating laws given in control schemes (C_1), (C_2) and (C_3), it is found that two additional terms are appended to (9.72) and (9.73). These two terms, $-\phi\hat{u}_{i,j}$ and $-\varphi\hat{\theta}_{i,j}$, are leakage terms to increase the robustness and ensure the boundedness of the learning algorithms.

The convergence property is summarized in the following theorem.

Theorem 9.4　Assume that Assumptions 9.1–9.3 hold for the multi-agent system (9.1). The closed loop system consisting of (9.1) and the control update algorithms (9.71)–(9.73), can ensure that: (1) the tracking error $\bar{e}_i(t)$ converges to the ζ-neighborhood of zero asymptotically in the sense of L_T^2-norm within finite iterations, where

$$\zeta \triangleq \frac{2TN\bar{\delta}\eta}{\mu_m\sigma_{\min}^2(H)} + \frac{(N+1)NT\varphi}{2p_m\mu_m\sigma_{\min}^2(H)}\theta^T\theta + \frac{\epsilon}{\sigma_{\min}^2(H)}, \tag{9.74}$$

with $T, N, \varphi, \mu_m, p_m, \sigma_{\min}(H)$ being the iteration length, amount of agents, leakage gain, minimum value of parameter μ_j in the stabilizing function, minimum value of the learning gain in the parameter updating law, minimum singular value of H, respectively, and $\epsilon > 0$ being an arbitrary small constant; (2) the iterative updating parameters $\hat{u}_{i,j}$ and $\hat{\theta}_{i,j}$ are bounded in the sense of L_T^2-norm, $\forall j = 1, \cdots, N$, $\forall i$.

Remark 9.11　From the expression of ζ in (9.74), any prespecified nonzero bound of tracking error can be obtained by tuning the design parameters η, φ, ϵ, p_j, and μ_j. More clearly, it is seen the magnitude of ζ is proportional to η, φ, and ϵ, where η denotes the compensation error when using the smooth hyperbolic tangent function, φ is the leakage gain in parameter updating, and ϵ can be predefined and sufficiently small.

Proof of Theorem 9.4: We still apply the BCEF given in (9.22)–(9.25) and check the difference of $E_i(T)$ first.

Part I. Difference of $E_i(t)$

By similar steps to the proof of Theorem 9.3, we can with little difficulty obtain

$$\Delta V^1_{i,j}(T) \le \int_0^T \left[-\lambda_{i,j}\gamma_{i,j}(\varepsilon_j + d^{in}_j)\tilde{\theta}^T_{i,j}\xi_{i,j} + \lambda_{i,j}\gamma_{i,j}(\varepsilon_j + d^{in}_j)b_j\hat{u}_{i,j} - \mu_j\gamma^2_{i,j} + 2\bar{\delta}\eta \right] d\tau. \qquad (9.75)$$

Next, let us consider $\Delta V^2_{i,j}(T)$, which is derived as

$$\begin{aligned}
\Delta V^2_{i,j}(T) =& \frac{\varepsilon_j + d^{in}_j}{2p_j} \int_0^T (\hat{\theta}_{i,j} - \theta)^T(\hat{\theta}_{i,j} - \theta)d\tau - \frac{\varepsilon_j + d^{in}_j}{2p_j} \int_0^T (\hat{\theta}_{i-1,j} - \theta)^T(\hat{\theta}_{i-1,j} - \theta)d\tau \\
=& -\frac{\varepsilon_j + d^{in}_j}{2p_j} \int_0^T (\hat{\theta}_{i,j} - \hat{\theta}_{i-1,j})^T(\hat{\theta}_{i,j} - \hat{\theta}_{i-1,j})d\tau \\
& + \frac{\varepsilon_j + d^{in}_j}{p_j} \int_0^T (\hat{\theta}_{i,j} - \theta)^T(\hat{\theta}_{i,j} - \hat{\theta}_{i-1,j})d\tau \\
=& -\frac{\varepsilon_j + d^{in}_j}{2p_j} \int_0^T (\hat{\theta}_{i,j} - \hat{\theta}_{i-1,j})^T(\hat{\theta}_{i,j} - \hat{\theta}_{i-1,j})d\tau + (\varepsilon_j + d^{in}_j) \int_0^T \lambda_{i,j}\gamma_{i,j}\tilde{\theta}_{i,j}\xi_{i,j}d\tau \\
& - \frac{(\varepsilon_j + d^{in}_j)\varphi}{p_j} \int_0^T \tilde{\theta}^T_{i,j}\hat{\theta}_{i,j}d\tau.
\end{aligned} \qquad (9.76)$$

Meanwhile, the following equality is true:

$$-\tilde{\theta}^T_{i,j}\hat{\theta}_{i,j} = -\frac{1}{2}\tilde{\theta}^T_{i,j}\tilde{\theta}_{i,j} - \frac{1}{2}\hat{\theta}^T_{i,j}\hat{\theta}_{i,j} + \frac{1}{2}\theta^T\theta. \qquad (9.77)$$

Then combining (9.76) and (9.77) we have

$$\Delta V^2_{i,j}(T) \le (\varepsilon_j + d^{in}_j) \int_0^T \lambda_{i,j}\gamma_{i,j}\tilde{\theta}_{i,j}\xi_{i,j}d\tau + \frac{(\varepsilon_j + d^{in}_j)T\varphi}{2p_j}\theta^T\theta. \qquad (9.78)$$

For the third component of $E_{i,j}(T)$, we have

$$\begin{aligned}
\Delta V^3_{i,j}(T) =& \frac{\varepsilon_j + d^{in}_j}{2q_j}b_j \left[\int_0^T \hat{u}^2_{i,j}d\tau - \int_0^T \hat{u}^2_{i-1,j}d\tau \right] \\
=& -\frac{\varepsilon_j + d^{in}_j}{2q_j} \int_0^T b_j(\hat{u}_{i,j} - \hat{u}_{i-1,j})^2 d\tau + \frac{\varepsilon_j + d^{in}_j}{q_j} \int_0^T b_j\hat{u}_{i,j}(\hat{u}_{i,j} - \hat{u}_{i-1,j})d\tau \\
=& -\frac{\varepsilon_j + d^{in}_j}{2q_j} \int_0^T b_j(\hat{u}_{i,j} - \hat{u}_{i-1,j})^2 d\tau - (\varepsilon_j + d^{in}_j) \int_0^T b_j\hat{u}_{i,j}\lambda_{i,j}\gamma_{i,j}d\tau \\
& - \frac{(\varepsilon_j + d^{in}_j)\phi}{q_j} \int_0^T b_j(\hat{u}_{i,j})^2 d\tau \\
\le& -(\varepsilon_j + d^{in}_j) \int_0^T b_j\hat{u}_{i,j}\lambda_{i,j}\gamma_{i,j}d\tau.
\end{aligned} \qquad (9.79)$$

Therefore, combing (9.75), (9.78), and (9.79) gives

$$\Delta E_{i,j}(T) \le -\int_0^T \mu_j\gamma^2_{i,j}d\tau + 2T\bar{\delta}\eta + \frac{(\varepsilon_j + d^{in}_j)T\varphi}{2p_j}\theta^T\theta, \qquad (9.80)$$

and consequently,

$$\Delta E_i(T) \leq - \sum_{j=1}^{N} \int_0^T \mu_j \gamma_{i,j}^2 d\tau + \kappa, \tag{9.81}$$

where $\kappa \triangleq 2TN\bar{\delta}\eta + \frac{(N+1)NT\varphi}{2p_m}\theta^T\theta > 0$, and $p_m = \min_j p_j$. Due to the existence of $\kappa > 0$, it is impossible to derive that $\Delta E_i(T)$ is negative even after sufficient iterations. Instead, we will show that the tracking error will enter the neighborhood of zero within finite iterations.

Part II. Bounded convergence of extended observation errors and tracking errors

By similar steps to the proof of Theorem 9.3, it is still true that $E_1(T)$ is finite. Let μ_m denote the smallest value of μ_j, $j = 1, \cdots, N$, that is, $\mu_m = \min_j \mu_j$. Then, for any finite sum of $\Delta E_i(T)$, we have

$$E_i(T) = E_1(T) + \sum_{k=2}^{i} \Delta E_k(T)$$

$$\leq E_1(T) - \sum_{k=2}^{i} \left[\sum_{j=1}^{N} \int_0^T \mu_j \gamma_{k,j}^2 d\tau - \kappa \right]$$

$$\leq E_1(T) - \sum_{k=2}^{i} \left[\mu_m \sum_{j=1}^{N} \int_0^T \gamma_{k,j}^2 d\tau - \kappa \right]$$

$$= E_1(T) - \sum_{k=2}^{i} \mu_m \left[\sum_{j=1}^{N} \int_0^T z_{k,j}^2 d\tau - \frac{\kappa}{\mu_m} \right]$$

$$= E_1(T) - \sum_{k=2}^{i} \mu_m \left[\int_0^T \|\bar{z}_k\|_2^2 d\tau - \frac{\kappa}{\mu_m} \right]. \tag{9.82}$$

Due to the positiveness of $E_i(T)$, we can show the boundedness and convergence of $\int_0^T \|\bar{z}_i\|_2^2 d\tau$ from (9.82).

(a) If $\int_0^T \|\bar{z}_i\|_2^2 d\tau$ tends to infinity at the ith iteration, then the RHS of (9.82) will diverge to infinity owing to the finiteness of κ/μ_m. This contradicts the positiveness of $E_i(T)$.

(b) For any given $\epsilon > 0$, there is a finite integer $i_0 > 0$ such that $\int_0^T \|\bar{z}_i\|_2^2 d\tau < \kappa/\mu_m + \epsilon$ for $i \geq i_0$. Otherwise, $\int_0^T \|\bar{z}_i\|_2^2 d\tau > \kappa/\mu_m + \epsilon$ holds for $i \to \infty$. Thus the RHS of (9.82) will approach $-\infty$, which again contradicts the positiveness of $E_i(T)$.

Hence, the extended observation error $\int_0^T \|\bar{z}_i\|_2^2 d\tau$ will enter the specified bound $\kappa/\mu_m + \epsilon$ within finite iterations.

Note the relationship between extended observation errors and tracking errors, that is, (9.6). We have

$$\int_0^T \|\bar{e}_i\|_2^2 d\tau = \int_0^T \|H^{-1}\bar{z}_i\|_2^2 d\tau$$

$$\leq \frac{1}{\sigma_{min}^2}(H) \int_0^T \|\bar{z}_i\|_2^2 d\tau \tag{9.83}$$

whence, the tracking error $\int_0^T \|\bar{e}_i\|_2^2 d\tau$ will enter the specified bound $\sigma_{min}^{-2}(H)(\kappa/\mu_m + \epsilon)$ within finite iterations.

Part III. Boundedness of $\hat{\theta}_{i,j}$ and $\hat{u}_{i,j}$

Recall the parameter updating law (9.73), which further yields

$$
\hat{\theta}_{i,j}^T \hat{\theta}_{i,j} = \frac{1}{(1+\varphi)^2}(\hat{\theta}_{i-1,j} + p_j \lambda_{i,j}\gamma_{i,j}\xi_{i,j})^T(\hat{\theta}_{i-1,j} + p_j \lambda_{i,j}\gamma_{i,j}\xi_{i,j})
$$
$$
= \frac{1}{(1+\varphi)^2}\left(\hat{\theta}_{i-1,j}^T \hat{\theta}_{i-1,j} + 2p_j \lambda_{i,j}\gamma_{i,j}\hat{\theta}_{i-1,j}^T\xi_{i,j} + (p_j\lambda_{i,j}\gamma_{i,j})^2\xi_{i,j}^T\xi_{i,j}\right). \tag{9.84}
$$

Using Young's inequality, we have

$$
2p_j\lambda_{i,j}\gamma_{i,j}\hat{\theta}_{i-1,j}^T\xi_{i,j} \le c\hat{\theta}_{i-1,j}^T\hat{\theta}_{i-1,j} + \frac{1}{c}(p_j\lambda_{i,j}\gamma_{i,j})^2\xi_{i,j}^T\xi_{i,j},
$$

where $0 < c < \varphi$. Substituting the above inequality in (9.84) yields

$$
\int_0^T \|\hat{\theta}_{i,j}\|_2^2 d\tau = \frac{1+c}{(1+\varphi)^2}\int_0^T \|\hat{\theta}_{i-1,j}\|_2^2 d\tau + \frac{1+1/c}{(1+\varphi)^2}\int_0^T (p_j\lambda_{i,j}\gamma_{i,j})^2\|\xi_{i,j}\|_2^2 d\tau. \tag{9.85}
$$

It is observed that $\lambda_{i,j}$ is bounded due to the definition of the BLF and constraint conditions. $\xi_{i,j}$ is a continuous function of $z_{i,j}$, thus it is also bounded provided that $z_{i,j}$ is bounded. Meanwhile, $\gamma_{i,j} = z_{i,j}$. In conclusion, the last term on the RHS of (9.85) is bounded for all iterations since the boundedness of $z_{i,j}$ has been proved in Part II. Denote the upper bound by ϑ.

Then (9.84) yields

$$
\int_0^T \|\hat{\theta}_{i,j}\|_2^2 d\tau = \sum_{k=1}^i \left(\frac{1+c}{(1+\varphi)^2}\right)^{i-k} \frac{1+1/c}{(1+\varphi)^2}\int_0^T (p_j\lambda_{k,j}\gamma_{k,j})^2\|\xi_{k,j}\|_2^2 d\tau
$$
$$
\le \sum_{k=1}^i \left(\frac{1+c}{(1+\varphi)^2}\right)^{i-k} \frac{1+1/c}{(1+\varphi)^2}\vartheta
$$
$$
\le \frac{1}{1-\frac{1+c}{(1+\varphi)^2}}\frac{1+1/c}{(1+\varphi)^2}\vartheta = \frac{1+1/c}{(1+\varphi)^2-1-c}\vartheta. \tag{9.86}
$$

The boundedness of $\hat{\theta}_{i,j}$ is obtained. By similar steps, we can further prove the boundedness of $\hat{u}_{i,j}$ for all iterations and for all agents. This completes the proof. ∎

Remark 9.12　As can be seen from the proof, the leakage term introduced to the iterative updating law would not affect the convergence property (i.e. convergence to the neighborhood of zero asymptotically). In other words, the leakage term could be removed from (9.72) and (9.73) while the bounded convergence of tracking errors is still established.

9.3.5　Practical Dead-Zone Based Algorithms

Several facts can be observed from the last subsection. First, it is hard to show the satisfaction of the state constraint since the L_T^2-norm of the state is only proved to converge into some neighborhood of zero. In practice, we may achieve better learning performance such as point-wise or uniform convergence so that the state constraint is fulfilled; however, in theory only L_T^2 convergence is guaranteed. As a result, it is impossible to analyze the specific behavior of the state in the time domain. Moreover, the leakage terms used in (9.72) and (9.73) aim only to ensure the boundedness of the learning process, since it is a kind of 'forgetting factor' type of mechanism. Furthermore, it is easy to see from (9.82) that the BLF is bounded before the tracking error enters the given neighborhood and thus the state satisfies constraints in these iterations.

Based on these observations, we further propose the following practical dead-zone based algorithms to ensure state constraint requirements:

$$(C_5): \quad u_{i,j} = \hat{u}_{i,j} - \frac{1}{b_{\min}} \hat{\theta}_{i,j}^T \xi_{i,j} \tanh\left(\frac{\lambda_{i,j}\gamma_{i,j}\hat{\theta}_{i,j}^T \xi_{i,j}}{\eta}\right)$$

$$- \frac{1}{b_{\min}(\varepsilon_j + d_j^{in})} \sigma_{i,j} \tanh\left(\frac{\lambda_{i,j}\gamma_{i,j}\sigma_{i,j}}{\eta}\right) \tag{9.87}$$

with iterative updating laws

$$\hat{u}_{i,j} = \begin{cases} \hat{u}_{i-1,j} - q_j \lambda_{i,j}\gamma_{i,j}, & \int_0^T z_{i-1,j}^2 d\tau > \varsigma', \\ \hat{u}_{i-1,j} & otherwise, \end{cases} \tag{9.88}$$

$$\hat{\theta}_{i,j} = \begin{cases} \hat{\theta}_{i-1,j} + p_j \lambda_{i,j}\gamma_{i,j}\xi_{i,j}, & \int_0^T z_{i-1,j}^2 d\tau > \varsigma', \\ \hat{\theta}_{i-1,j}, & otherwise, \end{cases} \tag{9.89}$$

where $q_j > 0, p_j > 0, \forall j = 1, \cdots, N$ are design parameters. The initial values of the iterative update laws are set to be zero, that is, $\hat{u}_{0,j} = 0, \hat{\theta}_{0,j} = 0, \forall j = 1, \cdots, N$. The parameter ς' denotes the bound of the convergent neighborhood, which is defined later.

Remark 9.13 The essential mechanism of the proposed algorithms is that learning processes of $\hat{u}_{i,j}$ and $\hat{\theta}_{i,j}$ will stop updating whenever the extended observation error enters the predefined neighborhood of zero so that the control system repeats the same tracking performance. Consequently, the boundedness of the learning algorithms (9.88)–(9.89) and the state constraints condition are fulfilled naturally so long as the bounded convergence is finished within finite iterations.

Then we have the following result.

Theorem 9.5 Assume that Assumptions 9.1–9.3 hold for the multi-agent system (9.1). The closed loop system consisting of (9.1) and the control update algorithms (9.87)–(9.89), can ensure that: (1) the tracking error $\bar{e}_i(t)$ converges to the predefined ς-neighborhood of zero asymptotically in the sense of L_T^2-norm within finite iterations, where

$$\varsigma \triangleq \frac{2TN\bar{\delta}\eta}{\mu_m \sigma_{\min}^2(H)} + \frac{\epsilon}{\sigma_{\min}^2(H)}, \tag{9.90}$$

with $T, N, \mu_m, \sigma_{\min}(H)$ being the iteration length, amount of agents, minimum value of parameter μ_j in the stabilizing function, minimum singular value of H, respectively, and $\epsilon > 0$ being an arbitrary small constant; (2) the system state $x_{i,j}$ is bounded by a predefined constraint, that is, $|x_{i,j}| < k_s$, will always be guaranteed for all iterations i and all agents j provided that $|\gamma_{1,j}| < k_{b_j}$ over $[0, T]$ for $j = 1, \cdots, N$; (3) the iterative updating parameters $\hat{u}_{i,j}$ and $\hat{\theta}_{i,j}$ are bounded in the sense of L_T^2-norm, $\forall j = 1, \cdots, N, \forall i$.

Remark 9.14 Here we assume that the upper bound of control gain, that is, $b_{\max}$, is also known so that the $\bar{\delta}$ can be calculated. Note that H could be defined by the topology relationship, and μ_m is a design parameter. Therefore, for any given ς, one can select η and ϵ sufficiently small and μ_m sufficiently large such that (9.90) is achieved. For the learning algorithms, the upper bound ς in (9.90) is satisfied as long as we define ς' as

$$\varsigma' \triangleq \frac{\sigma_{\min}^2(H)}{N}\varsigma = \frac{2T\bar{\delta}\eta}{\mu_m} + \frac{\epsilon}{N}.$$

Proof of Theorem 9.5: First, to show the finite iteration convergence to a desired neighborhood of zero, it is sufficient to use the same steps as in the proof of theorem 9.4, removing the last term on the RHS of (9.76). Thus the proof of item (1) is completed.

Next, let us verify the state constraint. Since we have shown that the tracking error will enter a neighborhood of zero within finite iterations, where the magnitude of the neighborhood could be predefined, then for a given neighborhood bound, there is a finite integer, say i_1, such that the tracking error enters the given neighborhood for $i \geq i_1$. Then it is evident that $E_i(T)$ is bounded, $\forall i < i_1$. Thus $V_{i,j}^1$ is also bounded, whence the constraints are satisfied following similar steps of previous proofs.

On the other hand, for the iterations $i \geq i_1$, the tracking error will enter a prespecified neighborhood. Then it is found that the learning processes (9.88) and (9.89) stop updating and the control system repeats its tracking performance. As a result, the state constraint is still satisfied.

Last, whenever the tracking error enters the predefined neighborhood, the learning algorithms (9.88)–(9.89) stop updating and the boundedness is therefore guaranteed. This completes the proof. ∎

9.4 Illustrative Example

To illustrate the applications of the proposed algorithms, consider a group of four agents. The communication topology graph is demonstrated in Figure 9.1, where vertex 0 represents the desired reference or virtual leader and the dashed lines stand for the communication links between leader and followers. In other words, agents 1 and 2 can access the information from the leader. The solid lines stand for the communication links among the four agents.

The Laplacian matrix of this connection graph is

$$L = \begin{bmatrix} 2 & -1 & 0 & -1 \\ -1 & 3 & -1 & -1 \\ 0 & -1 & 1 & 0 \\ -1 & -1 & 0 & 2 \end{bmatrix}$$

and $D = \text{diag}\{1, 1, 0, 0\}$.

In the simulations, the agent dynamics are modeled by

$$\dot{x}_{i,j} = \theta(t)(\sin(t)\cos(t) + 0.5\sin(x_{i,j})) + (1 + 0.1\sin(5t))u_{i,j},$$

where $\theta(t) = 2\sin(t)$. The initial states for the first iteration of the four agents are set as 1, 0.5, 0.2, 0.8, respectively. Then the state trajectories of the four agents at the first iteration are different. The desired reference is given as

$$x_r(t) = \sin^3(t) + 0.2\sin(5t) - 0.005\cos(10t) + 0.005.$$

The iteration length is set to be $[0, \pi]$. The log-type and tan-type BLFs are both simulated and the results are almost the same. Thus we only present the results with log-type BLF (9.7) with $k_b = 1$. The simulations are run for 20 iterations for each control scheme.

Case 1: Control Scheme (C_1).

The parameters used in control scheme (C_1) (9.12)–(9.14) and (9.10) are set to be $b_{\min} = 0.9$, $q_j = 5$, $p_j = 5$, and $\mu_j = 10$.

The trajectories of all agents at the 1st iteration and 20th iteration are shown in Figure 9.2, where it can be seen that the trajectories at the 1st iteration do not match the desired reference, while those at the 20th iteration coincide with the desired reference. The tracking performance of each agent is shown in Figure 9.3, where we find that all agents converge to the leader's trajectory quickly. In addition, the state constraints are satisfied.

Denote the maximum error as $\max_t |x_{i,j} - x_r|$ for the jth agent at the ith iteration. The maximum error profiles of all agents along the iteration axis are shown in Figure 9.4. As one can see, the maximum tracking errors for all agents are reduced significantly during the first few iterations.

It has been remarked that the input files may be discontinuous due to the introduction of the sign function in the control scheme. This point is verified by Figure 9.5. The four subfigures depict that the ultimate input files exhibit a large amount of the chattering phenomenon.

Case 2: Control Scheme (C_2).

It can be seen that control scheme (C_2) is a technical alternative of control scheme (C_1). The performance (C_2) would not differ from (C_1) as long as we choose the bounds sufficiently large. The parameters used in control scheme (C_2) are set to be $b_{\min} = 0.9$, $q_j = 50$, $p_j = 50$, and $\mu_j = 30$. The upper bound of the saturation function for the input and theta are set to be 1.5 and 4, respectively.

The trajectories of all agents for the 1st iteration and 20th iteration are given in Figure 9.6. The maximum error profiles along the iteration axis for all agents are displayed in Figure 9.7. As can been, these behaviors are little different to those of control scheme (C_1).

The input files of all agents at the 1st iteration, 10th iteration, and 20th iteration are provide in Figure 9.8. From the subfigures, one can see that the chattering phenomena is similar to the case of control scheme (C_1).

Case 3: Control Scheme (C_3).

The parameters used in control scheme (C_3) are set to be $b_{\min} = 0.9$, $q_j = 5$, $p_j = 1$, and $\mu_j = 10$. The decreasing parameter in the tanh function is selected as $\eta_i = 1/i^3$. In this case, the finite summation requirement, that is, $\sum_{i=1}^{\infty} \eta_i < 2$, implies that η_i converges to zero very fast. As a result, the tanh function approximates the sign function after a small number of iterations. In other words, the tracking performance under control scheme (C_3) would be almost the same to control scheme (C_1). This fact is shown in the following.

The trajectories of all agents for the 1st iteration and 20th iteration are given in Figure 9.9. The maximum error profiles along the iteration axis for all agents are displayed in Figure 9.10. As can been, these behaviors are little different to those of the control scheme (C_1).

The input files of all agents at the 1st iteration, 10th iteration, and 20th iteration are provided in Figure 9.11. From the subfigures, one can see that the chattering phenomena is not eliminated completely, although some improvements are seen compared to Figure 9.5 for control scheme (C_1). The reason lies in the fast attenuation of parameter η_i.

Case 4: Control Scheme (C_4).

In this case, we consider the control scheme (C_4), where the smoothing parameter used in tanh function is set to be $\eta = 0.1$. The other parameters are selected as $b_{min} = 0.9$, $q_j = 15, p_j = 20$, and $\mu_j = 10$.

The trajectories of all agents at the 1st iteration and 20th iteration are shown in Figure 9.12. The tracking performance improvements of each agent for different iterations are also shown in Figure 9.13 for comparison. The tracking performance is also good under the smoothing function.

The maximum tracking error profiles for all agents along the iteration axis are displayed in Figure 9.14. It is evident that the reduction degree of this scheme is as large as the previous schemes. Meanwhile, the plot also shows a flat trend, which means that the maximum error might not keep decreasing when the iteration number increases. This confirms our theoretical analysis.

In addition, the input profiles under control scheme (C_4) are presented in Figure 9.15. The discontinuity of input profiles in the previous two schemes is avoided due to the proposed smoothing technique. These observations are consistent with our analysis in the controller design parts.

Case 5: Control Scheme (C_5).

As has been shown in the theoretical analysis, the control scheme (C_5) might behave similarly to (C_4) because the former scheme is also a practical alternative of the latter one. However, due to the practical dead-zone technique, the ultimate tracking error may be inferior to (C_4). In this case, the parameters for the algorithms are set to be $b_{min} = 0.9$, $b_{max} = 1.1, q_j = 50, p_j = 50, \mu_j = 10, \bar{\delta} = 0.0028, \epsilon = 0.00001$.

The trajectories of all agents for the 1st iteration and 20th iteration are given in Figure 9.16. The maximum error profiles along the iteration axis for all agents are displayed in Figure 9.17. The input profiles for all agents are given in Figure 9.18.

Figure 9.1 Communication graph among agents in the network.

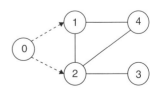

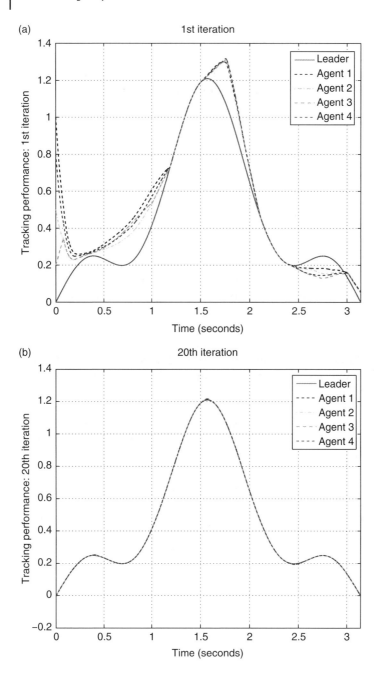

Figure 9.2 Trajectories of four agents at the 1st iteration and 20th iteration: (C_1) case.

(a)

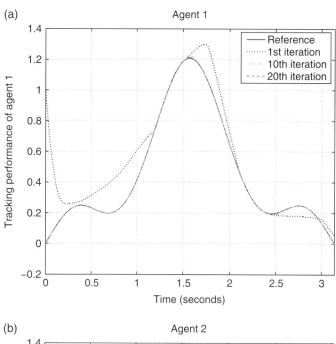

(b)

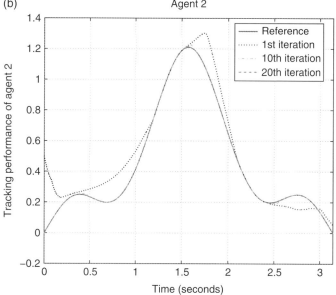

Figure 9.3 Tracking performance at the 1st iteration, 10th iteration, and 20th iteration for each agent: (C_1) case.

(c)

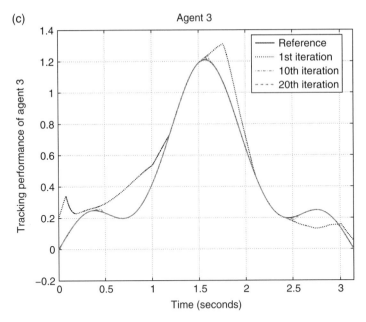

(d)

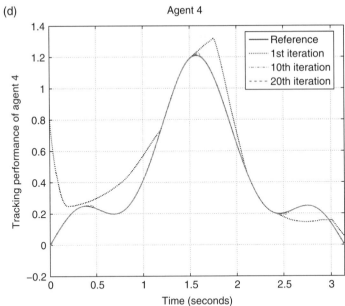

Figure 9.3 (Continued)

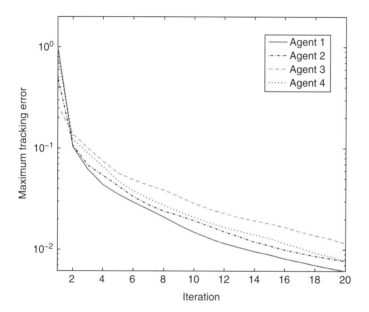

Figure 9.4 Maximum tracking error along iteration axis: (C_1) case.

(a)

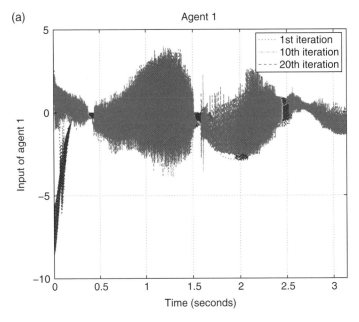

(b)

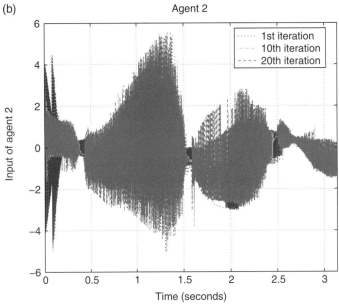

Figure 9.5 Input profiles at the 1st iteration, 10th iteration, and 20th iteration for each agent: (C_1) case.

(c)

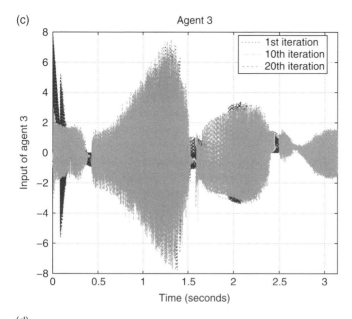

(d)

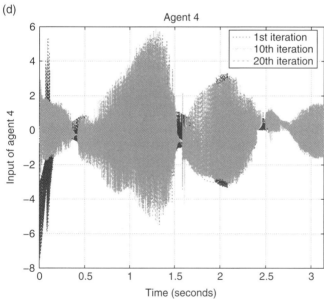

Figure 9.5 (Continued)

(a)

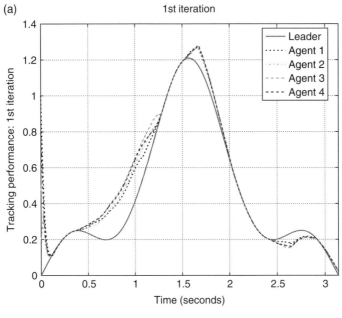

(b)

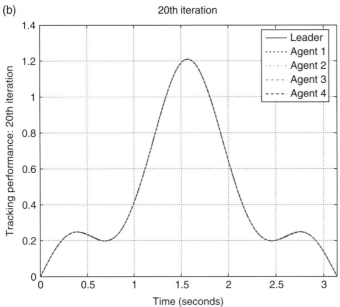

Figure 9.6 Trajectories of four agents at the 1st iteration and 20th iteration: (C_2) case.

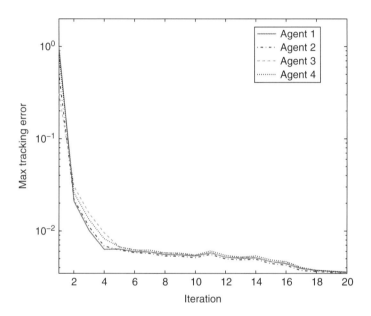

Figure 9.7 Maximum tracking error along iteration axis: (C_2) case.

(a)

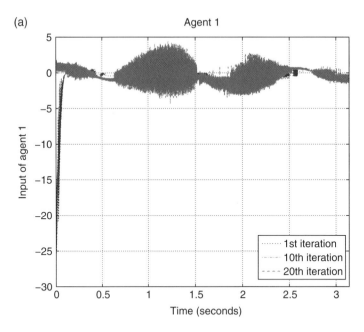

(b)

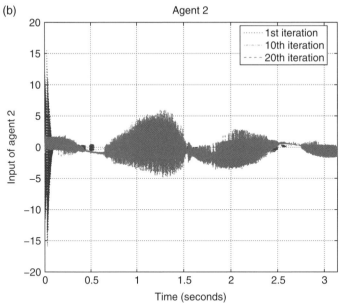

Figure 9.8 Input profiles at the 1st iteration, 10th iteration, and 20th iteration for each agent: (C_2) case.

(c)

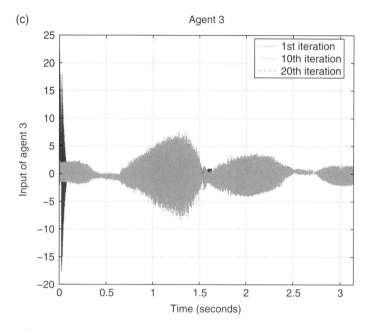

(d)

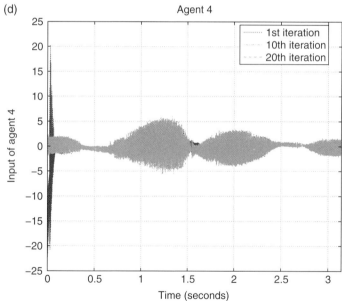

Figure 9.8 (Continued)

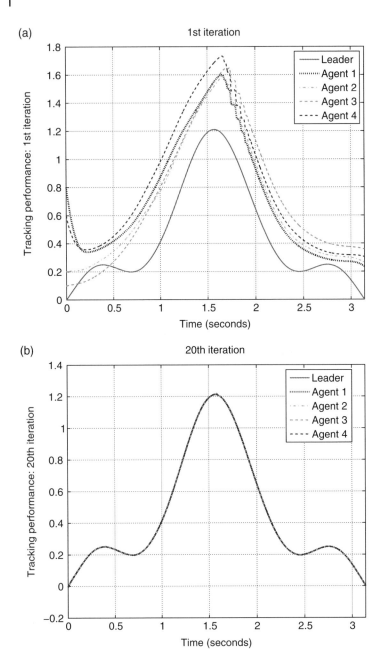

Figure 9.9 Trajectories of four agents at the 1st iteration and 20th iteration: (C_3) case.

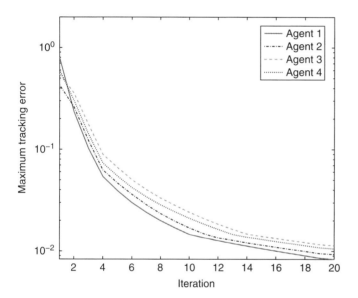

Figure 9.10 Maximum tracking error along iteration axis: (C_3) case.

(a)

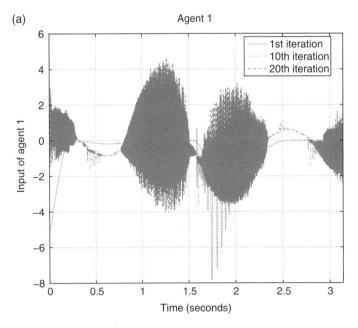

(b)

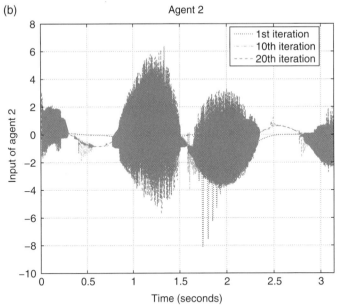

Figure 9.11 Input profiles at the 1st iteration, 10th iteration, and 20th iteration for each agent: (C_3) case.

(c)

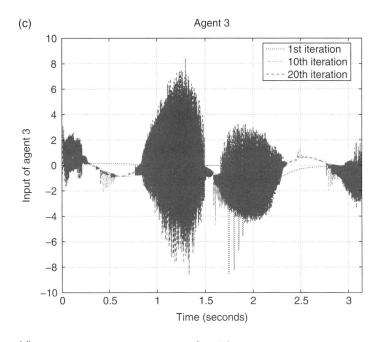

(d)

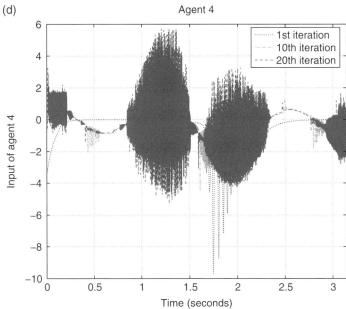

Figure 9.11 (Continued)

(a)

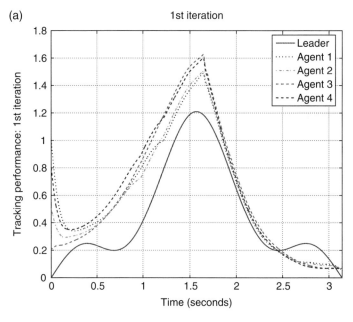

(b)

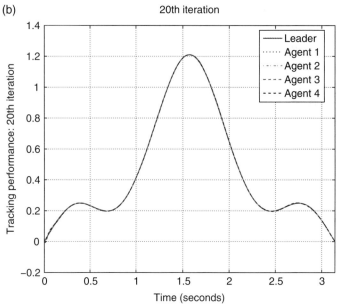

Figure 9.12 Trajectories of four agents at the 1st iteration and 20th iteration: (C_4) case.

(a)

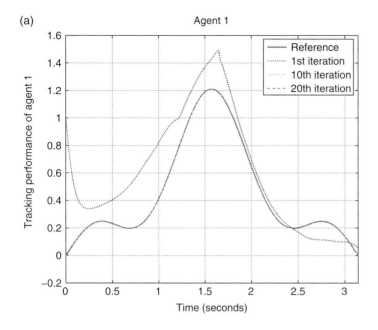

(b)

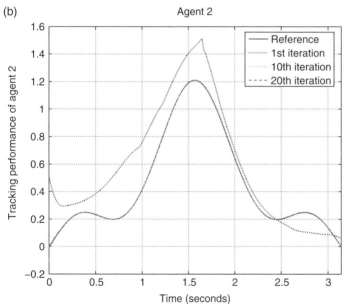

Figure 9.13 Tracking performance at the 1st iteration, 10th iteration, and 20th iteration for each agent: (C_4) case.

(c)

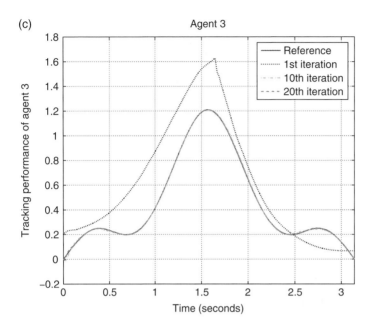

(d)

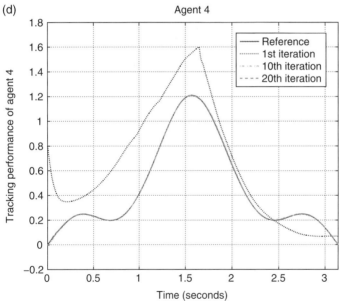

Figure 9.13 (Continued)

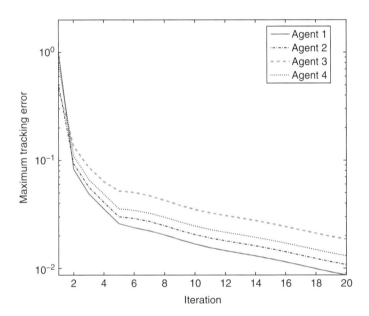

Figure 9.14 Maximum tracking error along iteration axis: (C_4) case.

(a)

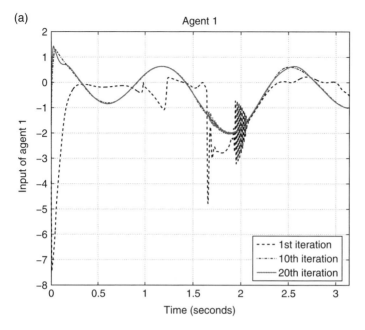

(b)

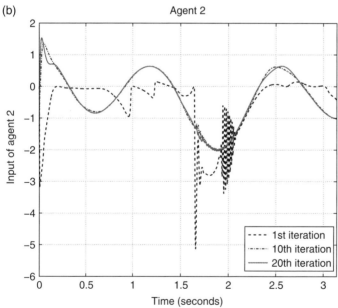

Figure 9.15 Input profiles at the 1st iteration, 10th iteration, and 20th iteration for each agent: (C_4) case.

(c)

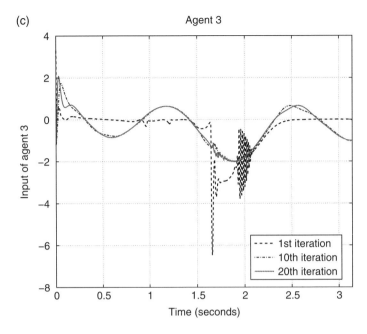

(d)

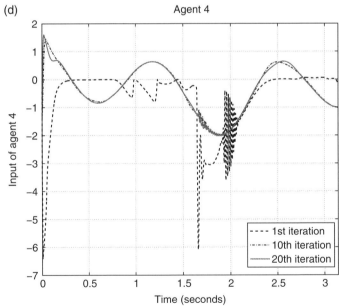

Figure 9.15 (Continued)

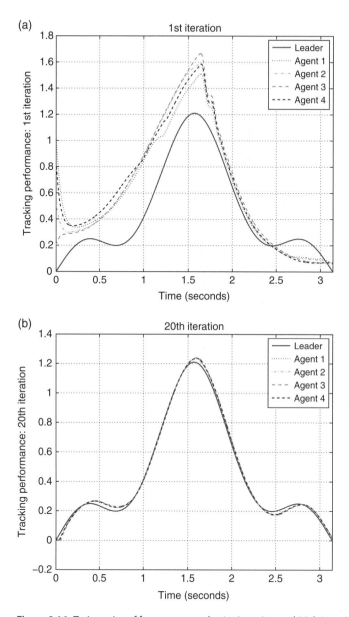

Figure 9.16 Trajectories of four agents at the 1st iteration and 20th iteration: (C_5) case.

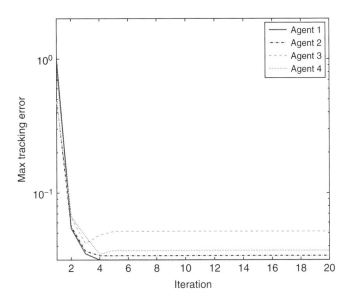

Figure 9.17 Maximum tracking error along iteration axis: (C_5) case.

(a)

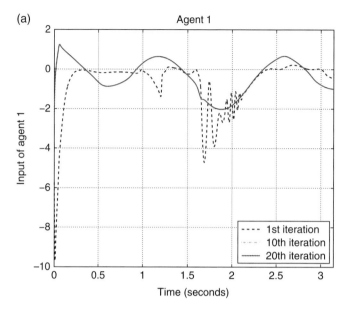

(b)

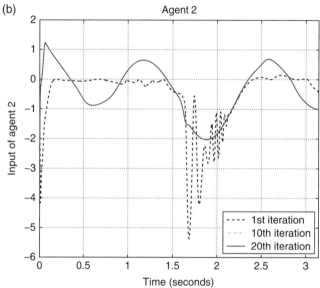

Figure 9.18 Input profiles at the 1st iteration, 10th iteration, and 20th iteration for each agent: (C_5) case.

(c)

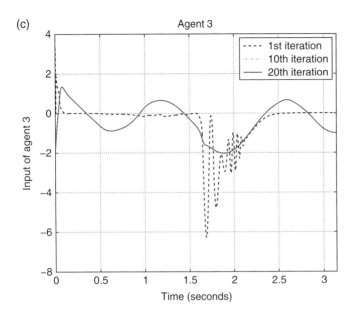

(d)

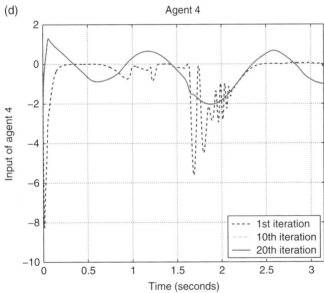

Figure 9.18 (Continued)

9.5 Conclusion

In this chapter, the nonlinear MAS with state constraints is considered under the framework of ILC. To ensure the state constraints while improving the tracking performance, a novel BLF is introduced. Five control schemes are designed in turn to address the consensus problem comprehensively from both theoretical and practical viewpoints. An original ILC scheme consisting of an iterative learning part, an unknown parameter learning part, and a robust compensating part is first investigated, where a sign function is involved to ensure zero-error convergence. Then a projection-based updating version is considered as an alternative for practical applications. The third scheme aims to provide a smooth function that approximates the sign function, anticipating improved performance, where the smooth parameter should satisfy a finite summation requirement. Next, the decreasing smooth parameter is considered to be fixed, which is much more practical for real applications. Finally, a dead-zone-like version is studied to complete the state constraints analysis. Consensus examples of nonlinear dynamic agents are presented to show the effectiveness of the developed algorithms.

10

Synchronization for Networked Lagrangian Systems under Directed Graphs

10.1   Introduction

Lagrangian systems are an important class of systems that can be used to model robotic manipulators, ground and underwater vehicles, helicopters, and satellites. Synchronization of networked Lagrangian systems has been reported in a number of publications. Leaderless synchronization algorithms, in which the final positions are constant and final velocities are zero, are presented in Hou *et al.* (2009), Hou *et al.* (2010), Ren (2009), and Min *et al.* (2011). In the leader-follower tracking scenario, the final velocity is usually time-varying, which complicates the control problem. In Zhang *et al.* (2014), an adaptive backstepping-based method is developed for followers to track a dynamic leader. However, the velocity and acceleration signals of the leader have to be available to all followers. A decentralized adaptive leader-follower control for a multi-manipulator with a similar information assumption is proposed in Cheng *et al.* (2008). In Mei *et al.* (2011), effective controllers are developed for two cases, namely, the leader having constant velocity and time-varying velocity. All the above mentioned works assume an undirected graph for communication among followers. A finite-time consensus tracking algorithm is developed in Khoo *et al.* (2009) under a directed graph when the system parameters are precisely known, by applying the terminal sliding mode technique. A similar tracking error definition to Khoo *et al.* (2009) is adopted in Chen and Lewis (2011) to synchronize a group of uncertain robot manipulators, and the universal approximation ability of a neural network is adopted to compensate for model uncertainties. Similarly, the neural network is used in Cheng *et al.* (2010) to track a dynamic leader, and a robust term is also included to counteract external disturbance and approximation error. They also explicitly discuss acyclic communication to avoid an implementation dead loop.

In this chapter, we consider the multi-agent synchronization problem by iterative learning control (ILC). Note that some robotic manipulator tracking algorithms are reported in the ILC literature (Tayebi, 2004; Sun *et al.*, 2006; Ouyang *et al.*, 2006). The problem formulation can be regarded as a single-leader-single-follower problem, where the information of the leader is known to the single follower. However, in the general MAS setup, the leader's information is usually only available to a small portion of the followers. It is currently not clear how the results in Tayebi (2004), Sun *et al.* (2006), and Ouyang *et al.* (2006) can be generalized to the networked synchronization problem. In Chapter 8, an ILC rule for the synchronization task is developed for parametric systems. However, the results rely on a symmetric graph Laplacian and do not apply to a directed graph.

Iterative Learning Control for Multi-agent Systems Coordination, First Edition.
Shiping Yang, Jian-Xin Xu, Xuefang Li, and Dong Shen.
© 2017 John Wiley & Sons Singapore Pte. Ltd. Published 2017 by John Wiley & Sons Singapore Pte. Ltd.

In this chapter, by fully utilizing the properties of Lagrangian systems, such as having a positive definite inertial matrix, being skew symmetric and linear in parameter properties, a distributed ILC rule is constructed to achieve the synchronization task for networked Lagrangian systems. The plant parameters are assumed to be unknown, and the systems dynamics are subject to bounded repeatable disturbances. The developed control rule contains three components, namely, one proportional-plus-derivative (PD) term and two learning terms. The PD term drives the tracking error to zero, one learning term compensates for the model uncertainties, and the other learning term is used for rejecting the unknown disturbance. In addition, the communication graph is directed and acyclic, which reduces the communication burden compared to an undirected graph.

The rest of this chapter is organized as follows. In Section 10.2, the problem formulation and some useful facts are presented. The learning controller design and performance analysis are conducted in Section 10.3 under the identical initialization condition. In Section 10.4, the results are generalized to the alignment condition, which is more practical. To demonstrate the efficacy of the proposed learning controller, a numerical example is presented in Section 10.5. Finally, conclusions are drawn in Section 10.6.

10.2 Problem Description

The dynamics of the followers are described by the following Lagrangian systems:

$$M_{i,j}(\mathbf{q}_{i,j}(t))\ddot{\mathbf{q}}_{i,j}(t) + C_{i,j}(\mathbf{q}_{i,j}(t), \dot{\mathbf{q}}_{i,j}(t))\dot{\mathbf{q}}_{i,j}(t) + G_{i,j}(\mathbf{q}_{i,j}(t)) = \boldsymbol{\tau}_{i,j}(t) + \mathbf{w}_j(t), \qquad (10.1)$$

where $j \in \mathcal{V}$ denotes the agent index, i represents the iteration index, t is the time argument and omitted in the following context for convenience, $\mathbf{q}_{i,j} \in \mathbb{R}^p$ is the vector of generalized coordinates, $M_{i,j}(\mathbf{q}_{i,j}) \in \mathbb{R}^{p \times p}$ is the inertial matrix, $C_{i,j}(\mathbf{q}_{i,j}, \dot{\mathbf{q}}_{i,j})\dot{\mathbf{q}}_{i,j} \in \mathbb{R}^p$ is the vector of Coriolis and centrifugal forces, $G_{i,j}(\mathbf{q}_{i,j})$ is the gravitational force, $\boldsymbol{\tau}_{i,j}$ is the control input for the jth agent, and $\mathbf{w}_j$ is the bounded and repeatable disturbance.

The system represented by (10.1) is very general, and can be used to model a large class of systems. The disturbance $\mathbf{w}_j$ is assumed to be repeatable. This is a reasonable assumption due to the fact that the disturbances are repeatable in many practical systems. For instance, the parasitic voltage ripple of power supply systems, disturbances and frictions in rotary systems, atmospheric drag and solar radiation on low earth orbit satellites. When the disturbance term is not repeatable, the sliding mode technique can be applied to reject it if it is bounded (Cheng *et al.*, 2010). As the focus of this book is on the learning control perspective, the non-repeatable disturbance is not considered in this chapter.

According to Spong *et al.* (2006), the system modeled in (10.1) has three interesting and useful properties:

1) Positive definiteness: $M_{i,j}(\mathbf{q}_{i,j})$ is uniformly positive definite for any $\mathbf{q}_{i,j}$, that is, there exist two positive constants α_j and β_j such that $0 < \alpha_j I \leq M_{i,j}(\mathbf{q}_{i,j}) \leq \beta_j I$.
2) Skew symmetric property: $\dot{M}_{i,j}(\mathbf{q}_{i,j}) - 2C_{i,j}(\mathbf{q}_{i,j}, \dot{\mathbf{q}}_{i,j})$ is skew symmetric.
3) Linear in parameters:

$$M_{i,j}(\mathbf{q}_{i,j})\mathbf{x} + C_{i,j}(\mathbf{q}_{i,j}, \dot{\mathbf{q}}_{i,j})\mathbf{y} + zG_{i,j}(\mathbf{q}_{i,j}) = Y_{i,j}(\mathbf{q}_{i,j}, \dot{\mathbf{q}}_{i,j}, \mathbf{x}, \mathbf{y}, z)\Theta_j,$$

where $\mathbf{x}, \mathbf{y} \in \mathbb{R}^p$, z is a scalar, the regressor $Y_{i,j} \in \mathbb{R}^{p \times m}$ is a known function, and $\Theta_j \in \mathbb{R}^m$ is an unknown constant matrix which represents the parameters of the system.

The leader's trajectory is given by $\mathbf{q}_0 \in C^2[0, T]$, and $\mathbf{q}_0(t)$ for $t \in (0, T]$ is only known to a few of the followers. The control task is to generate an appropriate $\tau_{i,j}$ such that perfect tracking is achieved, that is, $\lim_{i \to \infty} \left\| \mathbf{q}_{i,j} - \mathbf{q}_0 \right\| = 0$. In the multi-agent problem setup, the control input has to be constructed by the local information under the communication topology $\overline{\mathcal{G}}$, that is, $\tau_{i,j} = \tau_{i,j}(\ddot{\mathbf{q}}_{i,k}, \dot{\mathbf{q}}_{i,k}, \mathbf{q}_{i,k}), k \in \mathcal{N}_j$.

To restrict our discussion, the following assumptions about communication requirement and initialization conditions are imposed. These assumptions are extensively discussed in the previous chapters. Hence, they are not elaborated further.

Assumption 10.1   The communication graph $\overline{\mathcal{G}}$ is acyclic and contains a spanning tree with the leader being the root.

Assumption 10.2   The initial states of all followers are reset to the desired initial state after each iteration, that is, $\mathbf{q}_{i,j}(0) = \mathbf{q}_0(0), \dot{\mathbf{q}}_{i,j}(0) = \dot{\mathbf{q}}_0(0)$.

Assumption 10.3   The initial state of a follower at the current iteration is the final state of the previous iteration, namely, $\mathbf{q}_{i,j}(0) = \mathbf{q}_{i-1,j}(T), \dot{\mathbf{q}}_{i,j}(0) = \dot{\mathbf{q}}_{i-1,j}(T)$.

The following lemma is a useful result in multi-agent coordination.

Lemma 10.1 (Ren and Cao, 2011, pp. 9–10)   All eigenvalues of $L + B$ have positive real parts if and only if the graph $\overline{\mathcal{G}}$ contains a spanning tree with the leader being the root, where L is the Laplacian matrix of $\mathcal{G}$, $\mathcal{G}$ is the communication graph among followers, and $B = \mathrm{diag}(b_1, \ldots, b_N)$.

Remark 10.1   When the communication graph is undirected and connected, the graph Laplacian is symmetric positive semi-definite. This property is usually utilized to construct an appropriate Lyapunov function to facilitate controller design as we have done in Chapter 8. However, the Laplacian is asymmetric for directed graphs in general, which makes the controller design more challenging.

10.3   Controller Design and Performance Analysis

Define the actual tracking error $\mathbf{e}_{i,j} = \mathbf{q}_0 - \mathbf{q}_{i,j}$. As not all followers can obtain the tracking error, $\mathbf{e}_{i,j}$ cannot be used in the controller design. Following the convention in Khoo *et al.* (2009), the extended tracking errors are defined as follows:

$$\xi_{i,j} = \sum_{k \in \mathcal{N}_j} a_{j,k}(\mathbf{q}_{i,k} - \mathbf{q}_{i,j}) + b_j(\mathbf{q}_0 - \mathbf{q}_{i,j}), \tag{10.2}$$

$$\zeta_{i,j} = \sum_{k \in \mathcal{N}_j} a_{j,k}(\dot{\mathbf{q}}_{i,k} - \dot{\mathbf{q}}_{i,j}) + b_j(\dot{\mathbf{q}}_0 - \dot{\mathbf{q}}_{i,j}), \tag{10.3}$$

where $\boldsymbol{\xi}_{i,j}$ is the extended position tracking error, and $\boldsymbol{\zeta}_{i,j}$ is the extended velocity tracking error. In particular, $\dot{\boldsymbol{\xi}}_{i,j} = \boldsymbol{\zeta}_{i,j}$.

From (10.2) and (10.3), the extended tracking errors can be represented in compact forms as,

$$\boldsymbol{\xi}_i = ((L + B) \otimes I)\mathbf{e}_i, \tag{10.4}$$

$$\boldsymbol{\zeta}_i = ((L + B) \otimes I)\dot{\mathbf{e}}_i, \tag{10.5}$$

where $\boldsymbol{\xi}_i$, $\boldsymbol{\zeta}_i$, and $\mathbf{e}_i$ are column stack vectors of $\boldsymbol{\xi}_{i,j}$, $\boldsymbol{\zeta}_{i,j}$, and $\mathbf{e}_{i,j}$ for $j = 1, \dots, N$.

Based on Assumption 10.1 and Lemma 10.1, $L + B$ is of full rank since all the eigenvalues lie on the right half complex plane. Therefore, the minimum singular value $\underline{\sigma}(L + B) \neq 0$. As a result, it is straightforward to show that

$$\|\mathbf{e}_i\| \leq \frac{\|\boldsymbol{\xi}_i\|}{\underline{\sigma}(L + B)}, \text{ and } \|\dot{\mathbf{e}}_i\| \leq \frac{\|\boldsymbol{\zeta}_i\|}{\underline{\sigma}(L + B)},$$

by using the sub-multiplicative property of the matrix norm. Therefore, if $\boldsymbol{\xi}_i$ and $\boldsymbol{\zeta}_i$ converge to zero, $\mathbf{e}_i$ and $\dot{\mathbf{e}}_i$ converge to zero as well. Now the major issue is to design the distributed learning controllers, which drive $\boldsymbol{\xi}_i$ and $\boldsymbol{\zeta}_i$ to zero along the iteration axis. To facilitate the controller design, define the following auxiliary variable:

$$\mathbf{s}_{i,j} = \boldsymbol{\zeta}_{i,j} + \lambda\boldsymbol{\xi}_{i,j}, \tag{10.6}$$

where λ is a positive constant. Equation (10.6) can be treated as a stable filter with input $\mathbf{s}_{i,j}$. If $\|\mathbf{s}_{i,j}\| = 0$, $\boldsymbol{\xi}_{i,j}$ and $\boldsymbol{\zeta}_{i,j}$ converge to zero exponentially. In addition to $\|\mathbf{s}_{i,j}\| = 0$, if $\boldsymbol{\xi}_{i,j}(0) = 0$ and $\boldsymbol{\zeta}_{i,j}(0) = 0$, then $\boldsymbol{\xi}_{i,j}(t) = 0$ and $\boldsymbol{\zeta}_{i,j}(t) = 0$ for all $t \in [0, T]$.

From the system dynamics (10.1) and (10.6), the dynamics of the auxiliary variable $\mathbf{s}_{i,j}$, derived using the linear in parameter property, is given by

$$\begin{aligned} M_{i,j}&\dot{\mathbf{s}}_{i,j} + C_{i,j}\mathbf{s}_{i,j} \\ &= -(d_j + b_j)(\boldsymbol{\tau}_{i,j} + \mathbf{w}_j) + M_{i,j}\mathbf{x}_{i,j} + C_{i,j}\mathbf{y}_{i,j} + z_j G_{i,j} \\ &= -(d_j + b_j)(\boldsymbol{\tau}_{i,j} + \mathbf{w}_j) + Y_{i,j}\Theta_j \end{aligned} \tag{10.7}$$

where $\mathbf{x}_{i,j} = \sum_{k \in \mathcal{N}_j} a_{j,k}\ddot{\mathbf{q}}_{i,k} + b_j\ddot{\mathbf{q}}_0 + \lambda\boldsymbol{\zeta}_{i,j}$, $\mathbf{y}_{i,j} = \sum_{k \in \mathcal{N}_j} a_{j,k}\dot{\mathbf{q}}_{i,k} + b_j\dot{\mathbf{q}}_0 + \lambda\boldsymbol{\xi}_{i,j}$, and $z_j = d_j + b_j$.

Note that both $\mathbf{x}_{i,j}$ and $\mathbf{y}_{i,j}$ are distributed measurements.

The proposed controllers are

$$\boldsymbol{\tau}_{i,j} = \frac{1}{d_j + b_j}(K\mathbf{s}_{i,j} + Y_{i,j}\hat{\Theta}_{i,j}) - \hat{\mathbf{w}}_{i,j}, \tag{10.8}$$

$$\dot{\hat{\Theta}}_{i,j} = \gamma Y_{i,j}^T\mathbf{s}_{i,j}, \ \hat{\Theta}_{i,j}(0) = \hat{\Theta}_{i-1,j}(T), \ \hat{\Theta}_{0,j}(0) = 0, \tag{10.9}$$

$$\hat{\mathbf{w}}_{i,j} = \hat{\mathbf{w}}_{i-1,j} - \eta\mathbf{s}_{i,j}, \ \hat{\mathbf{w}}_{0,j} = 0, \tag{10.10}$$

where K is a positive definite matrix, γ and η are positive learning gains, $\hat{\Theta}_{i,j}$ is the estimate of constant unknown Θ_j, and $\hat{\mathbf{w}}_{i,j}$ is the estimate of time-varying but iteration-invariant disturbance $\mathbf{w}_j$. The controller (10.8) consists of three components, where $K\mathbf{s}_{i,j}$ is the proportional-plus-derivative term that is commonly used in robotics control (Spong *et al.*, 2006). $Y_{i,j}\hat{\Theta}_{i,j}$ compensates for the uncertainties in the system model, and $\hat{\mathbf{w}}_{i,j}$ is used for rejecting the repeatable disturbance. $\hat{\Theta}_{i,j}$ is updated by the differential updating rule (10.9), and $\hat{\mathbf{w}}_{i,j}$ is updated by the point-wise updating rule (10.10). As the communication graph in Assumption 10.1 contains a spanning tree with the leader being the root, $d_j + b_j \neq 0$. Therefore, the controller (10.8) is well defined.

Remark 10.2   The regressor $Y_{i,j}$ contains the acceleration signal. However, acceleration measurement is usually not available in most mechanical systems. On the other hand, the velocity signal can be easily measured by tachometers. Numerical differentiation of velocity generates large amount of noise, and should not be used for acceleration estimation. By using filtered differentiation (Slotine and Li, 1991) or the high gain observer (singular perturbation) (Khalil, 2002), a reasonable estimate of the acceleration signal can be obtained. Detailed examples can be found in Lee and Khalil (1997) and Islam and Liu (2010).

Remark 10.3   The differential updating rule (10.9) is applied since the system parameters are assumed to be constant. However, it is likely that the parameters are time-varying due to payload variations, mechanical wear, and aging. Similar to (10.10), a point-wise updating rule can be used to handle the time-varying parameters. Therefore, the point-wise updating rule for time-varying unknown parameters is not detailed in this chapter.

Theorem 10.1   Under Assumptions 10.1 and 10.2, the closed loop system consisting of (10.1) and controllers (10.8)–(10.10) can ensure that the actual tracking error $\mathbf{e}_{i,j}(t)$ converges to zero in the sense of $\mathcal{L}^2[0,T]$ norm for $j = 1, \ldots, N$ as iteration number i tends to infinity. Moreover, $\boldsymbol{\tau}_{i,j} \in \mathcal{L}^2[0,T]$ for any $j = 1, \ldots, N, i \in \mathbb{N}$.

Proof: Consider the composite energy function (CEF)

$$E_i(t) = \sum_{j \in \mathcal{V}} E_{i,j}(t), \tag{10.11}$$

where the individual energy function (EF) on agent j is defined as

$$E_{i,j}(t) = \underbrace{\frac{1}{2}\mathbf{s}_{i,j}^T M_{i,j}\mathbf{s}_{i,j} + \frac{1}{2\gamma}\tilde{\Theta}_{i,j}^T\tilde{\Theta}_{i,j}}_{V_{i,j}} + \underbrace{\frac{d_j + b_j}{2\eta}\int_0^t \tilde{\mathbf{w}}_{i,j}^T\tilde{\mathbf{w}}_{i,j}\, dr}_{U_{i,j}}, \tag{10.12}$$

where $\tilde{\Theta}_{i,j} = \Theta_j - \hat{\Theta}_{i,j}$ and $\tilde{\mathbf{w}}_{i,j} = \mathbf{w}_j - \hat{\mathbf{w}}_{i,j}$.

From the definition of E_i, it is easy to see $E_i \geq 0$. For clarity, the proof is divided into three parts. In Part I, it is shown $E_i(T)$ is non-increasing along the iteration axis. The convergence of tracking error is shown in Part II, and in Part III, the boundedness of control input energy is proven.

Part I: Calculate ΔE_i

Let $\Delta E_{i,j} = E_{i,j} - E_{i-1,j}$. As such, $\Delta E_i = \sum_{j \in \mathcal{V}} \Delta E_{i,j}$, and $\Delta E_{i,j}$ can be written as

$$\Delta E_{i,j} = V_{i,j} - V_{i-1,j} + U_{i,j} - U_{i-1,j}. \tag{10.13}$$

Together with the error dynamics (10.7), controller (10.8), and the skew symmetric property, $V_{i,j}$ can written as

$$
\begin{aligned}
V_{i,j} &= \int_0^t \dot{V}_{i,j}\, dr + V_{i,j}(0) \\
&= \int_0^t \left\{ \frac{1}{2}\mathbf{s}_{i,j}^T \dot{M}_{i,j}\mathbf{s}_{i,j} + \mathbf{s}_{i,j}^T M_{i,j}\dot{\mathbf{s}}_{i,j} + \frac{1}{\gamma}\tilde{\Theta}_{i,j}^T \dot{\tilde{\Theta}}_{i,j} \right\} dr + \frac{1}{2\gamma}\tilde{\Theta}_{i,j}(0)^T \tilde{\Theta}_{i,j}(0) \\
&= \int_0^t \left\{ -\mathbf{s}_{i,j}^T(K\mathbf{s}_{i,j} - Y_{i,j}\tilde{\Theta}_{i,j} + (d_j + b_j)\tilde{\mathbf{w}}_{i,j}) + \frac{1}{\gamma}\tilde{\Theta}_{i,j}^T \dot{\tilde{\Theta}}_{i,j} \right\} dr \\
&\quad + \frac{1}{2\gamma}\tilde{\Theta}_{i,j}^T(0)\tilde{\Theta}_{i,j}(0).
\end{aligned} \tag{10.14}
$$

By using the following equality:

$$
(\mathbf{a} - \mathbf{b})^T(\mathbf{a} - \mathbf{b}) - (\mathbf{a} - \mathbf{c})^T(\mathbf{a} - \mathbf{c}) = (\mathbf{c} - \mathbf{b})^T(2\mathbf{a} - \mathbf{b} - \mathbf{c}),
$$

the difference of $U_{i,j}$ between two consecutive iterations can be evaluated as

$$
\begin{aligned}
\Delta U_{i,j} &= U_{i,j} - U_{i-1,j} \\
&= \frac{d_j + b_j}{2\eta} \int_0^t \left\{ \tilde{\mathbf{w}}_{i,j}^T\tilde{\mathbf{w}}_{i,j} - \tilde{\mathbf{w}}_{i-1,j}^T\tilde{\mathbf{w}}_{i-1,j} \right\} dr \\
&= \frac{d_j + b_j}{2\eta} \int_0^t (\hat{\mathbf{w}}_{i-1,j} - \hat{\mathbf{w}}_{i,j})^T(2\mathbf{w}_j - \hat{\mathbf{w}}_{i,j} - \hat{\mathbf{w}}_{i-1,j}) dr \\
&= \frac{d_j + b_j}{2\eta} \int_0^t (\hat{\mathbf{w}}_{i-1,j} - \hat{\mathbf{w}}_{i,j})^T(2\mathbf{w}_j - 2\hat{\mathbf{w}}_{i,j} + \hat{\mathbf{w}}_{i,j} - \hat{\mathbf{w}}_{i-1,j}) dr \\
&= \frac{d_j + b_j}{2\eta} \int_0^t (\hat{\mathbf{w}}_{i-1,j} - \hat{\mathbf{w}}_{i,j})^T(2\tilde{\mathbf{w}}_{i,j} + \hat{\mathbf{w}}_{i,j} - \hat{\mathbf{w}}_{i-1,j}) dr \\
&\leq \frac{d_j + b_j}{\eta} \int_0^t (\hat{\mathbf{w}}_{i-1,j} - \hat{\mathbf{w}}_{i,j})^T\tilde{\mathbf{w}}_{i,j}\, dr.
\end{aligned} \tag{10.15}
$$

Using Equations (10.9), (10.10), (10.13), (10.14) and (10.15), $\Delta E_i(T)$ becomes

$$
\begin{aligned}
\Delta E_i(T) &= \sum_{j\in\mathcal{V}} \left\{ -\int_0^T \mathbf{s}_{i,j}^T K\mathbf{s}_{i,j}\, dr \right. \\
&\quad \left. + \frac{1}{2\gamma}\tilde{\Theta}_{i,j}^T(0)\tilde{\Theta}_{i,j}(0) - \frac{1}{2\gamma}\tilde{\Theta}_{i-1,j}^T(T)\tilde{\Theta}_{i-1,j}(T) - \frac{1}{2}\mathbf{s}_{i-1,j}(T)M_{i-1,j}\mathbf{s}_{i-1,j}(T) \right\} \\
&\leq -\sum_{j\in\mathcal{V}} \int_0^T \mathbf{s}_{i,j}^T K\mathbf{s}_{i,j}\, dr \leq 0.
\end{aligned} \tag{10.16}
$$

Therefore, $E_i(T)$ is non-increasing along the iteration axis.

Part II: Convergence of tracking error

Taking derivative of $E_{1,j}$ and substituting in the controllers (10.8)–(10.10) yields

$$\dot{E}_{1,j} = -\mathbf{s}_{1,j}^T(K\mathbf{s}_{1,j} - Y_{1,j}\tilde{\Theta}_{1,j} + (d_j + b_j)\tilde{\mathbf{w}}_{1,j}) + \frac{1}{\gamma}\tilde{\Theta}_{1,j}^T\dot{\Theta}_{1,j} + \frac{d_j + b_j}{2\eta}\tilde{\mathbf{w}}_{1,j}^T\tilde{\mathbf{w}}_{1,j}$$

$$= -\mathbf{s}_{1,j}^T K\mathbf{s}_{1,j} + \mathbf{s}_{1,j}^T Y_{1,j}\tilde{\Theta}_{1,j} + \frac{1}{\gamma}\tilde{\Theta}_{1,j}^T\dot{\Theta}_{1,j} - (d_j + b_j)\mathbf{s}_{1,j}^T\tilde{\mathbf{w}}_{1,j} + \frac{d_j + b_j}{2\eta}\tilde{\mathbf{w}}_{1,j}^T\tilde{\mathbf{w}}_{1,j}$$

$$\leq -(d_j + b_j)\mathbf{s}_{1,j}^T(\mathbf{w}_j + \eta\mathbf{s}_{1,j}) + \frac{d_j + b_j}{2\eta}(\mathbf{w}_j + \eta\mathbf{s}_{1,j})^T(\mathbf{w}_j + \eta\mathbf{s}_{1,j})$$

$$\leq \frac{d_j + b_j}{2\eta}\mathbf{w}_j^T\mathbf{w}_j. \tag{10.17}$$

As $\mathbf{w}_j$ is bounded, the boundedness of $\dot{E}_{1,j}$ can be concluded from (10.17). Furthermore, $E_{1,j}(t)$ can be calculated by integrating $\dot{E}_{1,j}$ over $[0, t]$, where $t \leq T$. Therefore, the boundedness of $E_{1,j}(t)$ is ensured for $t \in [0, T]$. $E_i(T)$ can be written as

$$E_i(T) = E_1(T) + \sum_{k=2}^{i}\sum_{j\in\mathcal{V}}\Delta E_{k,j}$$

$$= E_1(T) - \sum_{k=2}^{i}\sum_{j\in\mathcal{V}}\int_0^T \mathbf{s}_{k,j}^T K\mathbf{s}_{k,j}\, dr. \tag{10.18}$$

The finiteness of $E_1(T)$ and positiveness of $E_i(T)$ leads to $\lim_{i\to\infty}\int_0^T \mathbf{s}_{i,j}^T\mathbf{s}_{i,j}\, dr = 0$. Therefore, $\mathbf{e}_{i,j}$ converges to zero in the sense of $\mathcal{L}^2[0, T]$ norm.

Part III: Boundedness property

Equation (10.18) implies that $E_i(T)$ is finite. As such, $E_{i,j}(T)$ is bounded for all agents, and there exist two finite constants P_j^1 and P_j^2 such that

$$\frac{1}{2\gamma}\tilde{\Theta}_{i-1,j}^T(0)\tilde{\Theta}_{i-1,j}(0) = \frac{1}{2\gamma}\tilde{\Theta}_{i,j}^T(T)\tilde{\Theta}_{i,j}(T) \leq P_j^1,$$

$$\frac{d_j + b_j}{2\eta}\int_0^t \tilde{\mathbf{w}}_{i,j}^T\tilde{\mathbf{w}}_{i,j}\, dr \leq \frac{d_j + b_j}{2\eta}\int_0^T \tilde{\mathbf{w}}_{i,j}^T\tilde{\mathbf{w}}_{i,j}\, dr \leq P_j^2.$$

From Part I, it can be shown that

$$\Delta E_{i,j}(t) \leq \frac{1}{2\gamma}\tilde{\Theta}_{i,j}^T(0)\tilde{\Theta}_{i,j}(0) - \frac{1}{2\gamma}\tilde{\Theta}_{i-1,j}^T(t)\tilde{\Theta}_{i-1,j}(t)$$

$$- \frac{1}{2}\mathbf{s}_{i-1,j}M_{i-1,j}\mathbf{s}_{i-1,j}.$$

Therefore, $E_{i,j}(t)$ can be expressed as

$$E_{i,j}(t) = E_{i-1,j}(t) + \Delta E_{i,j}$$

$$\leq \frac{1}{2\gamma}\tilde{\Theta}_{i,j}^T(0)\tilde{\Theta}_{i,j}(0) - \frac{1}{2\gamma}\tilde{\Theta}_{i-1,j}^T(t)\tilde{\Theta}_{i-1,j}(t)$$

$$- \frac{1}{2}\mathbf{s}_{i-1,j}M_{i-1,j}\mathbf{s}_{i-1,j} + \frac{1}{2}\mathbf{s}_{i-1,j}^T M_{i-1,j}\mathbf{s}_{i-1,j}$$

$$+ \frac{1}{2\gamma} \tilde{\Theta}_{i-1,j}^T(t) \tilde{\Theta}_{i-1,j}(t) + \frac{d_j + b_j}{2\eta} \int_0^t \tilde{\mathbf{w}}_{i-1,j}^T \tilde{\mathbf{w}}_{i-1,j} \, dr$$

$$\leq \frac{1}{2\gamma} \tilde{\Theta}_{i,j}^T(0) \tilde{\Theta}_{i,j}(0) + \frac{d_j + b_j}{2\eta} \int_0^t \tilde{\mathbf{w}}_{i-1,j}^T \tilde{\mathbf{w}}_{i-1,j} \, dr$$

$$\leq P_j^1 + P_j^2.$$

Hence, $E_{i,j}(t)$ is bounded for all $t \in [0, T]$. From the definition of $E_{i,j}(t)$ in (10.12), it can be concluded that $\mathbf{s}_{i,j}$ and $\hat{\Theta}_{i,j}$ are bounded, and $\hat{\mathbf{w}}_j \in \mathcal{L}^2[0, T]$. Therefore, $\tau_{i,j} \in \mathcal{L}^2[0, T]$. ∎

Remark 10.4 In today's control systems, most controllers are realized by digital computers. In the controller design, the regressor $Y_{i,j}$ requires the acceleration signals of agents in the neighborhood $\mathcal{N}_j$ at the current time instance, which are not measurable. Therefore, the acceleration signals have to be estimated from velocity measurements. For example, pass the measured velocity through the filter $\frac{\alpha p}{p+\alpha}$, where p is the Laplace variable, and $\alpha \gg 1$ is a design parameter. Applying the zero-order hold operation, the discrete implementation of the differentiation is (Slotine and Li, 1991, p.202)

$$v_{i,j}((k+1)h) = a_1 v_{i,j}(kh) + a_2 \dot{\mathbf{q}}_{i,j}(kh),$$
$$\psi((k+1)h) = \alpha(\dot{\mathbf{q}}_{i,j}(kh) - v_{i,j}((k+1)h)),$$

where k is the discrete-time index, h is the sampling time, $v_{i,j}$ is the internal state, $\psi_{i,j}$ is the estimation of $\ddot{\mathbf{q}}_{i,j}$, $a_1 = e^{-\alpha h}$, and $a_2 = 1 - a_1$.

If the communication graph contains cycles, the acceleration signal may form closed loop dynamics, and is then likely to be unstable. This observation can be verified by numerical studies. However, stability analysis of such a sampled-data system is extremely difficult due to the nonlinear terms in the closed loop systems. To rule out such kinds of situation, the directed acyclic graph is required for communication.

Remark 10.5 In case the communication graph contains cycles, some of the edges have to be removed in order to make the graph acyclic. Alternatively, an agent simply does not use the data received from certain channels so that Assumption 10.1 is fulfilled.

If the upper and lower bounds of $\mathbf{w}_j$ are known, the projection operator can be used to achieve a stronger convergence result. The projection operator can be defined as

$$\text{proj}(z) = \begin{cases} \underline{z}^* & \text{if } z < \underline{z}^* \\ z & \underline{z}^* \leq z \leq \overline{z}^*, \\ \overline{z}^* & z > \overline{z}^* \end{cases}$$

where $\overline{z}^*$ and $\underline{z}^*$ are the upper and lower bounds of z. When the argument is a vector, the projection operator is defined element-wise. By applying the projection operator to (10.10), the controller becomes

$$\hat{\mathbf{w}}_{i,j} = \text{proj}(\hat{\mathbf{w}}_{i-1,j}) - \eta \mathbf{s}_{i,j}, \quad \hat{\mathbf{w}}_{0,j} = 0. \tag{10.19}$$

Corollary 10.1   Under Assumptions 10.1 and 10.2, the closed loop system consisting of (10.1) and controllers (10.8), (10.10), and (10.19) can ensure that the actual tracking error $\mathbf{e}_{i,j}(t)$ converges to zero uniformly for $j = 1, \dots, N$ as iteration number i tends to infinity. Moreover, $\boldsymbol{\tau}_{i,j}$ is bounded for any $j = 1, \dots, N, i \in \mathbb{N}$.

Proof: Following a similar procedure to that used in the proof of Theorem 10.1, it can be shown that $\mathbf{s}_{i,j}$ and $\hat{\Theta}_{i,j}$ are bounded. Therefore, the boundedness of $\hat{\mathbf{w}}_{i,j}$ can be concluded from (10.19). The controller (10.8) leads to the boundedness of control input $\boldsymbol{\tau}_{i,j}$. As such $\dot{\mathbf{s}}_{i,j}$ is finite. Together with (10.18), the uniform convergence of $\mathbf{s}_{i,j}$ and $\mathbf{e}_{i,j}$ can be concluded. ∎

Remark 10.6   The projection operation in (10.19) requires a priori information about the disturbance. When the upper and lower bounds of the disturbances are unknown, the upper and lower bounds can be chosen as arbitrarily large constants.

10.4 Extension to Alignment Condition

In addition to the Assumption 10.3, the desired trajectory has to be closed, that is, $\mathbf{q}_0(0) = \mathbf{q}_0(T)$ and $\dot{\mathbf{q}}_0(0) = \dot{\mathbf{q}}_0(T)$. The results in Section 10.3 are still valid when the alignment assumption is used. Although this assumption seems to be more stringent than the *iic*, it is more practical. For example, for a satellite orbiting the earth periodically, the position and speed at the beginning of the current cycle are identical to the ones at terminal time of the previous cycle.

Theorem 10.2   Under Assumptions 10.1 and 10.3, if the leader's trajectory is closed, the closed loop system consisting of (10.1) and controllers (10.8)–(10.10) can ensure that the actual tracking error $\mathbf{e}_{i,j}(t)$ converges to zero in the sense of $\mathcal{L}^2[0, T]$ norm for $j = 1, \dots, N$ as iteration number i tends to infinity. Moreover, $\boldsymbol{\tau}_{i,j} \in \mathcal{L}^2[0, T]$ for any $j = 1, \dots, N, i \in \mathbb{N}$.

Proof: Consider the same CEF as in (10.11). Since the *iic* is not assumed, $\mathbf{s}_{i,j}^T(0)M_{i,j}\mathbf{s}_{i,j}(0) \neq 0$. Similar to (10.14), $V_{i,j}$ can be expressed as

$$V_{i,j} = \int_0^t \left\{ -\mathbf{s}_{i,j}^T(K\mathbf{s}_{i,j} - Y_{i,j}\tilde{\Theta}_{i,j} + (d_j + b_j)\tilde{\mathbf{w}}_{i,j}) + \frac{1}{\gamma}\tilde{\Theta}_{i,j}^T\dot{\tilde{\Theta}}_{i,j} \right\} dr + \frac{1}{2\gamma}\tilde{\Theta}_{i,j}^T(0)\tilde{\Theta}_{i,j}(0)$$
$$+ \frac{1}{2}\mathbf{s}_{i,j}^T(0)M_{i,j}\mathbf{s}_{i,j}(0). \tag{10.20}$$

Assumption 10.3 implies that

$$\mathbf{s}_{i,j}^T(0)M_{i,j}\mathbf{s}_{i,j}(0) = \mathbf{s}_{i-1,j}^T(T)M_{i-1,j}\mathbf{s}_{i-1,j}(T).$$

Together with (10.16) and (10.20), $\Delta E_i(T)$ can be derived as

$$\Delta E_i(T) = -\sum_{j\in \mathcal{V}} \int_0^T \mathbf{s}_{i,j}^T K\mathbf{s}_{i,j} \, dr \leq 0.$$

Following the steps in the proof to Theorem 10.1, Theorem 10.2 can be concluded. ∎

Analogous to Corollary 10.1, the projection operation can also be applied under Assumption 10.3. As such, we have the corresponding Corollary 10.2.

Corollary 10.2 Under Assumptions 10.1 and 10.3, if the leader's trajectory is closed, the closed loop system consisting of (10.1) and controllers (10.8), (10.10), and (10.19) can ensure that the actual tracking error $\mathbf{e}_{i,j}(t)$ converges to zero uniformly for $j = 1, \dots, N$ as iteration number i tends to infinity. Moreover, $\boldsymbol{\tau}_{i,j}$ is bounded for any $j = 1, \dots, N$, $i \in \mathbb{N}$.

10.5 Illustrative Example

To demonstrate the effectiveness of the developed methods, consider four networked two-link robotic arms, which are modeled by the example in Slotine and Li (Slotine and Li, 1991, p.396) using

$$M_{i,j}(\mathbf{q}_{i,j})\ddot{\mathbf{q}}_{i,j} + C_{i,j}(\mathbf{q}_{i,j}, \dot{\mathbf{q}}_{i,j})\dot{\mathbf{q}}_{i,j} = \boldsymbol{\tau}_{i,j} + \mathbf{w}_j.$$

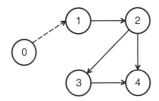

Figure 10.1 Directed acyclic graph for describing the communication among agents.

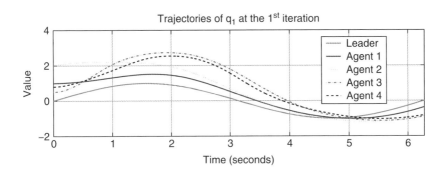

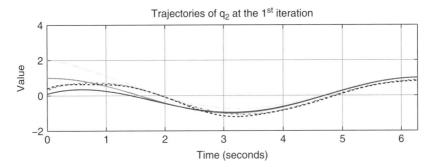

Figure 10.2 Trajectory profiles at the 1st iteration.

The detailed definitions of $M_{i,j}$ and $C_{i,j}$ are given in Slotine and Li (1991). The nominal values of the plant parameters are $m_1^0 = 1\,kg$, $l_1^0 = 1\,m$, $m_e^0 = 2\,kg$, $\delta_e = 30^\circ$, $I_1^0 = 0.12\,kg \cdot m^2$, $l_{c1}^0 = 0.5\,m$, $I_e^0 = 0.25\,kg \cdot m^2$, and $l_{ce}^0 = 0.6\,m$. To demonstrate the capability of the proposed method in dealing with heterogeneous agent systems, the actual parameters

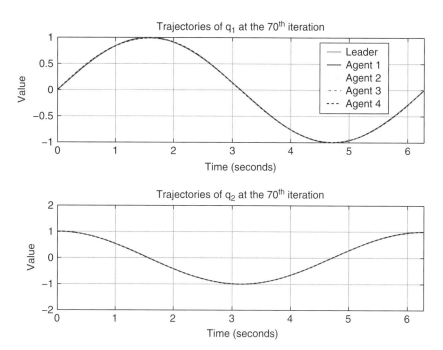

Figure 10.3 Trajectory profiles at the 70th iteration, all trajectories overlap with each other.

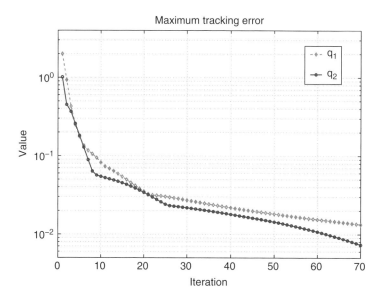

Figure 10.4 Maximum tracking error profile.

for agent j are set as follows: $\omega^j = (1 + 0.1j)\omega$, where $\omega \in \{m_1^0, l_1^0, m_e^0, I_1^0, l_{c1}^0, I_e^0, l_{ce}^0\}$. $\mathbf{w}_j$ can be used to model the iteration-invariant input disturbance, for instance the frictions in rotary motion systems. In this example, the input disturbance is set as $\mathbf{w}_j = \mathbf{1} \sin(j \cdot t)$. It is worth noting that all the parameters and disturbances are assumed to be unknown in the actual simulation. Therefore, the system parameters such as m_1^j, l_1^j, m_e^j, and so on are lumped into Θ_j, and estimated by $\hat{\Theta}_{i,j}$. Similarly, the unknown disturbance is estimated by $\hat{\mathbf{w}}_{i,j}$.

The leader's trajectory is a closed curve, and $\mathbf{q}_0 = [\sin(t), \cos(t)]^T$ for $t \in [0, 2\pi]$. The communication among the followers and the leader is described by a directed acyclic graph in Figure 10.1. Only the first follower has access to the leader's trajectory, and all other followers have to learn the desired trajectory from their neighbors. The alignment condition is assumed for the numerical study. The initial states of the followers are $\mathbf{q}_{1,1} = [1, 0.1]^T$, $\mathbf{q}_{1,2} = [2, 2]^T$, $\mathbf{q}_{1,3} = [0.5, 0.3]^T$, $\mathbf{q}_{1,4} = [0.8, 0.4]^T$, $\dot{\mathbf{q}}_{1,1} = [0.2, 0.6]^T$, $\dot{\mathbf{q}}_{1,2} = [0.4, 0.7]^T$, $\dot{\mathbf{q}}_{1,3} = [0.6, 0.3]^T$, and $\dot{\mathbf{q}}_{1,4} = [0.5, 0.5]^T$.

Controllers (10.8)–(10.10) are applied to the networked systems. There are four controller parameters to tune, namely, K, γ, η, and λ. Based on the theorems and corollaries in this chapter, if K is positive definite and all the other parameters are positive, individual tracking error asymptotically converges to zero in the sense of $\mathcal{L}^2[0, T]$ norm. However, the controller parameters directly affect the transient performance. Generally speaking, when the parameters are chosen with small values, the control profiles are smooth and the convergence rate is slow. Whereas, if the parameters have large magnitudes, the convergence rate is very fast. However, when the values go beyond certain limits, the control profiles become very oscillatory due to the fast learning in estimation of unknown terms. In the numerical study, the controller parameters are selected as $K = \text{diag}(7, 14)$, $\gamma = 1$, $\eta = 1$, and $\lambda = 1$.

The trajectory profiles of all agents at the 1st iteration are shown in Figure 10.2, where it can be seen that the followers' trajectories have large deviations from that of the leader. The developed controllers enable the followers to learn and improve their tracking performances from iteration to iteration. In contrast to Figure 10.2, the trajectory profiles at the 70th iteration are shown in Figure 10.3, where it can be seen that all the trajectories are indistinguishable. Define the maximum position errors at the ith iteration as $\max_{j \in \mathcal{V}} \|[\mathbf{e}_{i,j}]_1\|$ for the first component, and $\max_{j \in \mathcal{V}} \|[\mathbf{e}_{i,j}]_2\|$ for the second one. The maximum tracking error along the iteration profile is plotted in Figure 10.4. The maximum position errors of the first and second generalized coordinates at the 70th iteration have been respectively reduced to 0.67% and 0.72% of the ones at the 1st iteration. The initial control inputs at the 1st iteration are described in Figure 10.5. For each robotic arm, the input should reject the external disturbance and compensate for the model uncertainties. As the learning process evolves along the iteration domain, the control inputs gradually converge to the ones depicted in Figure 10.6, in which the trajectories are smooth.

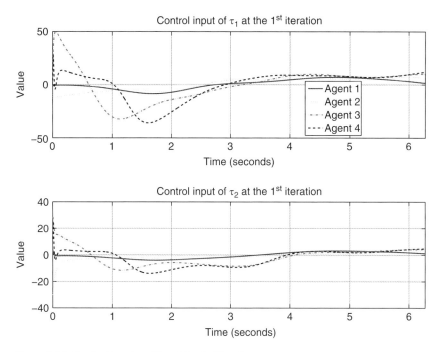

Figure 10.5 Control input profiles at the 1st iteration.

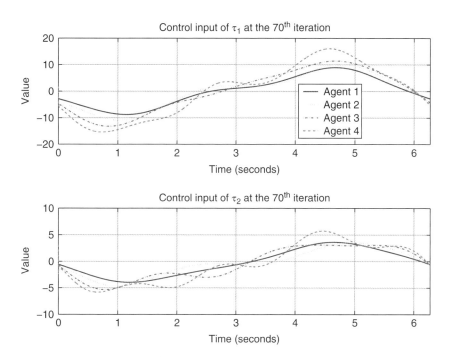

Figure 10.6 Control input profiles at the 70th iteration.

10.6 Conclusion

A leader-follower synchronization problem is formulated in the ILC framework for a group of Lagrangian systems with model uncertainties and repeatable external disturbances. By fully utilizing the properties of Lagrangian systems, learning controllers for the synchronization problem are developed under both the identical initialization condition and alignment condition. It is shown that the learning rules can effectively deal with both constant and time-varying unknowns. In contrast to many reported ILC techniques for the synchronization problem, that take advantage of the symmetric Laplacian for an undirected graph, it turns out that the directed acyclic graph is sufficient for communication among agents. A numerical study supports the theoretical results.

11

Generalized Iterative Learning for Economic Dispatch Problem in a Smart Grid

11.1 Introduction

The economic dispatch problem (EDP) is one of the important problems in power system operation (Wood and Wollenberg, 1996; Yalcinoz and Short, 1998; Liang, 1999; Madrigal and Quintana, 2000). Typically in a smart grid, there are multiple power generators and multiple loads. Each generator has a cost function and generation constraints, both of which are different from others. The objective of EDP is to find a power generation combination such that the overall cost is minimized provided that all power demands are satisfied. Traditionally this problem is solved by means of centralized optimization techniques. Many approaches have already been proposed in the literature. Wang *et al.* (1995) adopted an augmented Lagrangian relaxation method to perform short-term generation scheduling. Lee *et al.* (1998) proposed a Hopfield neural network. Bhattacharya and Chattopadhyay (2010) used a bio-geography-based optimization algorithm to solve both convex and non-convex EDP. These methods require a central station to gather the information such as power output values and cost coefficients of all the generators in order to perform the optimization. These approaches are susceptible to single point of failure, and the scalability is limited as they rely heavily on the central station's computing power and communication capability.

To solve the EDP, the equal incremental cost criterion—which means that the optimal solution without considering power line loss is achieved when the incremental costs of all generators are equal— is used as the guideline to find the optimal solution. The detailed criteria will be discussed in later sections. If we treat the incremental cost of each generator as the variable of interest, and allow generators to communicate and share their incremental costs with neighbors, then we can derive a distributed optimization method to solve EDP by applying the consensus algorithm. Thus, each generator will iteratively adjust its power generation until the equal incremental cost criterion is satisfied. The new learning mechanism can be regarded as a generalized iterative learning method.

A few approaches have been proposed in the literature following this idea. In Zhang and Chow (2012), a quadratic convex cost function is assumed for each generator. The equal incremental cost criterion is adopted to perform the optimization task. The paper uses a consensus algorithm to drive all the incremental costs to the same level. The approach is not completely distributed as it chooses leader generators to collect all the generators' power output in order to calculate the mismatch between the generation

Iterative Learning Control for Multi-agent Systems Coordination, First Edition.
Shiping Yang, Jian-Xin Xu, Xuefang Li, and Dong Shen.

and demand. Furthermore, it does not take power line loss into consideration. In Kar and Hug (2012), an innovative term is synthesized as the difference between local power generation and a fixed local reference. However, the algorithm only guarantees asymptotic convergence due to the adoption of a vanishing gain. If constant gain is selected, the equal incremental cost and demand constraints can not be satisfied simultaneously. Therefore, the algorithm only converges to the ϵ-neighborhood of the optimal dispatch.

In this chapter, we take power line loss calculation into account. The incremental cost criterion does not hold any more when considering power line loss. The consensus variable is now chosen as the product of a penalty factor and the incremental cost. There follows an overview of the algorithm structure.

Our approach is designed to have two levels of algorithm. The upper level constructs a set of distributed linear loss functions based on the previous "optimal power combination," which is obtained from the lower level algorithm. The lower level makes sure the consensus variable is driven to reach agreement among all agents while satisfying system constraints with the linear loss function, and generate a new "optimal power combination." Then, the new "optimal power combination" is fed back to the upper level algorithm. The whole process repeats until the "optimal power combination" converges. It turns out that the limit is the actual optimal power combination with loss consideration. Furthermore, each generator does not need to know the cost function parameters of other generators. The novelty of the proposed algorithm is that it can estimate the mismatch between demand, loss, and total generated power in a collective sense. With a tactical initialization, the local estimated mismatch may not equal the actual mismatch, but the summation of all the local estimated mismatches is preserved and exactly equal to the actual mismatch. The local estimated mismatches are used to adjust the power generation as if they were the true mismatch. The consensus variable is guaranteed to converge to the optimal value by the algorithm. In addition, the communication graph is assumed to be strongly connected, which is less restrictive than the bidirectional information exchange in Zhang and Chow (2011, 2012).

This chapter is organized as follows. In Section 11.2, several new terms of graph theory, discrete-time consensus results, and equal incremental cost criteria in traditional EDP are introduced. The problem description and main results for both constrained and unconstrained cases are presented in Section 11.3. Learning gain design is discussed in Section 11.4. To demonstrate the effectiveness of the proposed algorithms, numerical examples are presented in Section 11.5. Lastly, we conclude this chapter in Section 11.6.

11.2 Preliminaries

11.2.1 In-Neighbor and Out-Neighbor

In Appendix A we introduce some terminology of graph theory. In this chapter we need to distinguish the in-neighbor and out-neighbor depending on the information flow. Let $\mathcal{G} = (\mathcal{V}, \mathcal{E}, \mathcal{A})$ be a weighted directed graph with the set of vertices $\mathcal{V} = \{1, 2, ..., N\}$, and the set of edges $\mathcal{E} \subseteq \mathcal{V} \times \mathcal{V}$. $\mathcal{A}$ is the adjacency matrix. Let $\mathcal{V}$ also be the index set representing the generators in the smart grid. The in-neighbors of the ith generator are denoted by $N_i^+ = \{j \in \mathcal{V} | (j, i) \in \mathcal{E}\}$. Similarly, the out-neighbors of the ith generator are denoted by $N_i^- = \{j \in \mathcal{V} | (i, j) \in \mathcal{E}\}$. Physically, this means that a generator can obtain information from its in-neighbors, and send information to its out-neighbors. Since it is reasonable to assume that the generator i can obtain its own state information, we define that each vertex belongs to both its in-neighbors and out-neighbors, that is, $i \in N_i^+$ as well as $i \in N_i^-$. The in-degree and out-degree of vertex i are defined as $d_i^+ = |N_i^+|$ and $d_i^- = |N_i^-|$ respectively, where $|\cdot|$ denotes the cardinality of a set. It is easy to conclude that $d_i^+ \neq 0$ and $d_i^- \neq 0$ in a strongly connected graph.

11.2.2 Discrete-Time Consensus Algorithm

Let us define two matrices P, $Q \in \mathbb{R}^{N \times N}$ associated with a strongly connected graph $\mathcal{G} = (\mathcal{V}, \mathcal{E}, \mathcal{A})$:

$$p_{i,j} = \begin{cases} \frac{1}{d_i^+} & \text{if } j \in N_i^+ \\ 0 & \text{otherwise} \end{cases} \quad \forall i, j \in \mathcal{V},$$

similarly,

$$q_{i,j} = \begin{cases} \frac{1}{d_j^-} & \text{if } i \in N_j^- \\ 0 & \text{otherwise} \end{cases} \quad \forall i, j \in \mathcal{V}.$$

From the definition of P and Q, it is not difficult to verify that P is row stochastic, and Q is column stochastic (Horn and Johnson, 1985). Note that we actually have much freedom to choose the weights of P and Q, as long as P is row stochastic and Q is column stochastic, and satisfy the following assignments, $p_{i,j} > 0$ if $j \in N_i^+$, $p_{i,j} = 0$ otherwise, $q_{i,j} > 0$ if $i \in N_j^-$, $q_{i,j} = 0$ otherwise. The convergence result of our proposed algorithm is not affected by the weight selections.

Now consider the following two separate discrete-time systems:

$$\xi_i(k+1) = \sum_{j \in N_i^+} p_{i,j} \xi_j(k), \tag{11.1}$$

$$\xi_i'(k+1) = \sum_{j \in N_i^+} q_{i,j} \xi_j'(k), \tag{11.2}$$

where $\xi_i(k)$ and $\xi_i'(k)$ are state variables associated with vertex i in graph $\mathcal{G}$ at time step k. Systems (11.1) and (11.2) have the same structure but use two different sets of weights. They can be written in the compact form,

$$\xi(k+1) = P\xi(k), \tag{11.3}$$
$$\xi'(k+1) = Q\xi'(k), \tag{11.4}$$

where $\xi(k)$ and $\xi'(k)$ are the column stack vector of $\xi_i(k)$ and $\xi_i'(k)$. To investigate the asymptotic behavior of (11.3) and (11.4), the following theorem is needed.

Theorem 11.1 (Horn and Johnson, 1985, p.516) If $A \in \mathbb{R}^{N \times N}$ is a nonnegative and primitive matrix, then

$$\lim_{k \to \infty} (\rho(A)^{-1} A)^k = \mathbf{x}\mathbf{y}^T > 0$$

where $A\mathbf{x} = \rho(A)\mathbf{x}$, $\mathbf{y}^T A = \rho(A)\mathbf{y}^T$, $\mathbf{x} > 0$, $\mathbf{y} > 0$, and $\mathbf{x}^T \mathbf{y} = 1$.

The '>' symbol denotes that all the entries in a matrix or vector are greater than zero. Based on the definition, both P and Q are nonnegative and stochastic, so

$\rho(P) = \rho(Q) = 1$. Since they are derived from a strongly connected graph, and their diagonal entries are positive as well, then $P^{N-1} > 0$ and $Q^{N-1} > 0$, that is P and Q are primitive. From Theorem 11.1 we can derive the following two properties. Let $\mathbf{1}$ denote a vector of length N with all its elements being 1.

Property 11.1 $\lim_{k \to \infty} P^k = \mathbf{1}\omega^T$ where $\omega > 0$ and $\mathbf{1}^T \omega = 1$.

Property 11.2 $\lim_{k \to \infty} Q^k = \mu \mathbf{1}^T$ where $\mu > 0$ and $\mathbf{1}^T \mu = 1$.

By using Properties 11.1 and 11.2 we can get $\lim_{k \to \infty} \xi_i(k) = \omega^T \xi(0)$ in system (11.1), and $\lim_{k \to \infty} \xi_i'(k) = \mu_i \sum_{i=1}^{N} \xi_i'(0)$, where μ_i is the ith element of μ. In system (11.1), all state variables converge to a common value, which depends on the communication topology and initial state values. The algorithm in system (11.1) is the well known consensus algorithm for the first-order discrete-time system (Ren and Beard, 2008). In system (11.2), the state variables do not converge to a common value in general, but the summation of all state variable is preserved, that is, $\sum_{i=1}^{N} \xi_i'(k) = \sum_{i=1}^{N} \xi_i'(0), \forall k$.

These interesting properties will be utilized in the proposed algorithm design in Section 11.3.

11.2.3   Analytic Solution to EDP with Loss Calculation

A smart grid usually consists of multiple power generators. Assume that there are N power generators. The operating cost function of power generation is given by the following quadratic form:

$$C_i(x_i) = \frac{(x_i - \alpha_i)^2}{2\beta_i} + \gamma_i, \tag{11.5}$$

where x_i is the power generated by generator i, $\alpha_i \le 0$, $\beta_i > 0$, and $\gamma_i \le 0$.

The traditional EDP is to minimize the total generation cost

$$\min \sum_{i=1}^{N} C_i(x_i), \tag{11.6}$$

subject to the following two constraints,
Generator constraint:

$$\underline{x}_i \le x_i \le \bar{x}_i, \tag{11.7}$$

where $\underline{x}_i$ and $\bar{x}_i$ are the lower and upper bounds of the generator capability.
Demand constraint:

$$\sum_{i=1}^{N} x_i = D + L, \tag{11.8}$$

where D is the total demand and L is the total loss on the transmission lines satisfying $\sum_{i=1}^{N} \underline{x}_i < D + L < \sum_{i=1}^{N} \bar{x}_i$, that is, the problem is solvable.

The incremental cost for the generator i is $\frac{dC_i(x_i)}{dx_i} = \frac{x_i - \alpha_i}{\beta_i}$. The solution to traditional EDP with loss calculation is (Wood and Wollenberg, 1996):

$$
\begin{cases}
PF_i \frac{x_i - \alpha_i}{\beta_i} = \lambda^* & \text{for } \underline{x}_i < x_i < \bar{x}_i \\
PF_i \frac{x_i - \alpha_i}{\beta_i} < \lambda^* & \text{for } x_i = \bar{x}_i \\
PF_i \frac{x_i - \alpha_i}{\beta_i} > \lambda^* & \text{for } x_i = \underline{x}_i
\end{cases}
\tag{11.9}
$$

where λ^* is the consensus variable and PF_i is the penalty factor of unit i given by

$$
PF_i = \frac{1}{1 - \frac{\partial L}{\partial x_i}}
\tag{11.10}
$$

and $\frac{\partial L}{\partial x_i}$ is unit i incremental loss. The penalty factors are computed from losses traditionally represented using B coefficients:

$$
L = X^T B X + B_0^T X + B_{00},
\tag{11.11}
$$

where $X = [x_1, \ldots, x_N]^T$ is the vector of all generators' outputs, B is the square matrix, B_0^T is the vector of the same length as X and B_{00} is a constant.

Note that the parameter γ_i in the cost function does not affect the incremental cost.

Remark 11.1   The cost function in (11.5) is slightly different from the one in (11.12), which is commonly used by power engineers:

$$
C_i(x_i) = a_i x_i^2 + b_i x_i + c_i.
\tag{11.12}
$$

In fact, cost functions (11.5) and (11.12) are equivalent. It is not difficult to convert (11.12) to (11.5). Straightforward manipulation shows by setting $\alpha_i = -\frac{b_i}{2a_i}$, $\beta_i = \frac{1}{2a_i}$, and $\gamma_i = c_i - \frac{b_i^2}{4a_i}$, (11.5) and (11.12) are identical. The main motivation of using (11.5) is for notational simplicity in the next section.

11.3   Main Results

Let the communication topology among generators be the strongly connected graph $\mathcal{G}$ described in Section 11.2. Assume there is a command vertex which distributes the total demand D to a subset of $\mathcal{V}$. Denote the command vertex by vertex 0, and its out-neighbor set N_0^-. Recall that N_0^- is the set of vertices that can receive information from vertex 0. For simplicity, let the command vertex distribute the total demand equally among all the generators in N_0^-. By assumption, $1 \leq |N_0^-| \leq N$. In this section, we propose a two-level consensus algorithm that solves the economic dispatch problem distributively with loss consideration. The upper level algorithm estimates the power loss on the transmission lines and sends the loss information to the lower level algorithm. The lower level algorithm then learns and eliminates the mismatch information between power output, demand, and power loss in a distributed manner. After that, the results of the lower level computation are fed back to the upper level. This process repeats until there is no discrepancy between the two latest iterations.

11.3.1 Upper Level: Estimating the Power Loss

The loss on the transmission lines is computed traditionally using B coefficients:

$$L(X) = X^T B X + B_0^T X + B_{00}. \tag{11.13}$$

Note that the loss is in quadratic form of X. It is difficult to directly incorporate the nonlinear loss function to design a distributed algorithm that solves the EDP. The task of the upper level algorithm is to generate a set of distributed linear loss functions. To achieve this target, we approximate the loss function with a first-order Taylor approximation.

The gradient of $L(X)$ is

$$\mathbf{g} = [g_1, g_2, \ldots, g_N] = \left. \frac{\partial L}{\partial X} \right|_{X=X_{\text{new}}} = (2BX_{\text{new}} + B_0)^T,$$

where X_{new} is the current "optimal power combination" estimation obtained from the lower level algorithm.

The distributed linear loss function for generator i is

$$L_i(x_i) = g_i x_i + d,$$

where g_i is the ith component of $\mathbf{g}$, and

$$d = \frac{L(X_{\text{new}}) - \mathbf{g} X_{\text{new}}}{N}.$$

The loss function $L_i(x_i)$ depends only on x_i, and

$$\sum_{i=1}^{N} L_i(X_{\text{new}}^i) = L(X_{\text{new}}),$$

where X_{new}^i is the ith component of X_{new}. This implies that when each generator is working at the operation point X_{new}^i, the linear loss function is a perfect estimator of the quadratic loss function.

Then, the linear loss function is passed to the lower level algorithm, and the lower level algorithm would return a power combination X_{lower}. We update the new "optimal power combination" by the following equation:

$$X_{\text{new}} = wX_{\text{old}} + (1 - w)X_{\text{lower}}, \tag{11.14}$$

where w is the weighted value that determines how fast X_{new} is moving away from the X_{old}. If X_{new} converges, that is $X_{\text{new}} = X_{\text{old}} = X_{\text{lower}}$, then X_{new} is the actual optimal power combination with loss consideration, provided that all the constraints are satisfied.

With the linear loss function, the penalty factor in (11.10) becomes

$$PF_i = \frac{1}{1 - g_i}.$$

11.3.2 Lower Level: Solving Economic Dispatch Distributively

According to the incremental cost criterion (11.9), when all generators operate at the optimal configuration, equal incremental costs equal to the optimal value, that is,

$$PF_i \frac{x_i^* - \alpha_i}{\beta_i} = \lambda^*, \forall i \in \mathcal{V}. \tag{11.15}$$

Hence, the optimal power generation for each individual generator can be calculated if the optimal incremental cost λ^* is known, that is,

$$x_i^* = \frac{\beta_i \lambda^*}{PF_i} + \alpha_i, \ \forall i \in \mathcal{V}. \tag{11.16}$$

Equations (11.15) and (11.16) lead us to the following algorithm. Let $\lambda_i(k)$ be the estimation of optimal incremental cost by generator i, $x_i(k)$ be the corresponding power generation which is an estimation of optimal power generation, $l_i(k)$ be the corresponding loss information, and $y_i(k)$ be the collective estimation of the mismatch between demand and total power generation.

Initializations:

$$\begin{cases} \lambda_i(0) = \text{any fixed admissible value} \\ x_i(0) = \text{any fixed admissible value} \\ l_i(0) = 0 \\ y_i(0) = \begin{cases} \frac{D}{|N_0^-|} - (x_i(0) - l_i(0)) & \text{if } i \in N_0^- \\ -(x_i(0) - l_i(0)) & \text{otherwise} \end{cases} \end{cases}, \forall i \in \mathcal{V}.$$

We can now to state the main algorithm:

$$\lambda_i(k+1) = \sum_{j \in N_i^+} p_{i,j} \lambda_j(k) + \epsilon y_i(k), \tag{11.17a}$$

$$x_i(k+1) = \frac{\beta_i \lambda_i(k+1)}{PF_i} + \alpha_i, \tag{11.17b}$$

$$l_i(k+1) = g_i x_i(k+1) + d, \tag{11.17c}$$

$$y_i(k+1) = \sum_{j \in N_i^+} q_{i,j} y_j(k) + (l_i(k+1) - l_i(k)) - (x_i(k+1) - x_i(k)), \tag{11.17d}$$

where ϵ is a sufficiently small positive constant.

Remark 11.2 The iterative updating algorithm (11.17) only requires local information. Specifically, the updating rule for generator i only requires the information received from its in-neighbor set N_i^+. Hence, this updating rule is a complete distributed algorithm.

To analyze the properties and convergence of algorithm (11.17), rewrite it in the following matrix form:

$$\lambda(k+1) = P\lambda(k) + \epsilon \mathbf{y}(k), \tag{11.18a}$$

$$\mathbf{x}(k+1) = \mathbf{PF}\beta\lambda(k+1) + \alpha, \tag{11.18b}$$

$$\mathbf{l}(k+1) = \mathbf{G}\mathbf{x}(k+1) + d, \tag{11.18c}$$

$$\mathbf{y}(k+1) = Q\mathbf{y}(k) + (\mathbf{l}(k+1) - \mathbf{l}(k)) - (\mathbf{x}(k+1) - \mathbf{x}(k)), \tag{11.18d}$$

where $\mathbf{x}, \mathbf{y}, \mathbf{l}, \alpha, \lambda$ are the column stack vector of $x_i, y_i, l_i, \alpha_i, \lambda_i$ respectively, $\mathbf{PF} = \text{diag}([1/PF_1, 1/PF_2, \dots, 1/PF_N])$, $\mathbf{G} = \text{diag}([g_1, g_2, \dots, g_N])$, and $\beta = \text{diag}([\beta_1, \beta_2, \dots, \beta_N])$.

Equation (11.18d) preserves the summation of $x_i(k) + y_i(k) + l_i(k)$ over $\mathcal{V}$. This can be verified by premultiply both sides of (11.18d) by $\mathbf{1}^T$, and noticing that Q is column stochastic. We have

$$\mathbf{1}^T \mathbf{y}(k+1) = \mathbf{1}^T Q \mathbf{y}(k) + \mathbf{1}^T(\mathbf{l}(k+1) - \mathbf{l}(k)) - \mathbf{1}^T(\mathbf{x}(k+1) - \mathbf{x}(k)),$$
$$\mathbf{1}^T \mathbf{y}(k+1) = \mathbf{1}^T \mathbf{y}(k) + \mathbf{1}^T(\mathbf{l}(k+1) - \mathbf{l}(k)) - \mathbf{1}^T(\mathbf{x}(k+1) - \mathbf{x}(k)),$$
$$\mathbf{1}^T(\mathbf{y}(k+1) + \mathbf{x}(k+1)) = \mathbf{1}^T(\mathbf{l}(k+1) - \mathbf{l}(k)) + \mathbf{1}^T(\mathbf{y}(k) + \mathbf{x}(k)).$$

Noting the initialization of $x_i(0)$, $y_i(0)$ and $l_i(0)$, we can obtain $\sum_{i\in\mathcal{V}} x_i(0) + y_i(0) = D + L(0)$. The terms $\sum_{i\in\mathcal{V}}(l_i(k+1) - l_i(k))$ estimate the loss $L(k)$ on the transmission lines. Hence, $\mathbf{1}^T \mathbf{y}(k) = D + L(k) - \mathbf{1}^T \mathbf{x}(k)$ is the actual mismatch between sum of demand and loss and total power generation. The mismatch is obtained via a collective effort from all individual generators rather than a centralized method. That is the reason why $y_i(k)$ is called the collective estimation of the mismatch. The first term in the RHS of (11.18a) is the consensus part, it drives all $\lambda_i(k)$ to a common value. The second term $\epsilon y(k)$ provides a feedback mechanism to ensure $\lambda_i(k)$ converges to the optimal value λ^*. Equation (11.18b) just updates the estimated power generation $x_i(k)$ to the newest one.

Theorem 11.2 In algorithm (11.17), if the positive constant ϵ is sufficiently small, then the algorithm is stable, and all the variables converge to the solution to the traditional EDP, that is,

$$\lambda_i(k) \rightarrow \lambda^*, \ x_i(k) \rightarrow x_i^*, \ y_i(k) \rightarrow 0, \text{ as } k \rightarrow \infty, \forall i \in \mathcal{V}.$$

Proof: We use the eigenvalue perturbation approach to analyze the convergence properties. Replace $\mathbf{x}$ in (11.18c) with λ by using (11.18a) and (11.18b), we have

$$\mathbf{y}(k+1) = (Q - \epsilon \mathbf{PF}\beta + \epsilon \mathbf{GPF}\beta)\mathbf{y}(k) + (I - \mathbf{G})\mathbf{PF}\beta(I - P)\lambda(k), \tag{11.19}$$

where I is the identity matrix of appropriate dimension.

Writing (11.18a) and (11.19) in matrix form, we get the following composite system:

$$\begin{bmatrix} \lambda(k+1) \\ \mathbf{y}(k+1) \end{bmatrix} = \begin{bmatrix} P & \epsilon I \\ (I-\mathbf{G})\mathbf{PF}\beta(I-P) & Q - \epsilon \mathbf{PF}\beta + \epsilon \mathbf{GPF}\beta \end{bmatrix} \begin{bmatrix} \lambda(k) \\ \mathbf{y}(k) \end{bmatrix}. \tag{11.20}$$

Define

$$M \triangleq \begin{bmatrix} P & 0 \\ (I-\mathbf{G})\mathbf{PF}\beta(I-P) & Q \end{bmatrix}$$

and

$$E \triangleq \begin{bmatrix} 0 & I \\ 0 & -\mathbf{PF}\beta + \mathbf{GPF}\beta \end{bmatrix}.$$

The system matrix of (11.20) can be regarded as M perturbed by ϵE. M is a lower block triangular matrix; the eigenvalues of M are the union of the eigenvalues of P and Q. So M has two eigenvalues equal to 1, and the rest of its eigenvalues lie in the open unit disk on the complex plane. Denote these two eigenvalues by $\theta_1 = \theta_2 = 1$. It is easy to verify that $\mathbf{u}_1$, $\mathbf{u}_2$ and $\mathbf{v}_1^T$, $\mathbf{v}_2^T$ are the two linearly independent right and left eigenvectors of M,

$$U = [\mathbf{u}_1, \mathbf{u}_2] = \begin{bmatrix} \mathbf{0} & \mathbf{1} \\ \mu & -\eta\mu \end{bmatrix}, \tag{11.21}$$

where $\eta = [\sum_{i=1}^{N}(1 - a_i)][\sum_{i=1}^{N} \beta_i]$.

$$V^T = \begin{bmatrix} \mathbf{v}_1^T \\ \mathbf{v}_2^T \end{bmatrix} = \begin{bmatrix} \mathbf{1}^T(I - \mathbf{G})\mathbf{PF}\beta & \mathbf{1}^T \\ \boldsymbol{\omega}^T & \mathbf{0}^T \end{bmatrix}. \tag{11.22}$$

Furthermore, $V^T U = I$.

When ϵ is small, the variation of θ_1 and θ_2 perturbed by ϵE can be quantified by the eigenvalues of $V^T EU$:

$$V^T EU = \begin{bmatrix} 0 & 0 \\ \boldsymbol{\omega}^T \boldsymbol{\mu} & -\eta \boldsymbol{\omega}^T \boldsymbol{\mu} \end{bmatrix}.$$

The eigenvalues of $V^T EU$ are 0 and $-\eta \boldsymbol{\omega}^T \boldsymbol{\mu} < 0$. So $\frac{d\theta_1}{d\epsilon} = 0$ and $\frac{d\theta_2}{d\epsilon} = -\eta \boldsymbol{\omega}^T \boldsymbol{\mu} < 0$. That means θ_1 does not change against ϵ, and when $\epsilon > 0$, θ_2 becomes smaller. Let δ_1 be the upper bound of ϵ such that when $\epsilon < \delta_1$, $|\theta_2| < 1$. Since eigenvalues continuously depend on the entries of a matrix, in our particular case, the rest of the eigenvalues of $M + \epsilon E$ continuously depend on ϵ. Therefore, there exists an upper bound δ_2 such that when $\epsilon < \delta_2$, $|\theta_j| < 1, j = 3, 4, \ldots, 2N$. Hence, if we choose $\epsilon < \min(\delta_1, \delta_2)$, we can guarantee that the eigenvalue $\theta_1 = 1$ is simple, and all the rest of the eigenvalues lie in the open unit disk.

It can be verified that $\begin{bmatrix} 1 \\ 0 \end{bmatrix}$ is the eigenvector of the system matrix in (11.20) associated with $\theta_1 = 1$. Since all the rest of the eigenvalues are within the open unit disk,

$$\begin{bmatrix} \lambda(k) \\ \mathbf{y}(k) \end{bmatrix} \text{ converges to } span \begin{bmatrix} 1 \\ 0 \end{bmatrix}$$

as $k \rightarrow \infty$. That is $y_i(k) \rightarrow 0$. From (11.18c), we can derive $\mathbf{1}^T \mathbf{x}(k) = D$, that is, the demand constraint is satisfied. From (11.18a), $\lambda_i(k)$ converges to a common value, that is, the incremental cost criterion is satisfied. Therefore, we can conclude Theorem 11.2. ∎

11.3.3 Generalization to the Constrained Case

In order to take account of power generation constraints, define the following projection operators:

$$\phi_i(\lambda_i) = \begin{cases} \overline{x}_i & \text{if } \overline{\lambda}_i < \lambda_i \\ \frac{\beta_i \lambda_i}{PF_i} + \alpha_i & \text{if } \underline{\lambda}_i \leq \lambda_i \leq \overline{\lambda}_i \\ \underline{x}_i & \text{if } \lambda_i < \underline{\lambda}_i \end{cases} , \forall i \in \mathcal{V},$$

where $\underline{\lambda}_i = PF_i \frac{\underline{x}_i - \alpha_i}{\beta_i}$ and $\overline{\lambda}_i = PF_i \frac{\overline{x}_i - \alpha_i}{\beta_i}$. Now the distributed algorithm becomes

$$\lambda(k + 1) = P\lambda(k) + \epsilon \mathbf{y}(k), \tag{11.23a}$$

$$\mathbf{x}(k + 1) = \phi(\lambda(k + 1)), \tag{11.23b}$$

$$\mathbf{l}(k + 1) = \mathbf{Gx}(k + 1) + d, \tag{11.23c}$$

$$\mathbf{y}(k + 1) = Q\mathbf{y}(k) + (\mathbf{l}(k + 1) - \mathbf{l}(k)) - (\mathbf{x}(k + 1) - \mathbf{x}(k)), \tag{11.23d}$$

where

$$\phi(\lambda(k + 1)) = [\phi_1(\lambda_1(k + 1)), \phi_2(\lambda_2(k + 1)), \ldots, \phi_N(\lambda_N(k + 1))]^T.$$

The initial value of $\lambda_i(0)$ and $x_i(0)$ can be set to any admissible value. For simplicity, the initial value can be set as follows.

Initializations:

$$
\begin{cases}
\lambda_i(0) = \underline{\lambda}_i \\
x_i(0) = \underline{x}_i \\
l_i(0) = 0 \\
y_i(0) = \begin{cases} \dfrac{D}{|N_0^-|} - (x_i - l_i(0)) & \text{if } i \in N_0^- \\ -(x_i - l_i(0)) & \text{otherwise} \end{cases}
\end{cases}
, \forall i \in \mathcal{V}.
$$

Theorem 11.3 In algorithm (11.23), if the positive constant ϵ is sufficiently small, then the algorithm is stable, and all the variables converge to the solution to the traditional EDP.

Proof: By assumption, the total demand $\sum_{i=1}^{N} \underline{x}_i < D + L < \sum_{i=1}^{N} \bar{x}_i$: that means there is at least one generator is not saturated when the demand constraint is satisfied. If the smart grid operates in the linear region only, the rest of the proof follows the proof of Theorem 11.2 exactly. So we only consider the saturated case here.

The nonnegative matrix P in (11.23a) tends to map $\lambda(k)$ to the *span$\{1\}$*. Premultiplying $\mathbf{1}^T$ by both sides of (11.23a), we have

$$
\sum_i \lambda_i(k+1) = \sum_{i,j} p_{i,j}\lambda_j(k) + \epsilon e(k), \tag{11.24}
$$

where $e(k) = D + L - \sum_i x_i(k)$ is the mismatch between demand, loss, and total power generation, and ϵ can be treated as a proportional feedback gain. Without loss of generality, assume $e(k) > 0$, the overall level $\lambda_i(k)$ will increase and it approaches to the same value, and notice that the total power generation is a monotonically increasing function of incremental cost. Thus, the total power generation will increase. Therefore, the feedback mechanism in (11.24) will reduce the mismatch $e(k)$. In this process, some of the generators may reach their maximum capability. After some sufficiently long time K, if $x_i(K)$ is saturated, then $x_i(k)$ is always saturated for $k > K$. To investigate the transient behavior for $k > K$, algorithm (11.23) can be written in the following composite system:

$$
\begin{bmatrix} \lambda(k+1) \\ \mathbf{y}(k+1) \end{bmatrix} = \begin{bmatrix} P & \epsilon I \\ (I - \mathbf{G})\mathbf{PF}\tilde{\beta}(I - P) & Q - \epsilon\mathbf{PF}\tilde{\beta} + \epsilon\mathbf{GPF}\tilde{\beta} \end{bmatrix} \begin{bmatrix} \lambda(k) \\ \mathbf{y}(k) \end{bmatrix}, \tag{11.25}
$$

where $\tilde{\beta} = \text{diag}([\tilde{\beta}_1, \tilde{\beta}_2, \dots, \tilde{\beta}_N])$, and

$$
\tilde{\beta}_i = \begin{cases} 0 & \text{if } x_i(k) \text{ is saturated}, \\ \beta_i & \text{otherwise.} \end{cases}
$$

Based on our assumption, there is at least one nonzero $\tilde{\beta}_i$. Following a similar eigenvalue perturbation analysis, when ϵ is sufficiently small, the above system is stable. In addition, $\lambda(k) \to span\{1\}$, $\mathbf{y}(k) \to \mathbf{0}$, that is, it solves the traditional EDP. ∎

11.4 Learning Gain Design

Theorems 11.2 and 11.3 provide sufficient conditions for the proposed algorithms to work. However, they do not offer constructive methods to design the learning gain ϵ.

Therefore, in this section we present a systematic design method to select the appropriate learning gain.

Note that the learning gain is the same for all generators in (11.17a). However, the learning gains are not restricted to be identical. All the results in Section 11.3 are still valid when the learning gains are different from each other. Let the generators have distinct learning gains ϵ_i, then (11.17a) becomes

$$\lambda_i(k+1) = \sum_{j \in N_i^+} p_{i,j} \lambda_j(k) + \epsilon_i y_i(k).$$

Following a similar procedure as before, we can obtain the composite system below:

$$\begin{bmatrix} \lambda(k+1) \\ y(k+1) \end{bmatrix} = \begin{bmatrix} P & \Omega \\ (I - \mathbf{G})\mathbf{P}\mathbf{F}\beta(I - P) & Q - \Omega\mathbf{P}\mathbf{F}\beta + \Omega\mathbf{G}\mathbf{P}\mathbf{F}\beta \end{bmatrix} \begin{bmatrix} \lambda(k) \\ y(k) \end{bmatrix}, \quad (11.26)$$

where $\Omega = \text{diag}([\epsilon_1, \epsilon_2, \cdots, \epsilon_N])$. Denote the system matrix of (11.26) by H, that is,

$$H = \begin{bmatrix} P & \Omega \\ (I - \mathbf{G})\mathbf{P}\mathbf{F}\beta(I - P) & Q - \Omega\mathbf{P}\mathbf{F}\beta + \Omega\mathbf{G}\mathbf{P}\mathbf{F}\beta \end{bmatrix}.$$

Based on the our previous analysis, H has an eigenvalue of 1 with associated eigenvector $\pi_1 = \begin{bmatrix} 1 \\ 0 \end{bmatrix}$. If the modulus of the rest of the eigenvalues is less than 1, system (11.26) is stable, and solves the EDP. Thus, our ultimate task now is to find a suitable Ω such that H is stable.

By using the Gram–Schmidt orthonormalization process (Horn and Johnson, 1985, pp.15), together with π_1, we can generate a set of $2N - 1$ orthonormal bases $\{\pi_2, \pi_3, \cdots, \pi_{2N}\}$, where $\pi_j \in \mathbb{R}^{2N}, j = 2, 3, \cdots, 2N$. Then, we can construct a $2N \times (2N - 1)$ projection matrix

$$\Pi = \begin{bmatrix} \Pi_1 \\ \Pi_2 \end{bmatrix} = [\pi_2, \pi_3, \cdots, \pi_{2N}],$$

where $\Pi_1, \Pi_2 \in \mathbb{R}^{N \times (2N-1)}$. Applying the following projection, we have

$$\tilde{H} = \Pi^T H \Pi.$$

Hence, $\tilde{H}$ has the same set of eigenvalues as H except for the eigenvalue 1. Therefore, our task is to find an Ω such that $\rho(\tilde{H}) < 1$. This kind of problem is rather difficult to solve by the analytic method. From matrix analysis (Horn and Johnson, 1985), it is well known that

$$\rho(\tilde{H}) \leq \inf_S \|S\tilde{H}S^{-1}\|_2,$$

where S is a nonsingular matrix. Hence, if we could find an Ω so that $\inf_S \|S\tilde{H}S^{-1}\|_2 < 1$, then it is guaranteed that $\rho(\tilde{H}) < 1$. Inspired by the D-K iteration method in robust control literature (Mackenroth, 2004), we develop the following algorithm that determines an Ω that minimizes $\inf_S \|S\tilde{H}S^{-1}\|_2$, which is the greatest lower bound of $\rho(\tilde{H})$.

Step 1: *Initialize Ω, e.g., $\Omega = 0$;*

Step 2: *Find S that minimizes $\|S\tilde{H}S^{-1}\|_2$ by given Ω;*

Step 3: *Find new Ω that minimizes $\|S\tilde{H}S^{-1}\|_2$, where S is obtained from* Step 2. *Then, goto* Step 2.

The above algorithm is essentially an iterative numerical method which mimics the D-K iteration method. As remarked in Mackenroth (2004), up to now, it has not been possible to prove the convergence of the D-K iteration method. However, in practical applications, it takes only a few iterations to find a solution that is nearly optimal.

The minimization problem in Step 2 is equivalent to the generalized eigenvalue problem below in the field of linear matrix inequality (LMI):

$$\underset{S}{\text{minimize}} \quad t$$

$$\text{subject to} \quad 0 \prec S^T S,$$

$$\tilde{H}^T S^T S \tilde{H} \prec t S^T S.$$

The problem in Step 3 can be converted to the following LMI problem by using the Schur Complement:

$$\underset{\Omega}{\text{minimize}} \quad r$$

$$\text{subject to} \quad \begin{bmatrix} -r S^T S & \star \\ S(H_l + \Pi_1^T \Omega \Pi_2 - \Pi_2^T \Omega(I - G)\beta \Pi_2) & -I \end{bmatrix} \prec 0,$$

where $H_l = \Pi_1^T P \Pi_1 + \Pi_2^T (I - G)\beta(I - P)\Pi_1 + \Pi_2^T Q \Pi_2$, and $\star$ represents the corresponding symmetric component in the matrix.

These two LMI problems can be effectively handled by numerical software packages, for instance, MATLAB LMI control toolbox.

Remark 11.3 To design an appropriate learning gain Ω that ensures convergence, the proposed design method requires detailed information of communications P, Q, as well as the generator parameter β. P, Q, and β are fixed values, and they can be obtained by offline methods. Once the learning gain is calculated, P, Q, and β are no longer needed in the implementation. Similarly, a generator only requires the power generation information from its neighbors at the implementation stage. The time-varying global information, such as total power generation, is not required.

11.5 Application Examples

The communication topology is shown in Figure 11.1. We assume that there are four generators in this smart grid, labeled as vertices 1, 2, 3, and 4. The four generators are selected from the three types of generators in Table 11.1. Vertices 1 and 2 are Type A generators, vertex 3 is a Type B generator, and vertex 4 is a Type C generator. The communication among the four generators are denoted by solid lines, which is a strongly connected graph. Vertex 0 is the command vertex. The dashed line represents the communication between command vertex and power generators, that is the command vertex can send information to power generators 1 and 3. Based on Figure 11.1, matrices P and Q can be defined as

$$P = \begin{bmatrix} \frac{1}{2} & 0 & 0 & \frac{1}{2} \\ \frac{1}{3} & \frac{1}{3} & \frac{1}{3} & 0 \\ 0 & 0 & \frac{1}{2} & \frac{1}{2} \\ 0 & \frac{1}{2} & 0 & \frac{1}{2} \end{bmatrix}, \quad Q = \begin{bmatrix} \frac{1}{2} & 0 & 0 & \frac{1}{3} \\ \frac{1}{2} & \frac{1}{2} & \frac{1}{2} & 0 \\ 0 & 0 & \frac{1}{2} & \frac{1}{3} \\ 0 & \frac{1}{2} & 0 & \frac{1}{3} \end{bmatrix},$$

Figure 11.1 Communication topology among generators and command vertex in the network.

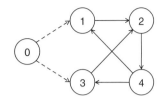

Table 11.1 Generator Parameters.

Generator Type	A (Coal-fired)	B (Oil-fired)	C (Oil-fired)
Range (MW)	[150, 600]	[100, 400]	[50, 200]
a ($\$/MW^2h$)	0.00142	0.00194	0.00482
b ($\$/MWh$)	7.2	7.85	7.97
c ($\$/h$)	510	310	78
α (MW)	−2535.2	−2023.2	−826.8
β ($MW^2h/\$$)	352.1	257.7	103.7
γ ($\$/h$)	−8616.8	−7631.0	−3216.7

respectively. The power lines loss matrix is defined as

$$B = \begin{bmatrix} 0.00676 & 0.00953 & -0.00507 & 0.00211 \\ 0.00953 & 0.0521 & 0.00901 & 0.00394 \\ -0.00507 & 0.00901 & 0.0294 & 0.00156 \\ 0.00211 & 0.00394 & 0.00156 & 0.00545 \end{bmatrix},$$

$$B_0 = \begin{bmatrix} -0.0766 \\ -0.00342 \\ 0.0189 \\ 0.0173 \end{bmatrix}, B_{00} = 0.0595.$$

The initial values in case studies 1–3 are given in Table 11.2, w is chosen to be 0.8, and total demand $D = 1300MW$.

11.5.1 Case Study 1: Convergence Test

In this case study, we want to verify the convergence property. The results are shown in Figure 11.2. The top subfigure shows the estimated power loss and actual power loss calculated by the upper level. The estimated power loss reaches a steady state and the value is equal to the actual calculated loss after 14 iterations. This demonstrates the convergence of the upper level algorithm. The other four subfigures at the bottom show the dynamics of the lower level algorithm. The collective estimated mismatch y_i goes to zero after 30 iterations. The estimated λ of all generators converge to the same value while meeting the power balance constraint. From the transient response, we notice that generators 1, 3, and 4 are saturated after the first 25 iterations, and they gradually increase to the final output as λ increases. Based on the results, $L = 76.02MW$, $\lambda^* = 1246.3$, $x_1^* = 600MW$, $x_2^* = 176.01MW$, $x_3^* = 400MW$, and $x_4^* = 200MW$. All the

Table 11.2 Initializations.

Variables	$i = 1$	$i = 2$	$i = 3$	$i = 4$
$x(0)$ (MW)	150	150	100	50
$y(0)$ (MW)	600	-150	650	-50
$l(0)(MW)$	0	0	0	0
$\lambda(0)$	736.29	957.38	906.46	871.24

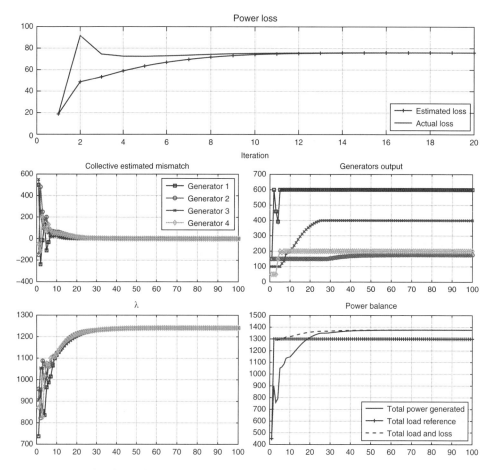

Figure 11.2 Results obtained with generator constraints.

power generations are within the generation ranges. No power generators exceed the operational ranges even in their transient responses.

11.5.2 Case Study 2: Robustness of Command Node Connections

Our proposed algorithm only requires that the communication between generators is strongly connected. If the command vertex has at least one edge to any one of the

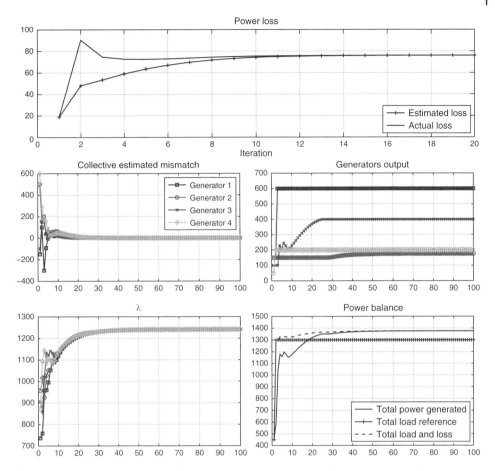

Figure 11.3 Robustness test when the command vertex is connected to generators 2 and 4.

generators, the algorithm works when the learning gain is appropriately chosen. This is a very flexible communication condition. In the previous case study, the command vertex is connected to generators 1 and 3. Now, change the command vertex connection to generators 2 and 4. Figure 11.3 shows the numerical results. The initial conditions for $x_i(0)$ are still the same as in Table 11.2. The initial conditions for $y_i(0)$ are changed to $y_1(0) = -150MW$, $y_2(0) = 500MW$, $y_3(0) = -100MW$, and $y_4(0) = 600MW$, and the final outcomes are $L = 76.02MW$, $\lambda^* = 1246.3$, $x_1^* = 600MW$, $x_2^* = 176.01MW$, $x_3^* = 400MW$, and $x_4^* = 200MW$. The results are identical to Case study 1, which indicates that the command vertex connection does not affect the final convergence results.

11.5.3 Case Study 3: Plug and Play Test

One of the most important features of a smart grid is its plug and play adaptability. In this case study, we add an additional generator to the grid. The four generators have already reached their optimal states before plugging in the fifth generator. The four generators' steady states are the same as shown in Case 1. The fifth generator is plugged in at time

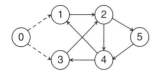

Figure 11.4 Communication topology with the fifth generator.

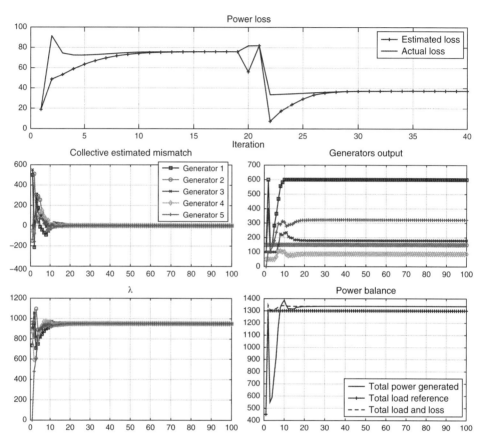

Figure 11.5 Results obtained with the fifth generator.

step $k = 20$. The fifth generator is a Type B generator. The initializations of the generators are the same as in Case 1. The new communication topology is shown in Figure 11.4. Thus matrices P and Q are changed to

$$P = \begin{bmatrix} \frac{1}{2} & 0 & 0 & \frac{1}{2} & 0 \\ \frac{1}{3} & \frac{1}{3} & \frac{1}{3} & 0 & 0 \\ 0 & 0 & \frac{1}{2} & \frac{1}{2} & 0 \\ 0 & \frac{1}{3} & 0 & \frac{1}{3} & \frac{1}{3} \\ 0 & \frac{1}{2} & 0 & 0 & \frac{1}{2} \end{bmatrix}, \quad Q = \begin{bmatrix} \frac{1}{2} & 0 & 0 & \frac{1}{3} & 0 \\ \frac{1}{2} & \frac{1}{3} & \frac{1}{2} & 0 & 0 \\ 0 & 0 & \frac{1}{2} & \frac{1}{3} & 0 \\ 0 & \frac{1}{3} & 0 & \frac{1}{3} & \frac{1}{2} \\ 0 & \frac{1}{3} & 0 & 0 & \frac{1}{2} \end{bmatrix},$$

respectively. After plugging in the fifth generator at $k = 100$, the output of generator 5 is set to $x_5(20) = 100MW$ and $y_5(20) = -100MW$. From the results in Figure 11.5, we can observe that the local estimated mismatch y_i goes to zero after a short disturbance. The other three generators reduce their outputs in order to accommodate the fifth generator output. Therefore, the incremental cost drops due to lower average output and lower loss on the transmission lines. Finally, the estimated incremental costs of all generators converge to the same value while meeting the power balance constraint. Based on the results, $L = 37.27MW$, $\lambda^* = 950.68$, $x_1^* = 600MW$, $x_2^* = 150MW$, $x_3^* = 178.94MW$, and $x_4^* = 86.80MW$, and $x_5^* = 321.53MW$. All the power generations are within the generation range. The optimization goal is fulfilled and the fifth generator is well adapted into the system.

11.5.4 Case Study 4: Time-Varying Demand

In this case study, the generators' setup is the same as in Case 1. In a practical situation, it is very likely that the demand is not a constant over time. Slight modification of our proposed algorithm can handle the demand change effectively. Let the initial demand

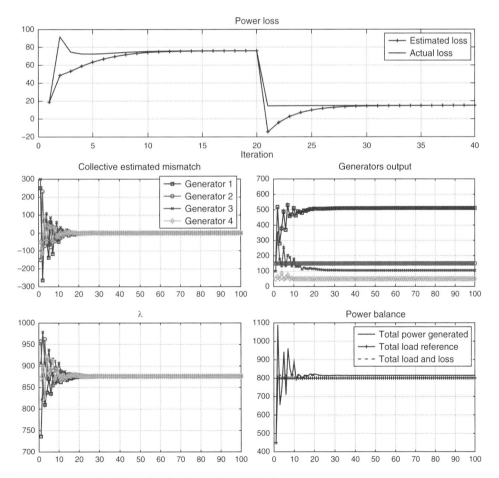

Figure 11.6 Results obtained with time-varying demand.

$D = 1300MW$ as usual,=. We then change it to $D' = 800MW$ at upper loop time step $k = 20$. The demand change is only known to generators 1 and 3 since they are connected to the command vertex. Thus, the algorithm needs to modify the local estimated mismatch at $k = 20$ before continuing to update the variables. Keep y_2, y_4 unchanged at $k = 20$, and update y_1, y_3 as follows at $k = 20$ before proceeding to the next updating iteration:

$$y_1(20) = y_1(20) + \frac{D' - D}{m},$$

$$y_3(20) = y_3(20) + \frac{D' - D}{m}.$$

Figure 11.6 shows the results. After the demand changes at $k = 20$, the algorithm asymptotically converges to the new optimal solution, that is, $\lambda^* = 875.89$, $x_1^* = 510.32MW$, $x_2^* = 150MW$, $x_3^* = 104.66MW$, $x_4^* = 50.00MW$, and $L = 14.98MW$. Compared to Case 1 results, all the outputs decrease since the demand is reduced.

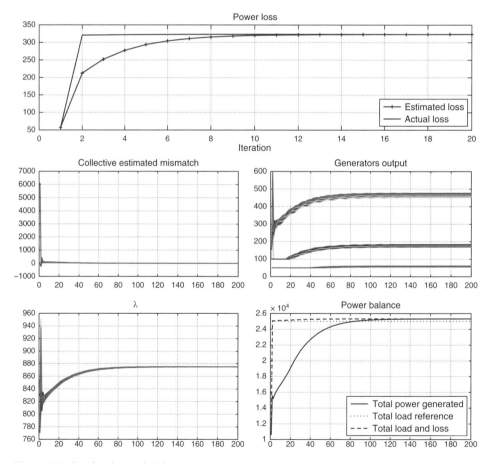

Figure 11.7 Results obtained with 100 generators.

11.5.5 Case Study 5: Application in Large Networks

This case study demonstrates the scalability of the proposed algorithm to very large networks. Consider a smart grid having 100 generators. The generators are randomly selected from the three types of generator in Table 11.1. The communication is defined as below. Let the out-neighbors of generator i be $N_i^- = \{\text{mod }(i + k, 100) \mid k = 0, 1, \dots, 20\}$, that is, the communication is a special circular graph, which is strongly connected. Let the command vertex 0 communicate with generators 25, 50, 75, and 100. In the case study, the learning gain $\epsilon = 2$. The initializations of $x_i(0), \lambda_i(0), y_i(0)$ are determined by the method in Theorem 11.3. The target demand $D = 25000MW$. Figure 11.7 shows the numerical results. The generator outputs are confined within the operational ranges. The collective estimated mismatches converge to zero. Total power generated by the smart grid converges to the target demand, and all the incremental costs converge to the same value, that is, the EDP is solved. In this particular case, the optimal incremental cost is $\lambda^* = 878.89$, Type A generator outputs are near $x_A = 480MW$, Type B generator outputs are $x_B = 180MW$, and Type C generator outputs are $x_C = 62MW$.

11.5.6 Case Study 6: Relation Between Convergence Speed and Learning Gain

The learning gain is the only design parameter that we can manipulate, and it plays a significant role in convergence speed and also the performance. If the learning gain is

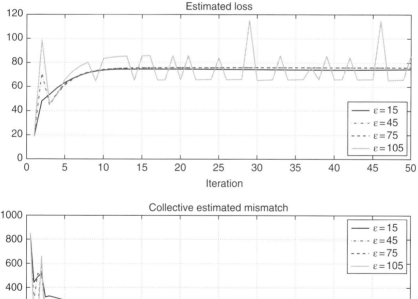

Figure 11.8 Relation between convergence speed and learning gain.

not properly selected, the results may oscillate, or even diverge. Thus, it is important to investigate the relation between convergence properties and learning gain. We select four sets of learning gains, that is, $\epsilon = 15, 45, 75$, and 105, and study the performance of the system with these learning gains. Let the initial conditions be the ones specified in Table 11.2. The mismatch between demand and total power generation is depicted in Figure 11.8. In general, when the learning gain is small, the convergence speed is relatively slow, and the transient response is smooth. A large gain learning results in a faster convergence speed. However, the performance may become oscillatory. In addition, if the learning goes beyond certain limit, the algorithm diverges, as in the experiment with $\epsilon = 105$.

11.6 Conclusion

In this chapter, a two-level consensus based iterative learning algorithm is proposed to solve the economic dispatch problem with loss calculation. The operating cost functions are modeled as quadratic functions. The loss information on the transmission lines is modeled using B coefficients. The upper level algorithm estimates the loss information and passes it to the lower level. The lower level algorithm drives the product of penalty factors and incremental cost to reach an agreement among all agents. Strongly connected communication is used to reduce the communication overhead. Several case studies show that the total operating cost can be minimized in a distributed manner while maintaining the power balance constraint. The proposed technique can also take care of demand variations and generator variations. In addition, the scalability of the technique is proved by solving a large network problem.

12

Summary and Future Research Directions

12.1 Summary

This book studies the multi-agent coordination and control problem from the iterative learning control (ILC) perspective. Consensus tracking is one of the most important coordination problems in multi-agent systems, and various coordination problems can be formulated and solved in the consensus framework, for example the formation control of multiple aerial vehicles, rendezvous of ground robots, sensor fusion, area coverage by multiple robots, and distributed optimization. Therefore, consensus tracking is chosen as the main problem studied in this book.

We follow the two main approaches in ILC, namely contraction-mapping (CM) and composite energy function (CEF)-based methods, to investigate the consensus tracking problems. First, the consensus tracking problem is formulated for a group of global Lipschitz continuous agents under the fixed communication assumption and perfect initialization condition. Distributed D-type iterative learning rules are developed for all of the followers to track a dynamic leader agent. Next, we relax the communication assumption by considering intermittent interactions among agents. It is shown that consensus tracking can be achieved under the very mild condition that the communication graph is uniformly strongly connected along the iteration axis. Furthermore, to remove the stringent initialization condition, we only assume that follower agents are reset to the same initial state at the beginning of every iteration, which may not be the same as the leader's initial state. It is shown that the previous D-type learning rule is still convergent. However, the final tracking trajectories may have large deviations from the leader's.

To improve the control performance under the imperfect initialization condition, a PD-type learning rule is developed for multi-agent systems. The PD-type learning rule has two advantages. On the one hand, it can ensure learning convergence. On the other hand, it offers the controller designer additional freedom to tune the final tracking performance. In addition, we introduce the idea of an input sharing mechanism into ILC. By sharing the learned information among agents in the network, the new learning controller not only improves convergence speed, but also smooths the transient performance. Then, we study an iterative switching formation problem. The geometric configurations between two consecutive iterations are governed by a high-order internal model (HOIM). By incorporating the HOIM into the learning rule, we successfully design learning rules to achieve the formation task.

The results summarized above are derived based on the CM method; hence, they are only applicable to linear or global Lipschitz nonlinear systems. The CM method for local

Iterative Learning Control for Multi-agent Systems Coordination, First Edition.
Shiping Yang, Jian-Xin Xu, Xuefang Li, and Dong Shen.
© 2017 John Wiley & Sons Singapore Pte. Ltd. Published 2017 by John Wiley & Sons Singapore Pte. Ltd.

Lipschitz systems has improved little over the past 30 years. The traditional λ-norm fails to construct a contraction-mapping for local Lipschitz systems as it is difficult to get a reasonable estimate of the upper bound of system state. By combining Lyapunov analysis and CM analysis methods, we can show that CM-based ILC can be applied to several classes of local Lipschitz systems such as stable systems with quadratic Lyapunov functions, exponentially stable systems, systems with bounded drift terms, and uniformly bounded energy bounded state systems under control saturation. This part of the discussion greatly complements the existing ILC literature, and also implies that the learning rules developed for multi-agent systems can be applied on a large scale to nonlinear systems.

Next, CEF-based ILC is adopted to study local Lipschitz systems whose dynamics can be linearly parameterized and full state information can be used for feedback. The controller design requires undirected communication. Initial rectifying action control is employed to deal with the imperfect initialization condition. The controller is applicable to high-order systems and able to reject state-dependent disturbances. This part of the results requires a symmetric Laplacian, namely the undirected communication graph.

Finally, we develop a set of distributed learning rules to synchronize networked Lagrangian systems under a directed acyclic graph. The learning rules fully utilize the properties of Lagrangian systems, and the CEF approach is adopted to analyze their convergence. Besides continuous-time systems, we also discuss a generalized iterative learning form of discrete-time systems for multi-agent coordination. The new learning algorithm takes advantage of the discrete-time consensus algorithm and is used to solve the economic dispatch problem in a smart grid environment.

To recap, a number of ILC design and analysis methods are introduced to deal with MAS under different types of nonlinearities and uncertainties. All important topics in this book are summarized in Table 12.1 for easy reference.

12.2 Future Research Directions

This book develops the general design principle for multi-agent coordination using the ILC approach, and investigates various problems in a systematic manner. Some specific problems like switching communication, imperfect initialization condition, and local Lipschitz continuous systems are discussed. Although many interesting results are presented, there are many other issues remaining unsolved not only in the multi-agent setup but also in ILC aspects. Here we list some of the open problems including both theoretical and applications issues.

12.2.1 Open Issues in MAS Control

1. PID-type ILC

In order to deal with imperfect initialization conditions, a PD-type learning rule is developed in Chapter 4. By using a similar idea and mathematical technique, it is possible to extend the PD-type learning rule to a PID-type learning rule. The PID-type learning rule offers more degrees of freedom to tune the tracking performance.

2. Extension of Input Sharing ILC

In Chapter 5, an input sharing mechanism is developed for a group of identical linear agent systems to perform leader-follower consensus tracking using the CM approach.

Table 12.1 The topics covered in this book.

Topics	Chapters
CM-based ILC	Chapters 2–7
CEF-based ILC	Chapters 8–10
Directed graph	Chapters 2–6, 10, 11
Undirected graph	Chapters 2, 8, 9
Iteration-varying graph	Chapters 3, 5
Identical initialization condition (*iic*)	Chapters 2–8, 10
Alignment condition	Chapters 8–10
Global Lipschitz condition (GLC)	Chapters 2–6
Local Lipschitz condition (LLC)	Chapters 7, 8
Input sharing	Chapter 5
D-type ILC	Chapters 2–6
P-type ILC	Chapter 7
PD-type ILC	Chapter 4
High-order internal model (HOIM)	Chapter 6
Linear matrix inequality (LMI)	Chapters 5, 11

The traditional ILC rule has only one learning resource, that is the correction term. The input sharing mechanism allows the followers to share their learned information with their neighbors. It is likely that additional learning resources will enhance the learning experience. In Chapter 5, numerical examples demonstrate that the input sharing mechanism can not only increase the convergence rate, but also smooth the transient performance. However, quantitative performance improvement is yet to be examined. Furthermore, it would be worthwhile to generalize the input sharing to the CEF-based ILC rule, which is applicable to local Lipschitz systems. In addition, the idea of input sharing is new and may lead to other novel controllers.

3. Leader's HOIM Not Accessible by All

In the problem description of Chapter 6, the leader's trajectory is assumed to be iteration-invariant. When the leader's trajectory changes from iteration to iteration, the existing learning rule does not work any more. Tracking an iteration-varying reference trajectory is still a challenging problem in the ILC literature. If the reference trajectories can be characterized by a high-order internal model (HOIM) over the iteration domain, the HOIM can be incorporated in the learning rule design, for example Liu *et al.* (2010). This approach has been applied in the multi-agent coordination problem in Chapter 6. However, all the followers are required to know the HOIM—the HOIM is regarded as global information. Such an approach conflicts with the spirit of distributed controller design because only a few of the followers can access the leader's information. A learning rule that does not require global access to the HOIM is desirable.

4. Extension to Local Lipschitzian

In Chapter 7, we show that the P-type learning rule can be applied to several local Lipschitz systems. One sufficiency condition says that if the unstable factor in the unforced systems satisfies the global Lipschitz condition, then a P-type rule is applicable to the system. At this stage, we do not know whether the controller converges or not when the unstable factor only satisfies the local Lipschitz condition. Based on our numerical study, a P-type rule may converge in certain cases. However, we are unable to identify a general result in this case. In addition, the D-type learning rule for local Lipschitz systems remains unknown and is worth investigation.

5. CEF-based ILC for Generic Graph

ILC is a kind of partial model-free control method. However, it is a waste if we do not use the inherent system properties when they are known. For example the Lagrangian system has three useful properties, linear in parameter, positive definiteness, and skewed symmetric property. It is possible that we can take advantages of these features to improve control performance. In Chapter 10 the three properties are fully utilized in the learning rule design. Unlike many existing results which rely on symmetric interaction topology, the controller works under a directed graph. However, the graph is required to be acyclic. Otherwise, it may lead to instability problems in the learning rule. It is an open problem to design a learning rule that works under a general directed graph using the CEF approach. The main challenge is to find a suitable energy function which facilitates distributed learning rule design.

6. Constrained MAS Control

There are some major challenges in essentially all kinds of control problems; one such challenge is the presence of physical system constraints. Physical systems always have practical limits in some or all states. For instance, in motion control systems, it is common that system outputs, like position and velocity, have to be bounded within certain ranges in order to ensure safety and performance. In electrical systems, designers usually introduce current limiters or general protective circuits to avoid overloading. In process control, for example, temperature or pressure in a batch reactor should be kept within a range.

Nearly all practical real world systems, including MAS, are subject to constraints in one way or the other. Violation of such constraints may result in degradation of MAS performance, or even hazards in the worst case. Therefore, the effective handling of constraints has become an important research topic. In Sun *et al.* (2010) and Xu *et al.* (2004), learning control with saturation on input control signals is discussed using the Composite Energy Function (CEF) approach, for parametric and nonparametric uncertainties. However, in some physical systems, it is commonly required that system outputs are bounded within certain ranges. State or output constrained problems have been discussed for other control topics such as adaptive control. In Ngo *et al.* (2005) and Tee *et al.* (2007, 2011, 2009), barrier Lyapunov functions (BLF) have been proposed to deal with such situations. A BLF is a Lyapunov-like function that grows to infinity as the arguments approach certain limits. By keeping the BLF bounded through analysis in the closed loop system, the constraint on the argument will not be violated. In the area of ILC, effective control schemes to ensure output boundedness have also been explored with the BLF approach (Jin and Xu, 2014, 2013; Xu and Jin, 2013). It would be of practical value to extend constrained ILC to MAS control problems. ILC with input constraints, state constraints, output constraints, and ILC based MAS, could be integrated, designed and analyzed under the framework of CEF.

Figure 12.1 The schematic diagram of an agent under networked control. $\mathbf{y}_i$, $\tilde{\mathbf{y}}_i$, and $\mathbf{y}_d$ are plant output, output signal received at the ILC side, reference output, respectively. $\mathbf{u}_i$ is the control profile generated by the ILC mechanism.

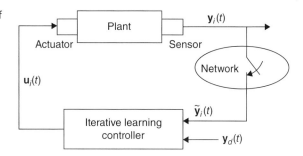

7. MAS with Networked Control

In most MAS explored hitherto, an agent is the integration of a physical plant and a controller. The communication graph describes the information exchange among controllers. In practical MAS, there is a possibility that the physical plant and the controller of an agent are separated and linked through a communication network, which is known as a networked control system. Such scenarios can be seen in cyber-physical network systems as well as any systems with remote control. A further generalization could include situations where sensors, actuators, controllers, and the physical plants are all distributed. Figure 12.1 demonstrates one simple case in which the plant is integrated with sensor and actuator, while the ILC mechanism is separated from the plant and the communication in between is carried out through a network.

In network control, two frequently encountered issues are data dropout and communication delays, which are causes of poor performance of network based communication. It is thus necessary to evaluate and compensate data dropout and time-delay factors (Yang, 2006; Hespanha, 2004; Gao and Wang, 2003, 2004). Since data dropout and delay are random and time-varying by nature, the existing control methods for deterministic data dropout and communication delays cannot be directly applied. Significant research efforts have been made on the control problems for networked systems with random data dropout and communication delays that are modeled in various ways in terms of the probability and characteristics of sources and destinations, for instance H_∞ control (Yang *et al.*, 2006), and ILC (Ahn *et al.*, 2008; Liu *et al.*, 2012, 2013; Bu *et al.*, 2016).

Now, we have two levels of network control in MAS as a new research problem, the individual agent level with random communication delays or data loss, and the coordination level with varying communication graph that has been well explored in this book.

8. Discrete or Sampled-Data MAS

The current trend toward digital rather than analog control of dynamic systems is mainly due to the availability of low-cost digital computers and the advantages found in working with digital signals rather than continuous-time signals. As a result, most practical control processes are handled as either discrete-time or hybrid processes consisting of interacting discrete and continuous components. These hybrid systems, in which the system to be controlled evolves in continuous-time and the controller evolves in discrete-time, are called sampled-data systems.

In this book, most theoretical results on MAS, ILC and the syntheses are given in continuous time. Deriving the discrete-time counterparts will be necessary when solving real-time MAS control tasks. Discrete-time MAS with ILC has been partially explored (Meng *et al.*, 2012, 2013c, 2014, 2015a,b; Meng and Moore, 2016).

On the one hand, ILC stability in discrete time becomes important, because otherwise the tracking error would diverge exponentially along the time axis, which is obviously not acceptable in any practical control applications. On the other hand, unlike in continuous-time problems, robust control approaches characterized by high feedback gain are no longer suitable in discrete-time implementations due to the inherent stability property. As a consequence, low gain profiles are essential in discrete-time control. To meet the control requirement such as precision tracking for MAS when agent model uncertainties are present, it is necessary to explore more subtle or smart control approaches that are based on the underlying characters of MAS model uncertainties. More efforts are required in this control area.

It is worth highlighting that ILC is a low gain control approach, thus is suitable for digital control. It is equally important to assure a monotonic tracking convergence in the iterative domain, when considering practical applications in MAS control. In Abidi and Xu (2011, 2015), ILC design and property analysis, which assures a monotonic tracking convergence along the iterative horizon, are present for linear discrete-time or sampled-data systems. Extension of Abidi and Xu (2011, 2015) to MAS with monotonic performance would be an interesting topic. As far as nonlinear dynamic MAS is concerned, how to guarantee monotonic convergence strictly in the iterative domain remains a challenging issue.

12.2.2 Applications

1. Transportation System

Freeway traffic flow is a typical MAS with constraints. As shown in Figure 12.2, within the limited capacity of the road, multiple agents (drivers) try to move at the maximum allowable speed while keeping a safe distance from the surrounding vehicles. Considering the fact that autonomous pilot systems are under development, the ability of coordination for each agent (now an autonomous vehicle) with other agents, becomes critical in order to achieve speed consensus along the freeway. This is a MAS-like traffic control problem at the micro level, which has been explored by researchers and now can be simulated by software such as PARAMICS with certain specific coordinated driving algorithms.

It is worth noting that macroscopic traffic flow patterns are in general repeated every day. For example, the traffic flow will start from a very low level at night, and increase gradually up to the first peak during morning rush hour, which is often from 7–9 AM, with a second one from 5–7 PM. This repeatability at the macroscopic level implies two features: 1) the macroscopic traffic model is invariant on a day/week or month axis, and 2) the macroscopic exogenous quantities (independent of the freeway states, for instance on-ramp flow, off-ramp flow, initial inflow on the mainstream, etc.) are invariant along the day/week or month axis also. Based on the traffic pattern repeatability, ILC can be adopted to learn and improve the traffic control. Both density control and flow control by ILC have be reported in Hou *et al.* (2008) and Zhao *et al.* (2014).

Clearly, the freeway or highway traffic flow imposes new challenges to MAS control, including the presence of physical constraints, the macroscopic interference from on-ramps and off-ramps, merging lanes, traffic lights, incidents, weather effects, as well as macro-level traffic control. Modeling and incorporating those factors into MAS will result in new problems and new outcomes.

Figure 12.2 Freeway traffic behaves as a multi-agent system, where each vehicle has to coordinate its location, speed, acceleration, and orientation with neighboring vehicles.

MAS for traffic network management in urban regions has also been explored in Choy *et al.* (2003), Fiosins *et al.* (2011), and Gokulan and Srinivasan (2010). Figure 12.3 shows a traffic network in a central business district (CBD). The section of the network could consist of one-way links, two-way links, major and minor priority lanes, signaled right and left turning movements, merge lanes, and so on. Each agent, denoted by the circular, square or triangular dots, is a traffic signal manager that determines the corresponding traffic light. The circle agent directly communicates with the neighboring square agents, and may be or may not be required to communicate with the triangle agents that are two blocks away. Clearly, the further away, the less important is the direct crosstalk in between two agents. The ultimate objective for traffic consensus could be equal travel time for each vehicle. This MAS problem has been dealt with by using algorithmic approaches. Considering the highly dynamic nature of traffic flow, and the repetitiveness over days, it would be interesting to seek a dynamic formulation of this problem and find an ILC based solution that could yield the desired consensus.

2. Task-Specific UAV

In both civil and military domains, unmanned aerial vehicle (UAV), unmanned ground vehicles (UGV), or autonomous underwater vehicles (AUV), have been studied as MAS in order to execute collaborative tasks that cannot be performed by a single UAV, UGV or AUV. Such collaborative tasks include surveillance in the battlefield or urban region, searching and exploration under an uncertain environment, tracking moving targets, and so on.

The challenges in this topic lie in the motion coordination among multiple agents with autonomy, the communication topology that decides the information exchange,

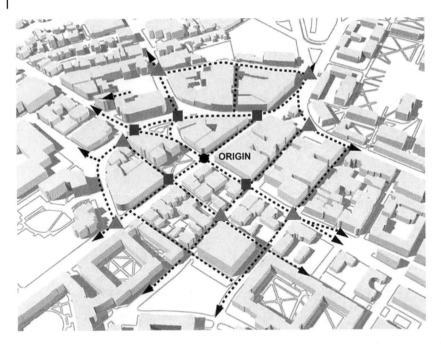

Figure 12.3 Layout of a road network in central business district (CBD). The circle, squares and triangles represent agents, and each controls the traffic signals at an intersection.

the environmental model uncertainties, and the task planning and decomposition. In this book, the first two issues on motion coordination and communication topology have been explored, which are, however, relatively easy jobs. In contrast, handling environmental model uncertainties and task planning still remain to be further investigated. Figure 4 provides an illustrative example. When three UAVs first enter the CBD region, they will start exploration of the region so that landmarks can be found to match the pre-stored digital map, to determine the locations and use the digital map for subsequent task execution. In general, exploration aims at building up an external world model. Searching could be performed next to identify targets, either static or mobile, to facilitate later surveillance and tracking. In this scenario, there are four targets to be identified: three static targets depicted by arrows in Figure 12.4 (an intersection and two specific room windows from two separate buildings) and a moving vehicle on the road. The surveillance task can be specified as to visit three static venues regularly within a certain time interval, and the tracking task is to trace the moving target.

In all three phases of exploration, searching, surveillance and tracking, tasks must be shared by three UAVs which are heterogenous, and information or knowledge should be effectively exchanged or shared within the MAS. Optimal task planning is important in MAS coordination and task execution (Tang and Ozguner, 2005), in order for a MAS to handle complex tasks subject to dynamic changes, randomness, environmental uncertainties, limitations of UAV capacity, and so on. More research efforts are required to develop hybrid approaches, to provide MAS with the desirable flexibility, adaptability, and distributed coordination capabilities.

We face many new problems that challenge existing MAS approaches and results. In new MAS problems, for instance the surveillance phase, the consensus may not be

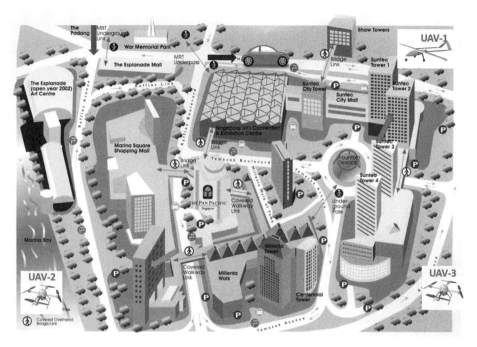

Figure 12.4 UAVs enter a CBD region to perform exploration, searching, surveillance, and tracking tasks. UAV-1 is fixed wing, UAV-2 and UAV-3 are rotary wing.

necessarily associated with motion quantities, but with other objectives such as the consensus on geometric coverage by multiple agents with equal visual sensing scope (Cortes *et al.*, 2004; Olfati-Saber, 2006; Huang *et al.*, 2010). Likewise, in the exploration phase where the main objective is to unveil an unknown region, the consensus for multiple agents could be the distributed detections that discover as much as possible of an uncertain frontier.

Although only UAVs were discussed briefly here, the ideas and methods based on MAS can be easily applied to UGVs or AUVs, where different kinds of communication media are used (Shaukat *et al.*, 2013; Teck *et al.*, 2014).

3. Quorum Sensing in Biological Systems

As a matter of fact, MAS was inspired by the biological world, in which "live" agents illustrate incredible collective behaviors, for instance the formation of birds or fish as shown in Figure 12.5. It is also observed that social insects, like ants and honey bees, are highly organized and cooperative, thus are typical MAS.

Recent research has further discovered that even single-celled bacteria also coordinate closely, monitoring cellular conditions and communicating with each other. From the viewpoint of MAS, the representative behaviors of bacteria or social insects are quorum sensing and quorum quenching.

Briefly speaking, quorum sensing is a system of stimuli and responses correlated to population density. Many species of bacteria use quorum sensing to coordinate gene expression according to the density of their local population. As shown in Figure 12.6, quorum sensing uses autoinducers, which are secreted by bacteria to communicate with other bacteria of the same kind, as signaling molecules. Autoinducers may be small,

(a) Formation of birds

(b) Formation of fish

Figure 12.5 Formations of live agents.

hydrophobic molecules, or they can be larger peptide-based molecules. Each type of molecule has a different mode of action. When only a few other bacteria of the same kind are in the vicinity, diffusion reduces the concentration of the inducer in the surrounding medium to almost zero, so that the bacteria produce little inducer. However, as the population grows, the concentration of the inducer passes a threshold, causing more inducer to be synthesized. This forms a positive feedback loop, and the receptor becomes fully activated. Activation of the receptor induces the up-regulation of other specific genes, causing all of the cells to begin transcription at approximately the same time.

When the population of bacteria is low, the autoinducers produced are low. Individual bacteria cannot sense the autoinducers, believe there to be a low population or a hostile environment, and hibernate. As the number of bacteria grow gradually, once the concentration of autoinducers exceeds a threshold, bacteria will be activated to exchange chemical signals, coordinate the release of toxins, and attack the host.

On the contrary, quorum quenching is a negative feedback process that stops diffusion or growth of the bacteria population. In a similar fashion, some social insects use quorum sensing to determine where to nest. In addition to its function in biological

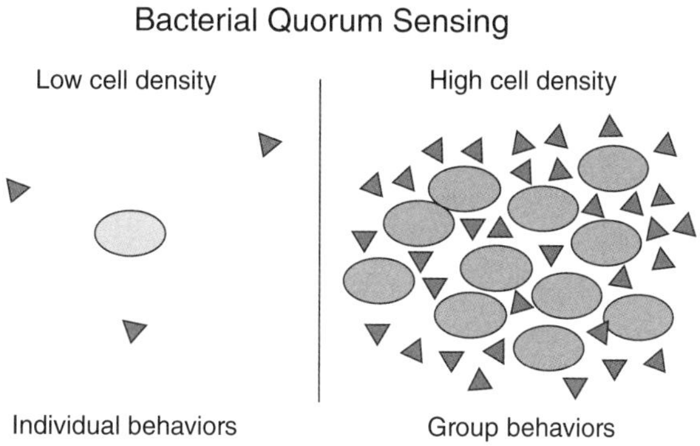

Figure 12.6 Bacteria quorum sensing with either individual or group behaviors, where the ellipses represent bacteria and triangles represent autoinducers.

Bacterial Quorum Sensing

Low cell density   High cell density

Individual behaviors   Group behaviors

systems, quorum sensing has several useful applications for computing and robotics. Quorum sensing can function as a decision-making process in any decentralized system, as long as individual components have: (a) a means of assessing the number of other components they interact with and (b) a standard response once a threshold number of components is detected (Waters and Bassler, 2005).

Despite the fact that quorum sensing/quenching finds more and more applications in microbiology and medicine (Wynendaele *et al.*, 2012, 2015a,b), there does not exist a dynamic MAS model that can quantitatively describe the collective behaviors of quorum sensing. Instead, works in this niche area focus on macroscopic models, for instance Painter and Hillen (2002). Since quorum sensing processes show similarity with respect to the same kind of bacteria, ILC based MAS could be a useful tool in modeling and analyzing quorum sensing.

4. Immunological System Modeling

In computer science and relevant research fields, the concept of the immune system is adopted to enhance MAS when dealing with a variety of practical problems, such as artificial immune systems for security detection (Harmer *et al.*, 2002), immune oriented MAS for biological image processing (Rodin *et al.*, 2004), resource sharing (Chingtham *et al.*, 2010), and so on. The immune system is chosen to model different aspects of MAS because it is compound, has autonomous entities, is able to cooperate, and has behaviors, receptors and means of action.

In general, the immune system shows highly specific and coherent characteristics that are observed in immunological responses, such as the link between innate immunity and adoptive immunity, the negative and positive selection mechanisms in the thymus to distinguish self and non-self, the cooperation between T cells and B cells within the immune system, the ability to detect the antigen by antigen presentation cells such as dentritic cells (first signal), the antigen presentation for the education of immune effector cells, the processes involving activation, differentiation, proliferation and programmed cellular death (apoptosis), the positive or negative co-stimuli by receptors and ligands (the second signal), the generation of danger signals by toll-like receptors (the third signal), and attacks from effector cells to bacteria, infected cells, mutated cells, and so on. Indeed, an immune cell is a typical live agent with specific functions.

It should be noted that each stage of the above mentioned immune response further consists of several interactions at the molecular protein level. For instance, Figure 12.7 shows the interactions between receptors and ligands that yield co-stimuli.

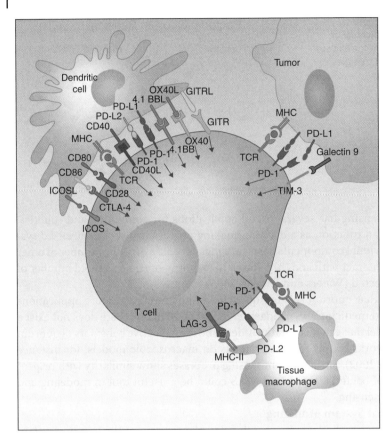

Figure 12.7 The receptors and associated ligands on T cell, dendritic cell, tissue macrophage cell, and tumor cell. Receptors and ligands binding together will generate positive/negative stimuli, and accordingly activate/inhibit the T cell.

It is surprising that the immune system itself has not been modeled as a MAS in a mathematical fashion, perhaps due to the problem complexity, for instance the high dimensionality, extremely large number of immune cells, multimodal interactions, randomness, and so on. Instead, immune systems have been explored as an agent based model (ABM) which is a kind of microscale model (Chiacchio *et al.*, 2014) that simulates the simultaneous operations and interactions of multiple agents in an attempt to re-create and predict the appearance of complex phenomena.

According to Chiacchio *et al.* (2014), immune system modeling approaches can be classified into top-down and bottom-up approaches. The top-down approach works by estimating the mean behavior at a macroscopic level, thus modeling populations instead of single entities. By using such an approach it is possible to model and represent a large number of entities. The top-down model is usually described by ordinary (Chrobak and Herrero, 2011; de Pillis *et al.*, 2008; Galante *et al.*, 2012; Li and Xu, 2015) or partial

(Painter, 2009) differential equations, where the latter are used when the space distribution is of importance for the problem.

The bottom-up approach works at a microscopic level. Entities (agents) and interactions are described and followed individually, and the general behavior of the system arises from the sum of the local behaviors of the involved entities. In this way, it is possible to describe local immunological processes with greater accuracy, avoiding rough approximations that are typical in top-down approaches (Dong *et al.*, 2010; Brown *et al.*, 2011; Pennisi *et al.*, 2009). Note that, as the bottom-up approach by ABM deals with entities that are followed individually, those modeling techniques demand significant computational effort in order to simulate the entire agent group, hence analytical exploration and design are hard to pursue.

Up to the 1990s, due to the absence or lack of intricate knowledge on the human immune system, analytical study was impossible. Agent based modeling and simulations were the only path to investigate the biological behaviors of immunity at a system level. Since then, the exponential increase in discoveries, new knowledge, and immunotherapy outcomes has opened the avenue for possible mathematical modeling, analysis, and therapeutical designs. Taking cancer therapy as an example, researchers now understand much more about various mechanism on how our immune system works, such as the mechanism of activation by dentritic cells, the checkpoint inhibition receptors and ligands, chemokines associated with tumor infiltrating lymphocytes, tumor associated macrophages, danger signals and danger model, recognition of tumor specific antigens and tumor associated antigens, suppressive or regulatory immune cells, and so on. As far as cancer therapy is concerned, there are cytokine based therapies using interleukin-2 (Rosenberg, 2014), macrophage activation factors (Inui *et al.*, 2013), interleukin-12 (Tugues *et al.*, 2015); antibody based therapies using ipilimumab (Fong and Small, 2008; Robert *et al.*, 2011), PD-1 (Flies and Chen, 2007; Mahoney *et al.*, 2015) or PD-L1 (Fehrenbacher *et al.*, 2016) as immune checkpoint inhibitors; immune cell based therapies including the latest chimeric antigen receptor T cell (Lipowska-Bhalla *et al.*, 2012; Porter *et al.*, 2011; Dai *et al.*, 2016), T cell receptor therapy (Stromnes *et al.*, 2015), NK (natural killer) cell therapy (Terme *et al.*, 2008); and all possible combinations of immunotherapy with chemotherapy (Antonia *et al.*, 2014), radiation therapy (Twyman-Saint Victor *et al.*, 2015), and so on.

With so many means available to monitor and partially control disease progression, we are in a position not only to model and simulate human immunity using MAS, but also to choose and design the most suitable therapy for individual patients in the framework of MAS. Further, with diversifying combined therapies, the current evidence-based clinical trial system will soon reach its limits in resources: typically the R&D cost of a new drug/therapy will soar to intolerably high levels if each has to go through the in vitro phase, animal model phase, and phases 1–3 of in vivo clinical trials. From a long term point of view, a MAS based immune system model could be used as a prognostic model to verify therapeutic designs and executions. For this purpose, the immunity oriented MAS should have an open and flexible architecture to incorporate any new discoveries in physiology and pathology at multiple levels, namely, whole body level, system level, organ and tissue level, cell level, protein level, DNA and RNA level. In addition, the fact of repeated cancer treatment from cycle to cycle may offer a chance for iteratively learning the optimal medication, such as varying treatment intervals, varying dosages and varying medicine combinations between cycles.

Appendix A

Graph Theory Revisit

Let $\mathcal{G} = (\mathcal{V}, \mathcal{E})$ be a weighted directed graph with the vertex set $\mathcal{V} = \{1, 2, \ldots, N\}$ and edge set $\mathcal{E} \subseteq \mathcal{V} \times \mathcal{V}$. Let $\mathcal{V}$ also be the index set representing the follower agents in the systems. A direct edge from k to j is denoted by an ordered pair $(k, j) \in \mathcal{E}$, which means that agent j can receive information from agent k. The neighborhood of the kth agent is denoted by the set $\mathcal{N}_k = \{j \in \mathcal{V} | (j, k) \in \mathcal{E}\}$. $\mathcal{A} = (a_{k,j}) \in \mathbb{R}^{N \times N}$ is the weighted adjacency matrix of $\mathcal{G}$. In particular, $a_{k,k} = 0$, $a_{k,j} = 1$ if $(j, k) \in \mathcal{E}$, and $a_{k,j} = 0$ otherwise.[1] The in-degree of vertex k is defined as $d_k^{in} = \sum_{j=1}^{N} a_{k,j}$, and the Laplacian of $\mathcal{G}$ is defined as $L = D - \mathcal{A}$, where $D = \text{diag}(d_1^{in}, \ldots, d_N^{in})$. The Laplacian of an undirected graph is symmetric, whereas the Laplacian of a directed graph is asymmetric in general. An undirected graph is said to be connected if there is a path between any two vertices.[2] A spanning tree is a directed graph whose vertices have exactly one parent except for one vertex, which is called the root and has no parent. We say that a graph contains or has a spanning tree if $\mathcal{V}$ and a subset of $\mathcal{E}$ can form a spanning tree.

Important Properties of a Laplacian Matrix:

- Zero is an eigenvalue of L and $\mathbf{1}$ is the associated eigenvector, namely, the row sum of a Laplacian matrix is zero.
- If $\mathcal{G}$ has a spanning tree, the eigenvalue 0 is algebraically simple and all other eigenvalues have positive real parts.
- If $\mathcal{G}$ is strongly connected, then there exists a positive column vector $\mathbf{w} \in \mathbb{R}^N$ such that $\mathbf{w}^T L = 0$.

Furthermore, if $\mathcal{G}$ is undirected and connected, then L is symmetric and has the following additional properties.

- $\mathbf{x}^T L \mathbf{x} = \frac{1}{2} \sum_{i,j=1}^{N} a_{ij} (x_i - x_j)^2$ for any $\mathbf{x} = [x_1, x_2, \ldots, x_N]^T \in \mathbb{R}^N$, and therefore L is positive semi-definite and all eigenvalues are positive except for one zero eigenvalue.
- The second smallest eigenvalue of L, which is denoted by $\lambda_2(L) > 0$, is called the algebraic connectivity of $\mathcal{G}$. It determines the convergence rate of the classic consensus algorithm.

---

1 An undirected graph is a special case of directed graph, satisfying $a_{k,j} = a_{j,k}$.
2 A path between vertices p and q is a sequence $(p = j_1, \ldots, j_l = q)$ of distinct vertices such that $(j_k, j_{k+1}) \in \mathcal{E}$, $\forall 1 \leq k \leq l - 1$.

Iterative Learning Control for Multi-agent Systems Coordination, First Edition.
Shiping Yang, Jian-Xin Xu, Xuefang Li, and Dong Shen.
© 2017 John Wiley & Sons Singapore Pte. Ltd. Published 2017 by John Wiley & Sons Singapore Pte. Ltd.

- The algebraic connectivity

$$\lambda_2(L) = \inf_{\substack{x \neq 0, 1^T x = 0}} \frac{x^T L x}{x^T x},$$

and therefore, if $1^T x = 0$, then $x^T L x \geq \lambda_2(L) x^T x$.

Appendix B

Detailed Proofs

B.1 HOIM Constraints Derivation

There are four constraints (6.6), (6.16), (6.21), (6.25) on selecting $\tilde{H}_{jk}$. To illustrate the derivation principles, we present step by step derivations for constraints (6.6) and (6.16). The same idea applies to the derivation of the constraints (6.21) and (6.25).

For (6.6), we have

$$
\begin{aligned}
\dot{z}^d_{i+1,1} &= \tilde{H}_1 \dot{z}^d_{i,1}, \\
\begin{bmatrix} \dot{x}^d_{i+1,1} \\ \dot{y}^d_{i+1,1} \\ \dot{\theta}^d_{i+1,1} \end{bmatrix}
&=
\begin{bmatrix} h_{11} & 0 & 0 \\ 0 & h_{12} & 0 \\ 0 & 0 & h_{13} \end{bmatrix}
\begin{bmatrix} \dot{x}^d_{i,1} \\ \dot{y}^d_{i,1} \\ \dot{\theta}^d_{i,1} \end{bmatrix}, \\
\begin{bmatrix} \dot{x}^d_{i+1,1} \\ \dot{y}^d_{i+1,1} \\ \dot{\theta}^d_{i+1,1} \end{bmatrix}
&=
\begin{bmatrix} h_{11}\dot{x}^d_{i,1} \\ h_{12}\dot{y}^d_{i,1} \\ h_{13}\dot{\theta}^d_{1,1} \end{bmatrix}.
\end{aligned}
\tag{B.1}
$$

From $\dot{z}^d_{i+1,1} = \tilde{H}_1 \dot{z}^d_{i,1}$, we can obtain $\dot{\theta}^d_{i+1,1} = h_{13}\dot{\theta}^d_{i,1}$. Since we know $\tan(\theta^d) = \frac{\dot{y}^d}{\dot{x}^d}$, then

$$
\begin{aligned}
\tan(\theta^d_{i+1,1}) &= \tan(h_{13}\theta^d_{i,1}) \\
&= \frac{\dot{y}^d_{i+1,1}}{\dot{x}^d_{i+1,1}} = \frac{h_{12}}{h_{11}} \frac{\dot{y}^d_{i,1}}{\dot{x}^d_{i,1}} = \frac{h_{12}}{h_{11}} \tan(\theta^d_{i,1}).
\end{aligned}
$$

For (6.16), we have

$$
\begin{aligned}
\dot{z}^d_{i+1,12} &= \tilde{H}_{12} \dot{z}^d_{i,12} \\
\dot{z}^d_{i+1,2} - \dot{z}^d_{i+1,1} &= \tilde{H}_{12}(\dot{z}^d_{i,2} - \dot{z}^d_{i,1}) \\
\dot{z}^d_{i+1,2} - \tilde{H}_1 \dot{z}^d_{i,1} &= \tilde{H}_{12}(\dot{z}^d_{i,2} - \dot{z}^d_{i,1}) \\
\dot{z}^d_{i+1,2} &= (\tilde{H}_1 - \tilde{H}_{12})\dot{z}^d_{i,1} - \tilde{H}_{12}\dot{z}^d_{i,2} \\
\begin{bmatrix} \dot{x}^d_{i+1,2} \\ \dot{y}^d_{i+1,2} \\ \dot{\theta}^d_{i+1,2} \end{bmatrix}
&=
\begin{bmatrix} (h_{11} - h_{121})\dot{x}^d_{i,1} + h_{121}\dot{x}^d_{i,2} \\ (h_{12} - h_{122})\dot{y}^d_{i,1} + h_{122}\dot{y}^d_{i,2} \\ (h_{13} - h_{123})\dot{\theta}^d_{i,1} + h_{123}\dot{\theta}^d_{i,2} \end{bmatrix}.
\end{aligned}
$$

Here, using the same technique in deriving (6.6) yields

$$\tan (\theta_{i+1,2}^d) = \tan ((h_{13} - h_{123})\theta_{i,1}^d + h_{123}\theta_{i,2}^d)$$

$$= \frac{\dot{y}_{i+1,2}^d}{\dot{x}_{i+1,2}^d} = \frac{(h_{12} - h_{122})\dot{y}_{i,1}^d + h_{122}\dot{y}_{i,2}^d}{(h_{11} - h_{121})\dot{x}_{i,1}^d + h_{121}\dot{x}_{i,2}^d}.$$

B.2 Proof of Proposition 2.1

By the Schur triangularization theorem (Horn and Johnson, 1985, pp. 79), there is a unitary matrix U and an upper triangular matrix Δ with diagonal entries being the eigenvalues of M, such that

$$\Delta = U^*MU,$$

where $*$ denotes the conjugate transpose.

Let $Q = \text{diag}(\alpha, \alpha^2, \ldots, \alpha^n)$, $\alpha \neq 0$, and set $S = QU^*$. So S is nonsingular. Now define a matrix norm $|\cdot|_S$ (Horn and Johnson, 1985, p.296) such that

$$|M|_S = |SMS^{-1}|,$$

where $|\cdot|$ can be any l_p vector norm induced matrix norm.

Compute SMS^{-1} explicitly, we can obtain

$$SMS^{-1} = \begin{bmatrix} \lambda_1 & \alpha^{-1}\delta_{1,2} & \alpha^{-2}\delta_{1,3} & \cdots & \alpha^{-n+1}\delta_{1,n} \\ 0 & \lambda_2 & \alpha^{-1}\delta_{2,3} & \cdots & \alpha^{-n+2}\delta_{2,n} \\ 0 & 0 & \lambda_3 & \cdots & \alpha^{-n+3}\delta_{3,n} \\ \vdots & \vdots & \vdots & \ddots & \vdots \\ 0 & 0 & 0 & 0 & \lambda_n \end{bmatrix},$$

where λ_i is an eigenvalue of M and $\delta_{i,j}$ is the (i,j)th entry of Δ.

Therefore, $|M|_S$ can be computed as below:

$$|M|_S = \max_{|\mathbf{z}|=1} |SMS^{-1}\mathbf{z}|$$

$$= \max_{|\mathbf{z}|=1} |M_0\mathbf{z} + E(\alpha)\mathbf{z}|$$

$$\leq \max_{|\mathbf{z}|=1} |M_0\mathbf{z}| + \max_{|\mathbf{z}|=1} |E(\alpha)\mathbf{z}|, \tag{B.2}$$

where $M_0 = \text{diag}(\lambda_1, \lambda_2, \ldots, \lambda_n)$ and

$$E(\alpha) = \begin{bmatrix} 0 & \alpha^{-1}\delta_{1,2} & \alpha^{-2}\delta_{1,3} & \cdots & \alpha^{-n+1}\delta_{1,n} \\ 0 & 0 & \alpha^{-1}\delta_{2,3} & \cdots & \alpha^{-n+2}\delta_{2,n} \\ 0 & 0 & 0 & \cdots & \alpha^{-n+3}\delta_{3,n} \\ \vdots & \vdots & \vdots & \ddots & \vdots \\ 0 & 0 & 0 & 0 & 0 \end{bmatrix}.$$

It is easy to verify that

$$\max_{|\mathbf{z}|=1} |M_0\mathbf{z}| = \max_{|\mathbf{z}|=1} \left| \begin{bmatrix} \lambda_1 z_1 & \lambda_2 z_2 & \cdots & \lambda_n z_n \end{bmatrix}^T \right|$$

$$\leq \max_{j=1,2,\ldots,n} |\lambda_j| \max_{|\mathbf{z}|=1} |\mathbf{z}| = \rho(M). \tag{B.3}$$

Define the last term in (B.2) as a function of α, $g(\alpha) = \max_{|\mathbf{z}|=1} |E(\alpha)\mathbf{z}|$. As $g(\alpha)$ is a continuous function of α and $\lim_{\alpha\to\infty} g(\alpha) = 0$, therefore, for any $\epsilon = (1 - \rho(M))/2$, there exists an α^* such that $g(\alpha) < \epsilon$ for all $\alpha > \alpha^*$. Substituting (B.3) in (B.2) and choosing $\alpha > \alpha^*$, we have

$$|M|_S \leq \rho(M) + \epsilon < (1 + \rho(M))/2 < 1.$$

Therefore, we can conclude that $\lim_{k\to\infty} \left(|M|_S \right)^k = 0.$ ∎

B.3 Proof of Lemma 2.1

The feasible region of the optimization problem

$$\min_{\gamma\in\mathbb{R}} \quad \max_{\alpha_1 \leq a \leq \sqrt{a^2+b^2} \leq \alpha_2} \quad |1 - \gamma(a+jb)|$$

can be classified into three regions according to γ.

1) $\gamma > 0$, denoting $J_1 \stackrel{\triangle}{=} \min_{\gamma>0} \max_{\alpha_1 \leq a \leq \sqrt{a^2+b^2} \leq \alpha_2} |1 - \gamma(a+jb)|$, the boundary of $a + jb$ can be seen in Figure B.1.

According to Proposition 2.4, $|1 - \gamma(a + jb)|$ reaches its maximum value at the boundary of the compact set (the shadow area in Figure B.1). The maximum value in chord $\overline{AB}$ is either at point $A = \left(\alpha_1, \sqrt{\alpha_2^2 - \alpha_1^2} \right)$ or at point $B = \left(\alpha_1, -\sqrt{\alpha_2^2 - \alpha_1^2} \right)$, because the imaginary parts of these two points reach the maximum value in $\overline{AB}$ while the real parts in $\overline{AB}$ retain the constant α_1:

Figure B.1 The boundary of complex parameter $a + jb$.

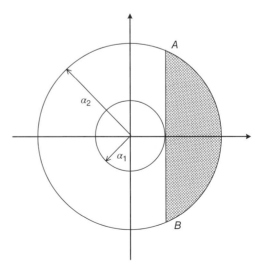

$$\max_{(a,b)\in \widehat{AB}} |1 - \gamma(a + jb)|$$

$$= \sqrt{(1 - \gamma\alpha_1)^2 + \gamma^2 \left(\sqrt{\alpha_2^2 - \alpha_1^2}\right)^2}$$

$$= \sqrt{1 - 2\gamma\alpha_1 + \gamma^2\alpha_2^2}. \tag{B.4}$$

Along the arc $\widehat{AB}$, $\forall \gamma > 0$, the real part $a = \alpha_2 \sin(\theta)$ and the imaginary part $b = \alpha_2 \cos(\theta)$, with $-\bar{\theta} \le \theta \le \bar{\theta}$ and $\bar{\theta} \triangleq \arccos\left(\dfrac{\alpha_1}{\alpha_2}\right)$. The maximum value can be calculated as

$$\max_{(a,b)\in \widehat{AB}} |1 - \gamma(a + jb)|$$

$$= \max_{-\bar{\theta}\le\theta\le\bar{\theta}} \sqrt{(1 - \gamma\alpha_2 \cos(\theta))^2 + \gamma^2\alpha_2^2 \sin^2(\theta)}$$

$$= \max_{-\bar{\theta}\le\theta\le\bar{\theta}} \sqrt{1 - 2\gamma\alpha_2 \cos(\theta) + \gamma^2\alpha_2^2}$$

$$= \sqrt{1 - 2\gamma\alpha_2\dfrac{\alpha_1}{\alpha_2} + \gamma^2\alpha_2^2}$$

$$= \sqrt{1 - 2\gamma\alpha_1 + \gamma^2\alpha_2^2}. \tag{B.5}$$

Consequently when $\gamma = \dfrac{\alpha_1}{\alpha_2^2}$ we have

$$J_1 = \min_{\gamma>0} \sqrt{1 - 2\gamma\alpha_1 + \gamma^2\alpha_2^2}$$

$$= \min_{\gamma>0} \sqrt{\alpha_2^2 \left(\gamma - \dfrac{\alpha_1}{\alpha_2^2}\right)^2 + 1 - \dfrac{\alpha_1^2}{\alpha_2^2}}$$

$$= \dfrac{\sqrt{\alpha_2^2 - \alpha_1^2}}{\alpha_2}. \tag{B.6}$$

2) $\gamma = 0, J_2 \triangleq \min_{\gamma=0} \max_{\alpha_1\le a\le\sqrt{a^2+b^2}\le\alpha_2} |1 - \gamma(a + jb)| = 1.$

3) $\gamma < 0, J_3 \triangleq \min_{\gamma<0} \max_{\alpha_1\le a\le\sqrt{a^2+b^2}\le\alpha_2} |1 - \gamma(a + jb)| > 1.$

Therefore, $\min_{\gamma\in\mathbb{R}} \max_{\alpha_1\le a\le\sqrt{a^2+b^2}\le\alpha_2} |1 - \gamma(a + jb)| = \min\{J_1, J_2, J_3\} = \dfrac{\sqrt{\alpha_2^2 - \alpha_1^2}}{\alpha_2}.$ ∎

B.4   Proof of Theorem 8.1

From Assumption 8.2 and Lemma 8.1, H is a symmetric positive definite matrix, and

$$\left(\min_j \underline{b}_j\right) I \leq B \leq (\max_j \overline{b}_j) I.$$

Therefore, $E_i(t)$ is a nonnegative function.

The proof consists of two parts. In Part A, the difference in E_i between two consecutive iterations is calculated; the convergence of tracking error is shown in Part B.

Part A: Difference of E_i

The difference of $E_i(t)$ is defined as

$$\Delta E_i(t) = E_i(t) - E_{i-1}(t),$$

$$= V_i(\mathbf{e}_i) - V_{i-1}(\mathbf{e}_{i-1}) + \frac{1}{2\kappa} \int_0^t \text{Trace}\left((\tilde{\Theta}_i(\tau))^T B \tilde{\Theta}_i(\tau)\right) d\tau$$

$$- \frac{1}{2\kappa} \int_0^t \text{Trace}\left((\tilde{\Theta}_{i-1}(\tau))^T B \tilde{\Theta}_{i-1}(\tau)\right) d\tau. \tag{B.7}$$

Assumption 8.3 indicates that $V_i(\mathbf{e}_i(0)) = 0$. The first term in (B.7) becomes

$$V_i(\mathbf{e}_i) = \int_0^t \dot{V}_i(\mathbf{e}_i)\, d\tau + V_i(\mathbf{e}_i(0))$$

$$= \int_0^t \dot{V}_i(\mathbf{e}_i)\, d\tau. \tag{B.8}$$

Noting that $\dot{V}_i(\mathbf{e}_i) = \mathbf{e}_i^T H \dot{\mathbf{e}}_i = \boldsymbol{\epsilon}_i^T \dot{\mathbf{e}}_i$, together with the closed loop error dynamics (8.13), yields

$$\dot{V}_i = \boldsymbol{\epsilon}_i^T B \tilde{\Theta}_i \boldsymbol{\xi}(t, \mathbf{x}_i) + \boldsymbol{\epsilon}_i^T \left(\boldsymbol{\eta}_d - \boldsymbol{\eta}(t, \mathbf{x}_i)\right) - \boldsymbol{\epsilon}_i^T \left(\gamma I + \left(\overline{\Phi}(\mathbf{x}_i)\right)^2\right) \boldsymbol{\epsilon}_i. \tag{B.9}$$

Since $\eta_j(t, x_{i,j})$ satisfies Assumption 8.1, noticing (8.4), we have

$$\left|\boldsymbol{\epsilon}_i^T \left(\boldsymbol{\eta}_d - \boldsymbol{\eta}(t, \mathbf{x}_i)\right)\right| \leq \sum_{j=1}^N \phi_j(x_d, x_{i,j})|\epsilon_{i,j}| \cdot |e_{i,j}| \leq \boldsymbol{\epsilon}_i^T \left(\overline{\Phi}(\mathbf{x}_i)\right)^2 \boldsymbol{\epsilon}_i + \frac{1}{4}\mathbf{e}_i^T \mathbf{e}_i$$

$$= \boldsymbol{\epsilon}_i^T \left(\overline{\Phi}(\mathbf{x}_i)\right)^2 \boldsymbol{\epsilon}_i + \frac{1}{4}\boldsymbol{\epsilon}_i^T H^{-2} \boldsymbol{\epsilon}_i$$

$$\leq \boldsymbol{\epsilon}_i^T \left(\overline{\Phi}(\mathbf{x}_i)\right)^2 \boldsymbol{\epsilon}_i + \frac{1}{4\underline{\sigma}(H)^2}\boldsymbol{\epsilon}_i^T \boldsymbol{\epsilon}_i. \tag{B.10}$$

Substituting (B.10) in (B.9), using the convergence condition (8.15) leads to

$$\dot{V}_i \leq \boldsymbol{\epsilon}_i^T B \tilde{\Theta}_i \boldsymbol{\xi}(t, \mathbf{x}_i) + \boldsymbol{\epsilon}_i^T \left(\overline{\Phi}(\mathbf{x}_i)\right)^2 \boldsymbol{\epsilon}_i + \frac{1}{4\underline{\sigma}(H)^2}\boldsymbol{\epsilon}_i^T \boldsymbol{\epsilon}_i - \boldsymbol{\epsilon}_i^T \left(\gamma I + \left(\overline{\Phi}(\mathbf{x}_i)\right)^2\right) \boldsymbol{\epsilon}_i$$

$$\leq -\alpha \boldsymbol{\epsilon}_i^T \boldsymbol{\epsilon}_i + \boldsymbol{\epsilon}_i^T B \tilde{\Theta}_i \boldsymbol{\xi}(t, \mathbf{x}_i). \tag{B.11}$$

Combining the third term and the fourth term in (B.7) yields

$$\text{Trace}\left(\tilde{\Theta}_i^T B\tilde{\Theta}_i\right) - \text{Trace}\left(\tilde{\Theta}_{i-1}^T B\tilde{\Theta}_{i-1}\right)$$

$$= \text{Trace}\left((\hat{\Theta}_{i-1} - \hat{\Theta}_i)^T B(2\Theta_i - \hat{\Theta}_i - \hat{\Theta}_{i-1})\right)$$

$$= \text{Trace}\left((\hat{\Theta}_{i-1} - \hat{\Theta}_i)^T B(2\Theta_i - 2\hat{\Theta}_i + \hat{\Theta}_i - \hat{\Theta}_{i-1})\right)$$

$$= \text{Trace}\left((\hat{\Theta}_{i-1} - \hat{\Theta}_i)^T B(2\tilde{\Theta}_i + \hat{\Theta}_i - \hat{\Theta}_{i-1})\right)$$

$$= -\text{Trace}\left((\hat{\Theta}_{i-1} - \hat{\Theta}_i)^T B(\hat{\Theta}_{i-1} - \hat{\Theta}_i)\right) + 2\text{Trace}\left((\hat{\Theta}_{i-1} - \hat{\Theta}_i)^T B\tilde{\Theta}_i\right). \quad \text{(B.12)}$$

From Equations (B.7), (B.8), (B.11), and (B.12), one has

$$\Delta E_i \le -\frac{1}{2}\mathbf{e}_{i-1}^T H\mathbf{e}_{i-1} + \int_0^t -\alpha\boldsymbol{\epsilon}_i^T\boldsymbol{\epsilon}_i d\tau + \int_0^t \boldsymbol{\epsilon}_i^T B\tilde{\Theta}_i\boldsymbol{\xi}(t, \mathbf{x}_i)d\tau$$

$$- \int_0^t \frac{1}{2\kappa}\text{Trace}\left((\hat{\Theta}_{i-1} - \hat{\Theta}_i)^T B(\hat{\Theta}_{i-1} - \hat{\Theta}_i)\right)d\tau$$

$$+ \int_0^t \frac{1}{\kappa}\text{Trace}\left((\hat{\Theta}_{i-1} - \hat{\Theta}_i)^T B\tilde{\Theta}_i\right)d\tau. \quad \text{(B.13)}$$

From the parameter updating rule (8.11), it can be shown that

$$\boldsymbol{\epsilon}_i^T B\tilde{\Theta}_i\boldsymbol{\xi}(t, \mathbf{x}_i) + \frac{1}{\kappa}\text{Trace}\left((\hat{\Theta}_{i-1} - \hat{\Theta}_i)^T B\tilde{\Theta}_i\right) = 0,$$

and $\text{Trace}\left((\hat{\Theta}_{i-1} - \hat{\Theta}_i)^T B(\hat{\Theta}_{i-1} - \hat{\Theta}_i)\right) \ge 0$.
Therefore, (B.13) becomes

$$\Delta E_i \le -\frac{1}{2}\mathbf{e}_{i-1}^T H\mathbf{e}_{i-1} \le 0. \quad \text{(B.14)}$$

Part B: Convergence of $e_{i,j}$
    If the boundedness of E_1 is proven, following the same steps as in Xu and Tan (2002a), we can show the point-wise convergence of $e_{i,j}$. Taking the derivative of E_1, together with (B.11), simple manipulations lead to

$$\dot{E}_1 \le -\alpha\boldsymbol{\epsilon}_1^T\boldsymbol{\epsilon}_1 + \boldsymbol{\epsilon}_1^T B(\Theta - \hat{\Theta}_1)\boldsymbol{\xi}(t, \mathbf{x}_1) + \frac{1}{2\kappa}\text{Trace}\left((\Theta - \hat{\Theta}_1)^T B(\Theta - \hat{\Theta}_1)\right)$$

$$\le \frac{1}{2\kappa}\text{Trace}\left(\Theta^T B\Theta\right).$$

Θ is a finite and continuous signal, hence, $\dot{E}_1$ is bounded in the interval $[0, T]$. Subsequently, E_1 is bounded in the finite-time interval $[0, T]$.   ∎

B.5   Proof of Corollary 8.1

The proof is completed by evaluating the CEF defined in (8.14) at the time $t = T$.

By using Assumption 8.4, $V_i(0) = V_{i-1}(T)$, the difference between $V_i(T)$ and $V_{i-1}(T)$ can be written as

$$\Delta V_i(T) = \int_0^T \dot{V}_i(\tau)d\tau + V_i(0) - V_{i-1}(T)$$

$$= \int_0^T \dot{V}_i(\tau)\,d\tau.$$

By following a similar approach to the proof of Theorem 8.1, eventually, we can obtain that

$$E_i(T) = E_1(T) + \sum_{k=2}^i \Delta E_k(T)$$

$$\leq E_1(T) - \sum_{k=2}^i \alpha \int_0^T \left(\mathbf{e}_k(\tau)\right)^T H^2 \mathbf{e}_k(\tau)\,d\tau.$$

Since $E_1(T)$ is bounded, $E_i(T)$ is nonnegative, and H^2 is positive definite, it follows that

$$\lim_{i \to \infty} \int_0^T \left(\mathbf{e}_i(\tau)\right)^T \mathbf{e}_i(\tau)\,d\tau = 0.$$

This completes the proof. ∎

Bibliography

Abidi, K. and Xu, J.X. (2011) Iterative learning control for sampled-data systems: From theory to practice. *IEEE Transactions on Industrial Electronics*, **58** (7), 3002–3015.

Abidi, K. and Xu, J.X. (2015) *Advanced Discrete-Time Control: Designs and Applications*, Springer-Verlag, Singapore. In Series of Studies in Systems, Decision and Control 23.

Ahn, H., Chen, Y., and Moore, K. (2008) Discrete-time intermittent iterative learning control with independent data dropouts, in *Proceedings of the 17th IFAC World Congress*, July 6–11, 2008, Seoul, Korea, pp. 12442–12447.

Ahn, H.S. and Chen, Y. (2009) Iterative learning control for multi-agent formation, in *ICROS-SICE International Joint Conference*, August 18–21, 2009, Fukuoka, Japan.

Ahn, H.S., Chen, Y., and Moore, K.L. (2007a) Iterative learning control: Brief survey and categorization. *IEEE Transactions on Systems, Man, and Cybernetics - Part C: Applications and Reviews*, **37** (6), 1099–1121.

Ahn, H.S., Moore, K.L., and Chen, Y. (2007b) *Iterative learning control: robustness and monotonic convergence for interval systems*, Springer Science & Business Media.

Ahn, H.S., Moore, K.L., and Chen, Y. (2010) Trajectory-keeping in satellite formation flying via robust periodic learning control. *International Journal of Robust and Nonlinear Control*, **20** (14), 1655–1666.

Altafini, C. (2013) Consensus problems on networks with antagonistic interactions. *IEEE Transactions on Automatic Control*, **58** (4), 935–946.

Angeli, D., Sontag, E.D., and Wang, Y. (2000a) A characterization of integral input to state stability. *IEEE Transactions on Automatic Control*, **45** (6), 1082–1097.

Angeli, D., Sontag, E.D., and Wang, Y. (2000b) Further equivalences and semiglobal versions of integral input to state stability. *Dynamics and Control*, **10** (2), 127–149.

Antonia, S.J., Larkin, J., and Ascierto, P.A. (2014) Immuno-oncology combinations: a review of clinical experience and future prospects. *Clinical Cancer Research*, **20** (24), 6258–6268.

Arimoto, S., Kawamura, S., and Miyazaki, F. (1984) Bettering operation of robots by learning. *Journal of Robotic Systems*, **1** (2), 123–140.

Bai, H. and Wen, J.T. (2010) Cooperative load transport: A formation-control perspective. *IEEE Transactions on Robotics*, **26** (4), 742–750.

Bhattacharya, A. and Chattopadhyay, P. (2010) Biogeography-based optimization for different economic load dispatch problems. *IEEE Transactions on Power Systems*, **25** (2), 1064–1077.

Bien, Z. and Xu, J.X. (1998) *Iterative learning control: analysis, design, integration and applications*, Kluwer Academic Publishers, Boston, USA.

Biggs, N. (1994) *Algebraic Graph Theory*, Cambridge University Press, 2nd edn..

Iterative Learning Control for Multi-agent Systems Coordination, First Edition.
Shiping Yang, Jian-Xin Xu, Xuefang Li, and Dong Shen.
© 2017 John Wiley & Sons Singapore Pte. Ltd. Published 2017 by John Wiley & Sons Singapore Pte. Ltd.

Bristow, D.A., Tharayil, M., and Alleyne, A.G. (2006) A survey of iterative learning control a learning-based method for high-performance tracking control. *IEEE Control Systems Magazine*, **26**, 96–114.

Brown, B.N., Price, I.M., Toapanta, F.R., DeAlmeida, D.R., Wiley, C.A., Ross, T.M., Oury, T.D., and Vodovotz, Y. (2011) An agent-based model of inflammation and fibrosis following particulate exposure in the lung. *Mathematical biosciences*, **231** (2), 186–196.

Bu, X., Hou, Z., Jin, S., and Chi, R. (2016) An iterative learning control design approach for networked control systems with data dropouts. *International Journal of Robust and Nonlinear Control*, **26** (1), 91–109.

Cao, M., Morse, A.S., and Anderson, B.D.O. (2005) Coordination of an asynchronous multi-agent system via averaging, in *Proceedings of the 16th IFAC World Congress*, July 3–8, 2005, Prague, Czech Republic.

Cao, Y., Yu, W., Ren, W., and Chen, G. (2013) An overview of recent progress in the study of distributed multi-agent coordination. *IEEE Transactions on Industrial Informatics*, **9** (1), 427–438.

Chen, G. and Lewis, F.L. (2011) Distributed adaptive tracking control for synchronization of unknown networked lagrangian systems. *IEEE Transactions on Systems, Man, and Cybernetics - Part B: Cybernetics*, **41** (3), 805–816.

Chen, Y. and Wen, C. (1999) *Iterative Learning Control: Convergence, Robustness and Applications*, Springer-Verlag, London.

Chen, Y., Wen, C., Gong, Z., and Sun, M. (1997) A robust high-order ptype iterative learning controller using current iteration tracking error. *International Journal of Control*, **68** (2), 331–342.

Chen, Y., Wen, C., Gong, Z., and Sun, M. (1999) An iterative learning controller with initial state learning. *IEEE Transaction on Automatic Control*, **44** (2), 371–375.

Cheng, L., Hou, Z.G., and Tan, M. (2008) Decentralized adaptive leader-follower control of multi-manipulator system with uncertain dynamics, in *Proceedings of The 34th Annual Conference of The IEEE Industrial Electronics Society*, November 10–13, 2008, Florida, USA, pp. 1608–1613.

Cheng, L., Hou, Z.G., Tan, M., Lin, Y., and Zhang, W. (2010) Neural-network-based adaptive leader-following control for multiagent systems with uncertainties. *IEEE Transactions on Neural Networks*, **21** (8), 1351–1358.

Chi, R., Hou, Z., and Xu, J.X. (2008) Adaptive ilc for a class of discrete-time systems with iteration-varying trajectory and random initial condition. *Automatica*, **44**, 2207–2213.

Chiacchio, F., Pennisi, M., Russo, G., Motta, S., and Pappalardo, F. (2014) Agent-based modeling of the immune system: Netlogo, a promising framework. *BioMed research international*, **2014**.

Chingtham, T.S., Sahoo, G., and Ghose, M.K. (2010) An artificial immune system model for multi agents resource sharing in distributed environments. *International Journal of Computer Science and Engineering*, **2** (5), 1813–1818.

Chow, T.W.S. and Fang, Y. (1998) An iterative learning control method for continuous-time systems based on 2-d system theory. *IEEE Transactions on Circuits and Systems¡ªI: Fundamental Theory and Applications*, **45** (4), 683–689.

Choy, M.C., Srinivasan, D., and Cheu, R.L. (2003) Cooperative, hybrid agent architecture for real-time traffic signal control. *IEEE Transactions on Systems, Man and Cybernetics, Part A: Systems and Humans*, **33** (5), 597–607.

Chrobak, J.M. and Herrero, H. (2011) A mathematical model of induced cancer-adaptive immune system competition. *Journal of Biological Systems*, **19** (03), 521–532.

Cortes, J., Martinez, S., Karatas, T., and Bullo, F. (2004) Coverage control for mobile sensing networks. *IEEE Transactions on Robotics and Automation*, **20** (2), 243–255.

Cortex, J. (2006) Finite-time convergent gradient flows with applications to network consensus. *Automatica*, **42** (11), 1993–2000.

Dai, H., Wang, Y., Lu, X., and Han, W. (2016) Chimeric antigen receptors modified t-cells for cancer therapy. *Journal of the National Cancer Institute*, **108** (7), djv439.

de Pillis, L.G., Fister, K.R., Gu, W., Head, T., Maples, K., Neal, T., Murugan, A., and Kozai, K. (2008) Optimal control of mixed immunotherapy and chemotherapy of tumors. *Journal of Biological Systems*, **16** (01), 51–80.

Dong, X., Foteinou, P.T., Calvano, S.E., Lowry, S.F., and Androulakis, I.P. (2010) Agent-based modeling of endotoxin-induced acute inflammatory response in human blood leukocytes. *PloS One*, **5** (2), e9249.

Duan, Z.S. and Chen, G.R. (2012) Does the eigenratio λ_2/λ_n represent the synchronizability of a complex network? *Chinese Physics B*, **21** (8), 080 506.

Fang, L. and Antsaklis, P.J. (2006) On communication requirements for multi-agent consensus seeking. *Networked Embedded Sensing and Control, Proceedings of Workshop NESC05*, pp. 53–68.

Fang, Y. and Chow, T.W.S. (2003) 2-d analysis for iterative learning controller for discrete-time systems with variable initial conditions. *IEEE Transactions on Circuits and Systems-I: Fundamental Theory and Applications*, **50** (5), 722–727.

Fehrenbacher, L., Spira, A., Ballinger, M., Kowanetz, M., Vansteenkiste, J., Mazieres, J., Park, K., Smith, D., Artal-Cortes, A., Lewanski, C. *et al.* (2016) Atezolizumab versus docetaxel for patients with previously treated non-small-cell lung cancer (poplar): a multicentre, open-label, phase 2 randomised controlled trial. *The Lancet*.

Fiosins, M., Fiosina, J., Müller, J.P., and Görmer, J. (2011) Agent-based integrated decision making for autonomous vehicles in urban traffic, in *Advances on Practical Applications of Agents and Multiagent Systems*, Springer-Verlag, Berlin, Heidelberg, pp. 173–178.

Flies, D.B. and Chen, L. (2007) The new b7s: playing a pivotal role in tumor immunity. *Journal of Immunotherapy*, **30** (3), 251–260.

Fong, L. and Small, E.J. (2008) Anti–cytotoxic t-lymphocyte antigen-4 antibody: the first in an emerging class of immunomodulatory antibodies for cancer treatment. *Journal of clinical oncology*, **26** (32), 5275–5283.

Galante, A., Tamada, K., and Levy, D. (2012) B7-h1 and a mathematical model for cytotoxic t cell and tumor cell interaction. *Bulletin of mathematical biology*, **74** (1), 91–102.

Gao, H. and Wang, C. (2003) Delay-dependent robust H_∞ and L_2-L_∞ filtering for a class of uncertain nonlinear time-delay systems. *IEEE Transactions on Automatic Control*, **48** (9), 1661–1666.

Gao, H. and Wang, C. (2004) A delay-dependent approach to robust H_∞ filtering for uncertain discrete-time state-delayed systems. *IEEE Transactions on Signal Processing*, **52** (6), 1631–1640.

Gokulan, B.P. and Srinivasan, D. (2010) Distributed geometric fuzzy multiagent urban traffic signal control. *IEEE Transactions on Intelligent Transportation Systems*, **11** (3), 714–727.

Harmer, P.K., Williams, P.D., Gunsch, G.H., and Lamont, G.B. (2002) An artificial immune system architecture for computer security applications. *IEEE transactions on Evolutionary computation*, **6** (3), 252–280.

Hatano, Y. and Mesbahi, M. (2005) Agreement over random networks. *IEEE Transactions on Automatic Control*, **50** (11), 1867–1872.

Hespanha, J.P. (2004) *Stochastic hybrid systems: application to communication networks*, Springer-Verlag, New York. Lecture Notes in Computer Science.

Hong, Y., Hu, J., and Gao, L. (2006) Tracking control for multi-agent consensus with an active leader and variable topology. *Automatica*, **42** (7), 1177–1182.

Horn, R.A. and Johnson, C.R. (1985) *Matrix Analysis*, Cambridge University Press.

Hou, Z., Xu, J.X., and Yan, J. (2008) An iterative learning approach for density control of freeway traffic flow via ramp metering. *Transportation Research Part C: Emerging Technologies*, **16** (1), 71–97.

Hou, Z.G., Cheng, L., and Tan, M. (2009) Decentralized robust adaptive control for the multiagent system consensus problem using neural networks. *IEEE Transactions on Systems, Man, And Cybernetics-Part B: Cybernetics*, **39** (3), 636–647.

Hou, Z.G., Cheng, L., Tan, M., and Wang, X. (2010) Distributed adaptive coordinated control of multi-manipulator systems using neural networks, in *Robot Intelligence: An Advanced Knowledge Processing Approach*, Springer-Verlag, London, chap. 3, pp. 44–69.

Huang, D., Xu, J.X., and Lum, K.Y. (2010) Surveillance for a simply connected region: A one-center disk-covering problem, in *Proceedings 8th IEEE International Conference on Control and Automation (ICCA)*, June 9–11, 2010, Xiamen, China, pp. 860–865.

Huang, J. (2011) Remarks on 'synchronized output regulation of linear networked systems'. *IEEE Transactions on Automatic Control*, **56** (3), 630–631.

Inui, T., Kuchiike, D., Kubo, K., Mette, M., Uto, Y., Hori, H., and Sakamoto, N. (2013) Clinical experience of integrative cancer immunotherapy with gcmaf. *Anticancer Research*, **33** (7), 2917–2919.

Islam, S. and Liu, P.X. (2010) Adaptive iterative learning control for robot manipulators without using velocity signals, in *Proceedings of IEEE/ASME Iternational Conference on Advanced Intelligent Mechatronics*, July 6–9, 2010, Montreal, Canada, pp. 1293–1298.

Jadbabaie, A., Lin, J., and Morse, A.S. (2003) Coordination of groups of mobile autonomous agents using nearest neighbor rules. *IEEE Transactions on Automatic Control*, **48** (6), 988–1001.

Jiang, Z.P. and Wang, Y. (2001) Input-to-state stability for discrete-time nonlinear systems. *Automatica*, **37** (6), 857–869.

Jin, X. and Xu, J.X. (2013) Iterative learning control for output-constrained systems with both parametric and nonparametric uncertainties. *Automatica*, **49** (8), 2508–2516.

Jin, X. and Xu, J.X. (2014) A barrier composite energy function approach for robot manipulators under alignment condition with position constraints. *International Journal of Robust and Nonlinear Control*, **24** (17), 2840–2851.

Johansson, B., Speranzon, A., Johansson, M., and Johansson, K.H. (2008) On decentralized negotiation of optimal consensus. *Automatica*, **44** (4), 1175–1179.

Kang, M.K., Lee, J.S., and Han, K.L. (2005) Kinematic path-tracking of mobile robot using iterative learning control. *Journal of Robotic Systems*, **22** (2), 111–121.

Kar, S. and Hug, G. (2012) Distributed robust economic dispatch in power systems: A consensus + innovations approach, in *IEEE Power and Energy Society General Meeting*, July 22–26, 2012, San Diego, California, USA, pp. 1–8.

Khalil, H.K. (2002) *Nonlinear Systems*, Prentice Hall, 3rd edn..

Khoo, S., Xie, L., and Man, Z. (2009) Robust finite-time consensus tracking algorithm for multirobot systems. *IEEE/ASME Transactions on Mechatronics*, **14** (2), 219–228.

Lee, K., Sode-Yome, A., and Park, J.H. (1998) Adaptive hopfield neural networks for economic load dispatch. *IEEE Transactions on Power Systems*, **13** (2), 519–526.

Lee, K.W. and Khalil, H.K. (1997) Adaptive output feedback control of robot manipulators using high-gain observer. *International Journal of Control*, **67**, 869–886.

Li, J. and Li, J. (2013) Adaptive iterative learning control for coordination of second-order multi-agent systems. *International Journal of Robust and Nonlinear Control*, **24** (18), 3282–3299.

Li, S., Du, H., and Lin, X. (2011) Finite-time consensus algorithm for multi-agent systems with double-integrator dynamics. *Automatica*, **47**, 1706–1712.

Li, X. and Xu, J.X. (2015) A mathematical model of immune response to tumor invasion incorporated with danger model. *Journal of Biological Systems*, **23** (3), 505–526.

Li, Y., Li, T., and Jing, X. (2014) Indirect adaptive fuzzy control for input and output constrained nonlinear systems using a barrier lyapunov function. *International Journal of Adaptive Control and Signal Processing*, **28** (2), 184–199.

Li, Z., Duan, Z., Chen, G., and Huang, L. (2010) Consensus of multiagent systems and synchronization of complex networks: A unified viewpoint. *IEEE Transactions on Circuits and Systems - I*, **57** (1), 213–224.

Liang, R.H. (1999) A neural-based redispatch approach to dynamic generation allocation. *IEEE Transactions on Power Systems*, **14** (4), 1388–1393.

Lipowska-Bhalla, G., Gilham, D.E., Hawkins, R.E., and Rothwell, D.G. (2012) Targeted immunotherapy of cancer with car t cells: achievements and challenges. *Cancer Immunology, Immunotherapy*, **61** (7), 953–962.

Liu, C., Xiong, R., Xu, J., and Wu, J. (2013) On iterative learning control for remote control systems with packet losses. *Journal of Applied Mathematics*, **2013**, 1–14.

Liu, C., Xu, J., and Wu, J. (2012) Iterative learning control for remote control systems with communication delay and data dropout. *Mathematical Problems in Engineering*, **2012**, 1–14.

Liu, C., Xu, J.X., and Wu, J. (2010) On iterative learning control with high-order internal models. *International journal of Adaptive Control and Signal Processing*, **24** (9), 731–742.

Liu, Y. and Jia, Y. (2012) An iterative learning approach to formation control of multi-agent systems. *Systems & Control Letters*, **61** (1), 148–154.

Liu, Y.J. and Tong, S. (2016) Barrier lyapunov functions-based adaptive control for a class of nonlinear pure-feedback systems with full state constraints. *Automatica*, **64**, 70–75.

Longman, R.W. (2000) Iterative learning control and repetitive control for engineering practice. *International Journal of Control*, **73** (10), 930–954.

Ma, C.Q. and Zhang, J.F. (2010) Necessary and sufficient conditions for consensusability of linear multi-agent systems. *IEEE Transactions on Automatic Control*, **55** (5), 1263–1268.

Mackenroth, U. (2004) *Robust Control Systems Theory and Case Studies*, Springer-Verlag Berlin Heidelberg.

Madrigal, M. and Quintana, V. (2000) An analytical solution to the economic dispatch problem. *IEEE Power Engineering Review*, **20** (9), 52–55.

Mahoney, K.M., Rennert, P.D., and Freeman, G.J. (2015) Combination cancer immunotherapy and new immunomodulatory targets. *Nature reviews Drug discovery*, **14** (8), 561–584.

Mei, J., Ren, W., and Ma, G. (2011) Distributed coordinated tracking with a dynamic leader for multiple euler-lagrange systems. *IEEE Transactions on Automatic Control*, **56** (6), 1415–1421.

Meng, D. and Jia, Y. (2012) Iterative learning approaches to design finite-time consensus protocols for multi-agent systems. *Systems & Control Letters*, **61** (1), 187–194.

Meng, D. and Jia, Y. (2014) Formation control for multi-agent systems through an iterative learning design approach. *International Journal of Robust and Nonlinear Control*, **24** (2), 340–361.

Meng, D., Jia, Y., and Du, J. (2013a) Coordination learning control for groups of mobile agents. *Journal of the Franklin Institute*, **350** (8), 2183–2211.

Meng, D., Jia, Y., and Du, J. (2013b) Multi-agent iterative learning control with communication topologies dynamically changing in two directions. *IET Control Theory & Applications*, **7** (2), 260–271.

Meng, D., Jia, Y., and Du, J. (2015a) Robust consensus tracking control for multiagent systems with initial state shifts, disturbances, and switching topologies. *IEEE Transactions on Neural Networks and Learning Systems*, **26** (4), 809–824.

Meng, D., Jia, Y., Du, J., and Yu, F. (2012) Tracking control over a finite interval for multi-agent systems with a time-varying reference trajectory. *Systems & Control Letters*, **61** (7), 807–818.

Meng, D., Jia, Y., Du, J., and Yu, F. (2013c) Tracking algorithms for multiagent systems. *IEEE Transactions On Neural Networks And Learning Systems*, **24** (10), 1660–1676.

Meng, D., Jia, Y., Du, J., and Zhang, J. (2014) On iterative learning algorithms for the formation control of nonlinear multi-agent systems. *Automatica*, **50** (1), 291–295.

Meng, D., Jia, Y., Du, J., and Zhang, J. (2015b) High-precision formation control of nonlinear multi-agent systems with switching topologies: A learning approach. *International Journal of Robust and Nonlinear Control*, **25** (13), 1993–2018.

Meng, D., Jia, Y., Du, J., Zhang, J., and Li, W. (2013d) Formation learning algorithms for mobile agents subject to 2-d dynamically changing topologies, in *IEEE American Control Conference*, June 17–19, 2013, Washington, DC, USA, pp. 5165–5170.

Meng, D. and Moore, K.L. (2016) Learning to cooperate: Networks of formation agents with switching topologies. *Automatica*, **64**, 278–293.

Min, H., Sun, F., Wang, S., and Li, H. (2011) Distributed adaptive consensus algorithm for networked euler-lagrange systems. *IET Control Theory & Application*, **5** (1), 145–154.

Moore, K.L. (1993) *Iterative learning control for deterministic systems*, Springer-Verlag. Advances in Industrial Control.

Moore, K.L., Chen, Y., and Ahn, H.S. (2006) Iterative learning control: A tutorial and big picture, in *Proceedings of the 45th IEEE Conference on Decision & Control*, December 13–15, 2006, San Diego, CA, USA, pp. 2352–2357.

Moreau, L. (2005) Stability of multiagent systems with time-dependent communication links. *IEEE Transactions on Automatic Control*, **50** (2), 169–182.

Moyle, P.B. and Cech, J.J. (2003) *Fishes: An Introduction to Ichthyology*, Benjamin Cummings, 5th edn..

Ngo, K.B., Mahony, R., and Jiang, Z.P. (2005) Integrator backstepping using barrier functions for systems with multiple state constraints, in *Proceedings 44th IEEE Conference on Decision and Control*, December 12–15, 2005, Seville, Spain, pp. 8306–8312.

Norrlof, M. and Gunnarsson, S. (2002) Time and frequency domain convergence properties in iterative learning control. *International Journal of Control*, **75** (14), 1114–1126.

Olfati-Saber, R. (2006) Flocking for multi-agent dynamic systems: Algorithms and theory. *IEEE Transactions on Automatic Control*, **51** (3), 401–420.

Olfati-Saber, R., Fax, J.A., and Murray, R.M. (2007) Consensus and cooperation in networked multi-agent systems. *Proceedings of the IEEE*, **95** (1), 215–233.

Olfati-Saber, R. and Murray, R.M. (2004) Consensus problems in networks of agents with switching topology and time-delays. *IEEE Transactions on Automatic Control*, **49** (9), 1520–1533.

Ouyang, P.R., Zhang, W.J., and Gupta, M.M. (2006) An adaptive switching learning control method for trajectory tracking of robot manipulators. *Mechatronics*, **16**, 51–61.

Painter, K.J. (2009) Continuous models for cell migration in tissues and applications to cell sorting via differential chemotaxis. *Bulletin of Mathematical Biology*, **71** (5), 1117–1147.

Painter, K.J. and Hillen, T. (2002) Volume-filling and quorum-sensing in models for chemosensitive movement. *Canadian Applied Mathematics Quarterly*, **10** (4), 501–543.

Park, K.H. (2005) An average operator-based pd-type iterative learning control for variable initial state error. *IEEE Transactions on Automatic Control*, **50** (6), 865–869.

Park, K.H., Bien, Z., and Hwang, D.H. (1999) A study on the robustness of a pid-type iterative learning controller against initial state error. *International Journal of Systems Science*, **30** (1), 49–59.

Pennisi, M., Pappalardo, F., and Motta, S. (2009) Agent based modeling of lung metastasis-immune system competition, in *International Conference on Artificial Immune Systems*, August 9–12, 2009, York, UK, pp. 1–3.

Polycarpous, M. and Ioannouq, P. (1996) A robust adaptive nonlinear control design. *Automatica*, **32** (3), 423–427.

Porter, D.L., Levine, B.L., Kalos, M., Bagg, A., and June, C.H. (2011) Chimeric antigen receptor–modified t cells in chronic lymphoid leukemia. *New England Journal of Medicine*, **365** (8), 725–733.

Qiu, Z., Liu, S., and Xie, L. (2016) Distributed constrained optimal consensus of multi-agent systems. *Automatica*, **68**, 209–215.

Ren, B., Ge, S.S., Tee, K.P., and Lee, T.H. (2010) Adaptive neural control for output feedback nonlinear systems using a barrier lyapunov function. *IEEE Transactions on Neural Networks*, **21** (8), 1339–1345.

Ren, W. (2008a) On consensus algorithms for double integrator dynamics. *IEEE Transactions on Automatic Control*, **53** (6), 1503–1509.

Ren, W. (2008b) Synchronization of coupled harmonic oscillators with local interaction. *Automatica*, **44** (12), 3195–3200.

Ren, W. (2009) Distributed leaderless consensus algorithms for networked euler-lagrange systems. *International Journal of Control*, **82** (11), 2137–2149.

Ren, W. and Beard, R.W. (2008) *Distributed Consensus in Multi-vehicle Cooperative Control*, Communication and Control Engineering Series, Springer-Verlag, London.

Ren, W., Beard, R.W., and Atkins, E.M. (2007) Information consensus in multivehicle cooperative control. *IEEE Control Systems*, **27** (2), 71–82.

Ren, W. and Cao, Y. (2011) *Distributed Coordination of Multi-agent Networks*, Communication and Control Engineering Series, Springer-Verlag, London.

Robert, C., Thomas, L., Bondarenko, I., O'Day, S., Weber, J., Garbe, C., Lebbe, C., Baurain, J.F., Testori, A., Grob, J.J. *et al.* (2011) Ipilimumab plus dacarbazine for previously untreated metastatic melanoma. *New England Journal of Medicine*, **364** (26), 2517–2526.

Rodin, V., Benzinou, A., Guillaud, A., Ballet, P., Harrouet, F., Tisseau, J., and Le Bihan, J. (2004) An immune oriented multi-agent system for biological image processing. *Pattern Recognition*, **37** (4), 631–645.

Rogers, E., Galkowski, K., and Owens, D.H. (2007) *Control systems theory and applications for linear repetitive processes*, vol. 349, Springer Science & Business Media.

Rosenberg, S.A. (2014) Il-2: the first effective immunotherapy for human cancer. *The Journal of Immunology*, **192** (12), 5451–5458.

Saab, S.S. (1994) On the p-type learning control. *IEEE Transactions on Automatic Control*, **39** (11), 2298–2302.

Sahoo, S.K., Panda, S.K., and Xu, J.X. (2004) Iterative learning-based high-performance current controller for switched reluctance motors. *IEEE Transactions on Energy Conversion*, **19** (3), 491–498.

Shaukat, M., Chitre, M., and Ong, S.H. (2013) A bio-inspired distributed approach for searching underwater acoustic source using a team of auvs, in *Proceedings 2013 MTS/IEEE OCEANS-Bergen*, June 10–13, 2013, Bergen, Norway, pp. 1–10.

Shi, J., He, X., Wang, Z., and Zhou, D. (2014) Iterative consensus for a class of second-order multi-agent systems. *Journal of Intelligent & Robotic Systems*, **73** (1–4), 655–664.

Slotine, J.J.E. and Li, W. (1991) *Applied Nonlinear Control*, Prentice Hall.

Sontag, E.D. (2006) *Input to state stability: Basic concepts and results*, Springer.

Spong, M.I., Marino, R., Peresada, S.M., and Taylor, D.G. (1987) Feedback linearizing control of switched reluctance motors. *IEEE Transactions on Automatic Control*, **32** (5), 371–379.

Spong, M.W., Hutchinson, S., and Vidyasagar, M. (2006) *Robot Modeling and Control*, John Wiley & Sons Inc.

Stromnes, I.M., Schmitt, T.M., Hulbert, A., Brockenbrough, J.S., Nguyen, H.N., Cuevas, C., Dotson, A.M., Tan, X., Hotes, J.L., Greenberg, P.D. *et al.* (2015) T cells engineered against a native antigen can surmount immunologic and physical barriers to treat pancreatic ductal adenocarcinoma. *Cancer cell*, **28** (5), 638–652.

Sun, M., Ge, S.S., and Mareels, I.M.Y. (2006) Adaptive repetitive learning control of robotic manipulators without the requirement for initial repositioning. *IEEE Transactions on Robotics*, **22** (3), 563–568.

Sun, M. and Wang, D. (2002) Iterative learning control with initial rectifying action. *Automatica*, **38** (7), 1177–1182.

Sun, M., Wang, D., and Chen, P. (2010) Repetitive learning control of nonlinear systems over finite intervals. *Science in China Series F: Information Sciences*, **53** (1), 115–128.

Syrmos, V.L., Abdallah, C.T., Dorato, P., and Grigoriadis, K. (1997) Static output feedback - a survey. *Automatica*, **33** (2), 125–137.

Tahbaz-Salehi, A. and Jadbabaie, A. (2008) A necessary and sufficient condition for consensus over random networks. *IEEE Transactions on Automatic Control*, **53** (3), 791–795.

Tang, Z. and Ozguner, U. (2005) Motion planning for multitarget surveillance with mobile sensor agents. *IEEE Transactions on Robotics*, **21** (5), 898–908.

Tayebi, A. (2004) Adaptive iterative learning control of robot manipulators. *Automatica*, **40**, 1195–1203.

Tayebi, A. and Islam, S. (2006) Adaptive iterative learning control of robot manipulators: Experimental results. *Control Engineering Practice*, **14**, 843–851.

Teck, T.Y., Chitre, M., and Hover, F.S. (2014) Collaborative bathymetry-based localization of a team of autonomous underwater vehicles, in *2014 IEEE International Conference on Robotics and Automation (ICRA)*, May 31–June 7, 2014, Hong Kong, China, pp. 2475–2481.

Tee, K.P., Ge, S.S., and Tay, E.H. (2007) Adaptive control of a class of uncertain electrostatic microactuators, in *Proceedings of American Control Conference*, July 9–13, New York, USA, pp. 3186–3191.

Tee, K.P., Ge, S.S., and Tay, E.H. (2009) Barrier lyapunov functions for the control of output-constrained nonlinear systems. *Automatica*, **45** (4), 918–927.

Tee, K.P., Ren, B., and Ge, S.S. (2011) Control of nonlinear systems with time-varying output constraints. *Automatica*, **47** (11), 2511–2516.

Terme, M., Ullrich, E., Delahaye, N.F., Chaput, N., and Zitvogel, L. (2008) Natural killer cell–directed therapies: moving from unexpected results to successful strategies. *Nature immunology*, **9** (5), 486–494.

Tugues, S., Burkhard, S., Ohs, I., Vrohlings, M., Nussbaum, K., Vom Berg, J., Kulig, P., and Becher, B. (2015) New insights into il-12-mediated tumor suppression. *Cell Death & Differentiation*, **22** (2), 237–246.

Twyman-Saint Victor, C., Rech, A.J., Maity, A., Rengan, R., Pauken, K.E., Stelekati, E., Benci, J.L., Xu, B., Dada, H., Odorizzi, P.M. *et al.* (2015) Radiation and dual checkpoint blockade activate non-redundant immune mechanisms in cancer. *Nature*, **520** (7547), 373–377.

Wang, D., Ye, Y., and Zhang, B. (2014) *Practical Iterative Learning Control with Frequency Domain Design and Sampled Data Implementation*, Springer.

Wang, L. and Xiao, F. (2010) Finite-time consensus problems for networks of dynamic agents. *IEEE Transactions on Automatic Control*, **55** (4), 950–955.

Wang, S.J., Shahidehpour, S.M., Kirschen, D., Mokhtari, S., and Irisarri, G. (1995) Short-term generation scheduling with transmission and environmental constraints using an augmented lagrangian relaxation. *IEEE Transactions on Power Systems*, **10** (3), 1294–1301.

Wang, X. and Hong, Y. (2008) Finite-time consensus for multi-agent networks with second-order agent dynamics, in *Proceedings of the 17th IFAC World Congress*, July 6–11, 2008, Seoul, Korea, pp. 15 185–15 190.

Wang, Y., Gao, F., and III, F.J.D. (2009) Survey on iterative learning control, repetitive control, and run-to-run control. *Journal of Process Control*, **19**, 1589–1600.

Waters, C.M. and Bassler, B.L. (2005) Quorum sensing: cell-to-cell communication in bacteria. *Annu. Rev. Cell Dev. Biol.*, **21**, 319–346.

Wieland, P., Sepulchre, R., and Allgower, F. (2011) An internal model principle is necessary and sufficient for linear output synchronization. *Automatica*, **47** (3), 1068–1074.

Wood, A.J. and Wollenberg, B.F. (1996) *Power generation, operation & control*, Wiley, New York, 2nd edn..

Wu, C.W. (2006) Synchronization and convergence of linear dynamics in random directed networks. *IEEE Transactions on Automatic Control*, **51** (7), 1207–1270.

Wynendaele, E., Pauwels, E., Van De Wiele, C., Burvenich, C., and De Spiegeleer, B. (2012) The potential role of quorum-sensing peptides in oncology. *Medical Hypotheses*, **78** (6), 814–817.

Wynendaele, E., Verbeke, F., DHondt, M., Hendrix, A., Van De Wiele, C., Burvenich, C., Peremans, K., De Wever, O., Bracke, M., and De Spiegeleer, B. (2015a) Crosstalk between the microbiome and cancer cells by quorum sensing peptides. *Peptides*, **64**, 40–48.

Wynendaele, E., Verbeke, F., Stalmans, S., Gevaert, B., Janssens, Y., Van De Wiele, C., Peremans, K., Burvenich, C., and De Spiegeleer, B. (2015b) Quorum sensing peptides selectively penetrate the blood-brain barrier. *PloS One*, **10** (11), e0142071.

Xiang, J., Wei, W., and Li, Y. (2009) Synchronized output regulation of linear networked systems. *IEEE Transactions on Automatic Control*, **54** (6), 1336–1341.

Xiao, L. and Boyd, S. (2004) Fast linear iterations for distributed averaging. *Systems & Control Letters*, **53** (1), 65–78.

Xie, G. and Wang, L. (2005) Consensus control for a class of networks of dynamic agents: Fixed topology, in *Proceedings of the 44th IEEE Conference on Decision and Control and the European Control Conference 2005*, December 12–15, 2005, Seville, Spain, pp. 96–101.

Xu, J.X. (2011) A survey on iterative learning control for nonlinear systems. *International Journal of Control*, **84** (7), 1275–1294.

Xu, J.X., Chen, Y., Lee, T.H., and Yamamoto, S. (1999) Terminal iterative learning control with an application to rtpcvd thickness control. *Automatica*, **35** (9), 1535–1542.

Xu, J.X. and Jin, X. (2013) State-constrained iterative learning control for a class of mimo systems. *IEEE Transactions on Automatic Control*, **58** (5), 1322–1327.

Xu, J.X., Panda, S.K., and Lee, T.H. (2008) *Real-time Iterative Learning Control: Design and Applications*, Springer-Verlag, London.

Xu, J.X. and Qu, Z. (1998) Robust iterative learning control for a class of nonlinear systems. *Automatica*, **34** (8), 983–988.

Xu, J.X. and Tan, Y. (2002a) A composite energy function based learning control approach for nonlinear systems with time varying parametric uncertainties. *IEEE Transaction on Automatic Control*, **47** (11), 1940–1945.

Xu, J.X. and Tan, Y. (2002b) Robust optimal design and convergence properties analysis of iterative learning control approaches. *Automatica*, **38** (11), 1867–1880.

Xu, J.X. and Tan, Y. (2003) *Linear and Nonlinear Iterative Learning Control*, Springer-Verlag, Germany. In series of Lecture Notes in Control and Information Sciences.

Xu, J.X., Tan, Y., and Lee, T.H. (2004) Iterative learning control design based on composite energy function with input saturation. *Automatica*, **40** (8), 1371–1377.

Xu, J.X. and Xu, J. (2004) On iterative learning from different tracking tasks in the presence of time-varying uncertainties. *IEEE Transactions On Systems, Man, and Cybernetics – Part B: Cybernetics*, **34** (1), 589–597.

Xu, J.X. and Yan, R. (2005) On initial conditions in iterative learning control. *IEEE Transaction on Automatic Control*, **50** (9), 1349–1354.

Xu, J.X. and Yang, S. (2013) Iterative learning based control and optimization for large scale systems, in *13th IFAC Symposium on Large Scale Complex Systems: Theory and Applications*, July 7–10, 2012, Shanghai, China, pp. 74–81.

Xu, J.X., Zhang, S., and Yang, S. (2011) A hoim-based iterative learning control scheme for multi-agent formation, in *2011 IEEE International Symposium on Intelligent Control*, September 28–30, 2011, Denver, CO, USA, pp. 218–423.

Yalcinoz, T. and Short, M.J. (1998) Neural networks approach for solving economic dispatch problem with transmission capacity constraints. *IEEE Transactions on Power Systems*, **13** (2), 307–313.

Yang, F., Wang, Z., Hung, Y., and Gani, M. (2006) H_∞ control for networked systems with random communication delays. *IEEE Transactions on Automatic Control*, **51** (3), 511–518.

Yang, S., Tan, S., and Xu, J.X. (2013) Consensus based approach for economic dispatch problem in a smart grid. *IEEE Transactions on Power Systems*, **28** (4), 4416–4426.

Yang, S. and Xu, J.X. (2012) Adaptive iterative learning control for multi-agent systems consensus tracking, in *IEEE International Conference on Systems, Man, and Cybernetics*, October 14–17, 2012, COEX, Seoul, Korea, pp. 2803–2808.

Yang, S. and Xu, J.X. (2016) Leader-follower synchronization for networked lagrangian systems with uncertainties: A learning approach. *International Journal of Systems Science*, **47** (4), 956–965.

Yang, S., Xu, J.X., and Huang, D. (2012) Iterative learning control for multi-agent systems consensus tracking, in *The 51st IEEE Conference on Decision and Control*, December 10–13, 2012, Maui, Hawaii, USA, pp. 4672–4677.

Yang, S., Xu, J.X., Huang, D., and Tan, Y. (2014) Optimal iterative learning control design for multi-agent systems consensus tracking. *Systems & Control Letters*, **69**, 80–89.

Yang, S., Xu, J.X., Huang, D., and Tan, Y. (2015) Synchronization of heterogeneous agent systems by adaptive iterative learning control. *Asian Journal of Control*, **17** (6), 2091–2104.

Yang, T.C. (2006) Networked control system: a brief survey. *IEE Proceedings - Control Theory and Applications*, **153** (4), 403–412.

Yin, C., Xu, J.X., and Hou, Z. (2010) A high-order internal model based iterative learning control scheme for nonlinear systems with time-iteration-varying parameters. *IEEE Transaction on Automatic Control*, **55** (11), 2665–2670.

Yufka, A., Parlaktuna, O., and Ozkan, M. (2010) Formation-based cooperative transportation by a group of non-holonomic mobile robots, in *2010 IEEE International Conference on Systems Man and Cybernetics (SMC)*, October 10–13, 2010, Istanbul, Turkey, pp. 3300–3307.

Zhang, W., Wang, Z., and Guo, Y. (2014) Backstepping-based synchronization of uncertain networked lagrangian system. *International Journal of Systems Science*, **42** (2), 145–158.

Zhang, X.D. (2011) The laplacian eigenvalues of graphs: A survey, in *arXiv:math.OC/arXiv:1111.2897v1*.

Zhang, Y. and Tian, Y.P. (2009) Consentability and protocol design of multi-agent systems with stochastic switching topology. *Automatica*, **45**, 1195–1201.

Zhang, Z. and Chow, M.Y. (2011) Incremental cost consensus algorithm in a smart grid environment, in *IEEE Power and Energy Society General metting*, July 24–27, 2011, Michigan, USA, pp. 1–6.

Zhang, Z. and Chow, M.Y. (2012) Convergence analysis of the incremental cost consensus algorithm under different communication network topologies in a smart grid. *IEEE Transactions on Powers Systems*, **27** (4), 1761–1768.

Zhao, X., Xu, J., and Srinivasan, D. (2014) Novel and efficient local coordinated freeway ramp metering strategy with simultaneous perturbation stochastic approximation-based parameter learning. *IET Intelligent Transport Systems*, **8** (7), 581–589.

Zhou, K.M. and Doyle, J.C. (1998) *Essentials of Robust Control*, Prentice Hall, Upper Saddle River, New Jersy.

Index

Iterative Learning Control for Multi-agent Systems Coordination, First Edition.
Shiping Yang, Jian-Xin Xu, Xuefang Li, and Dong Shen.
© 2017 John Wiley & Sons Singapore Pte. Ltd. Published 2017 by John Wiley & Sons Singapore Pte. Ltd.